Undergraduate Lecture Notes in Physics

Undergraduate Lecture Notes in Physics (ULNP) publishes authoritative texts covering topics throughout pure and applied physics. Each title in the series is suitable as a basis for undergraduate instruction, typically containing practice problems, worked examples, chapter summaries, and suggestions for further reading.

ULNP titles must provide at least one of the following:

- An exceptionally clear and concise treatment of a standard undergraduate subject.
- A solid undergraduate-level introduction to a graduate, advanced, or non-standard subject.
- A novel perspective or an unusual approach to teaching a subject.

ULNP especially encourages new, original, and idiosyncratic approaches to physics teaching at the undergraduate level.

The purpose of ULNP is to provide intriguing, absorbing books that will continue to be the reader's preferred reference throughout their academic career.

Series Editors

Neil Ashby
University of Colorado, Boulder, CO, USA

William Brantley
Department of Physics, Furman University, Greenville, SC, USA

Matthew Deady
Physics Program, Bard College, Annandale-on-Hudson, NY, USA

Michael Fowler
Department of Physics, University of Virginia, Charlottesville, VA, USA

Morten Hjorth-Jensen
Department of Physics, University of Oslo, Oslo, Norway

Michael Inglis
SUNY Suffolk County Community College, Long Island, NY, USA

More information about this series at http://www.springer.com/series/8917

Martin Beech

Introducing the Stars

Formation, Structure and Evolution

 Springer

Martin Beech
Campion College
The University of Regina
Regina, SK, Canada

ISSN 2192-4791   ISSN 2192-4805   (electronic)
Undergraduate Lecture Notes in Physics
ISBN 978-3-030-11703-0   ISBN 978-3-030-11704-7   (eBook)
https://doi.org/10.1007/978-3-030-11704-7

Library of Congress Control Number: 2019931020

This Springer imprint is published by the registered company Springer Nature Switzerland AG
The registered company address is: Gewerbestrasse 11, 6330 Cham, Switzerland

Introducing the Stars

Study is like the heaven's glorious Sun
That will not be deep searched with saucy looks;
Small have continued plodders ever won
Save base authority from others' books.
These earthly godfathers of Heaven's lights
That give name to every fixed star
W. Shakespeare, *Love's Labour's Lost*

Introduction

It is sometimes simpler to say what something isn't rather than what it is. This text is concerned with the stars, but it is not a definitive guide to stellar properties, and it is not a scientific treatise concerned with the fundamental physics pertaining to stellar interiors. Nor is this a text that provides a refined and complete mathematical discourse or a comprehensive discussion of stellar evolutionary phenomena. Nor is it a definitive history of astrophysics.

So, why read on any further?

Firstly, this text is an introduction to the stars, and my intention is to follow the axiom that it is better to learn how to walk before one attempts to run. Secondly, the material in this text has evolved over many years through the hard-knock school of student scrutiny and requirements. The material that will appear in the following chapters is largely based upon a one-semester, 12-week course on the properties of stars for third and fourth year undergraduate students that have completed at least one course in each of the areas of introductory calculus, physics, and astronomy. The students that have successfully completed this class in the past are not all physics majors, and indeed, it is intended that the material should be accessible to any undergraduate working toward a typical science degree, be it physics, mathematics, computing, geology, engineering, or even biochemistry. It is not my aim in this text to provide a fully developed theory—dotting every *i* and crossing every *t*—of stellar structure and evolution, for indeed, there are many classic texts (see below) that do precisely this and do it very well. These latter texts, however, are not for the greenhorn; rather, they are for the beginning graduate student and those working toward a career in astrophysics. This text will introduce the first-time reader to the equations of stellar structure; it will outline the essential physics relating to star formation, stellar structure, and stellar evolution; and it will hopefully lead some initiate readers to the classic texts from which they can begin to hone their knowledge of the minutia, the more complex and the deeper, more subtle issues.

Although the approach in this text is exploratory and introductory, this is not to say that there won't be challenges to face, and on occasion, the going will get tough. But, my aim is to guide the reader toward what can be done, through the employment of mostly straightforward analytic techniques, to explain the observed properties of the stars from their birth to their death. Recourse will be made to computational models and detailed numerical results when appropriate, but broadly, the approach to be followed will be that of finding a physically plausible solution, rather than striving to extract a finely tuned mathematical explanation (remember we are learning to walk here—the running can come later).

The pioneering astrophysicist Arthur Eddington noted in 1927, "Proof is an ideal whom the pure mathematician tortures himself. In physics we are generally content to sacrifice before the lesser shrine of plausibility." In this text, we shall make many sacrifices to the lesser shrine—although it should also be pointed out that these sacrifices are not specifically wrong or acts of willful cheating, but rather they are sacrifices that allow for the development of ideas and analytic results that can be verified through more rigorous techniques and full computational simulations. I will not lead you astray in this text, but I will occasionally be strategically mute concerning the complexities in some situations. These complexities are the subject matter of the more advanced texts that the student can tackle once they have found their running feet.

The German physicist Heinrich Hertz, famous for his discovery of radio waves (and immortalized in the *SI* unit for frequency), often talked of developing his *innere scheinbilder*, or mental virtual images, of a physical system, and this book is a first step toward building such mental images of the stars. There is a well-known story often but not exclusively attributed to Martin Schwarzschild, one of the great pioneers of computational astrophysics, wherein once the money for the computer ran out, he had to make recourse to pen, paper, and thinking. Here, the idea is to start with the pen, paper, and thinking first and then worry about the full computational simulations later. Schwarzschild also commented on another occasion that it took at least one year of hard thinking to fully understand every twist and turn that the computational output revealed. Indeed, thinking is hard, and computation is easy (well, once the code has been written and tested), but the two must be used together to fully understand the stars, and it is also worth pointing out at this early stage that there are still many aspects, both on the theoretical front and with respect to computational complexity, about star formation, star structure, and star evolution that are far from being well understood.

The term "astrophysics" was introduced into the language of science by German astronomer Johann Zöllner in the mid-1860s, and it is a term describing the concatenation of astronomy and physics. Even though humanity has been aware of and has studied the stars for millennia, the idea that the structural properties of a star might be amenable to a detailed physical understanding is something that is altogether new and has only recently been added to the timeline of human endeavors. The first detailed text on stellar structure, *The Internal Constitution of the Stars*, was written by the indomitable Arthur Eddington and published in 1927 (not yet 100 years ago). This book—a classic that is still in print to this very day—is a beautiful disposition on the stars, but it is now beginning to look a little worn around the edges with respect to the physics underlying the stellar models. This, of course, is entirely to be expected given the advancements that have taken place over the past 90 years. For all this, however, Eddington has a wonderful turn of phrase (his *innere scheinbilder* were well developed), and in spite of its somewhat outdated results, I thoroughly encourage every student who is truly interested in the stars to find a copy of his text and read it through (see below).

The material to follow this introduction has been divided into six broad chapter headings. The first section sets out to describe what the observations tell us about star properties, while the second section begins to build up a picture of how the stars are envisioned to form. The third section begins to develop a physical picture of the material body of a star, while the fourth section begins to treat the energetic and thermal body. The fifth section looks at some aspects relating to stellar evolution and the manner in which stars end their days. The sixth chapter sees a change in approach and focuses on a number of selected review topics; this chapter is intended to make the reader think beyond the pure mathematical context of astronomy and to take a look at a number of philosophical and anthropomorphic issues where astronomy makes direct contact with humanity and with physical speculation.

As readers move through the various chapters and sections, they will encounter a number of exercises and problem sets. There are many reasons for including such challenges to the reader. Firstly, they are a good way to develop ones analytic and mental skills, and secondly, they save a lot of what would be otherwise tedious reading and printer's ink. The exercises and problems are described according to a three-level system that runs from *easy* to *straightforward*, to *requires some thought and is potentially very time consuming*. To help the reader identify the various levels, I will use the emoji code ☺, ☺, and ☹, with ☺ being the designation for easy. With this said, (☺) student exercise number 1 as indicated above is to find a copy of Eddington's text, and (☹) student exercise number 2 is to read Eddington's Chapter 1. Also presented under the exercises are a number of suggested topics for term papers, review essays, and computer programming studies. A list of these exercises is given below, and in order to avoid the author's old habit of leaving essay writing until the night before the due date (do what I say here—not what I did), it is suggested that these exercises be read through before even starting Chapter 1.

Exercise 1.6. The value of historical eclipse observations.
Exercise 1.24. Cepheid variable stars as standard candles.
Exercise 2.16. Max Planck's natural units.
Exercise 3.22. Develop a computer program to solve the Lane-Emden equation.

See also Exercises 3.24, 3.25, and 3.27.

Exercise 4.8. The OPAL opacity code.
Exercise 4.17. The solar neutrino problem.
Exercise 4.18. The Vogt-Russell theorem.
Exercise 4.19. A star in a fully reflective box.
Exercise 4.20. Homology relationships.
Exercise 4.23. Solving Eddington's quartic equation.
Exercise 4.24. The *mad star* web page.
Exercise 4.25. Energy generation in the CNO cycle.
Exercise 5.17. Hawaii District Court Case 1:2008cv00136.
Exercise 5.25. Becoming a red giant.
Exercise 6.1. The fallacy of the Copernican principle.
Exercise 6.2. The Crab Nebula.
Exercise 6.3. In the name of war.
Exercise 6.4. Extinction from space.
Exercise 6.5. The Gaia hypothesis.
Exercise 6.7. The history of η Carina.
Exercise 6.13. Not in my universe.
Exercise 6.13. The weak anthropic principle.
Exercise 6.14. A fine-tuned universe.

Other term paper topics will no doubt emerge from a reading of the text, and bear in mind Albert Einstein's wonderful comment that "Logic will take you from A to B. Imagination will take you everywhere."

Eddington's *The Internal Constitution of the Stars* is the first of the classic star-maker books. However, in the 65 years since Eddington's text first appeared, it was joined by at least five other classics. Each of these classic texts, just like Eddington's, is worth the read. The earlier texts may be somewhat dated now, but they were written by the founding giants of the field. The authors of these classic texts were all celebrated and honored practitioners, and their words still resonate with scientific passion and depth. Here is a list of the classic texts that I shall occasionally refer the reader to (there are numerous additional texts, all very worthy, that might have been included in the classics list, but I will leave it as (⊛) student exercise number 3 to find them):

1. *An Introduction to Stellar Structure*, by Subrahmanyan Chandrasekhar, published by University of Chicago in 1938
2. *Structure and Evolution of the Stars*, by Martin Schwarzschild, published by Princeton University Press in 1958
3. *Principles of Stellar Evolution and Nucleosynthesis*, by Donald Clayton, published by McGraw-Hill in 1968

4. *Principles of Stellar Structure*, Vol. I and II, by Jon Cox and Thomas Giuli, published by Gordon and Breech in 1968
5. *Stellar Structure and Evolution*, by Rudolf Kippenhahn and Alfred Weigert, published by Springer-Verlag in 1990

In the preface to his text, Eddington wrote that his study had been initiated "primarily in the hope that an understanding of the internal mechanisms will throw light on the external phenomena accessible to observations." His text is accordingly exploratory, and he strives to find the appropriate physics to develop analytic solutions that can be further constrained by direct comparison with the detailed observational data. Indeed, Eddington wrote, "I conceive the chief aim of the physicist in discussing a theoretical problem is to obtain *in sight*, to see which of the numerous factors are particularly concerned in any effect and how they work together to give it."

Subrahmanyan Chandrasekhar, writing just 13 years after Eddington, took an entirely different approach. Indeed, Chandrasekhar wrote, "an attempt is made to develop the theory of stellar structure from a consistent point of view, as far as possible, rigorously." While Eddington was "hoping" to find answers, Chandrasekhar was all about understanding the stars through mathematical and physical rigor. Chandrasekhar further noted that the goal of his studies was to "derive the complete march of the physical variables and the chemical composition through the entire [stellar] configuration." To this, he added, "the fundamental problem is to seek a theoretical relation of the kind F (L, M, R, *composition*) $= 0$." In this wonderful compaction, Chandrasekhar reduced the whole theory of stellar structure to the discovery of the function F parameterized according to the mass M, radius R, luminosity L, and composition of a star.

There is in fact no single analytic function F to satisfy Chandrasekhar's fundamental relation; rather, as we shall see in the chapters that follow, there are a series of four interconnected differential equations that can only be solved for using specialized numerical techniques on a computer. Indeed, Schwarzschild's classic book is all about the early attempts in the 1950s of taking the fundamental differential equations and solving them in a systematic manner. Almost inconceivably in the modern era, the early calculations were made by hand with the occasional help of a mechanical calculator. Between the publication of Eddington's text and that of Chandrasekhar, major developments took place in the field of quantum mechanics and in the understanding of the atom and nucleosynthesis. Between the publications of Chandrasekhar's text and that of Schwarzschild, the most important astrophysical developments were in those areas relating to the aging and evolution of the stars.

As we shall see later on, as a star ages, its internal composition changes, as a result of the fusion reactions taking place within its core. It is this internal alteration that drives the modification of the external and observable properties—the luminosity, radius, and temperature. Donald Clayton's text provides a detailed discussion of the nuclear physics appropriate to the interior of the stars, and he also offers a detailed analysis of the special numerical techniques required to construct stellar models. Between the publication of Schwarzschild's book and that of Clayton, the most important development to come about was the appearance of the first generally accessible—rather than military only—electronic computers. The accelerated speed of electronic calculation opened up the possibility to not only rapidly produce stellar models but to also develop detailed algorithms to deal with important and highly complex aspects of input physics—areas that were previously simplified for the sake of generating timely, hand-produced results.

Our final two classic texts are concerned with, in the case of Cox and Giuli, the development of highly detailed physical models and, in the case of Kippenhahn and Weigert, with a modern exposition on stellar evolutionary models by two of the pioneering investigators in the field. For those readers about to begin their exploratory walk through the following chapters, classic texts 3, 4, and 5 are not required reading, but I do recommend consultative visits. These latter texts are not for the beginner; rather, they are the domain of those who have found their running feet and who wish to embark on a research career in stellar astrophysics. Clayton indicates, for example, that the idea behind his text was to produce a tome that could be assimilated by a full-time graduate student over the course of one year. The two volumes by Cox and Giuli are additionally formidable and closely argued and amount to over 1000 pages of detailed physical reasoning and mathematical development.

In addition to the classic texts listed above, there is a very rich literature and history of research relating to the properties of the stars. Most of this latter material can be accessed through the SAO/NASA-sponsored ADS web server at www. adsabs.harvard.edu. Contemporary research papers on all astronomical topics can be accessed through the Cornell University library on its arXiv web server at www. arxiv.org. In terms of general information about the stars and star systems, I would highly recommend the internet websites maintained by Jim Kaler (www.stars.astro. illinois.edu) and that by the Swinburne University (www.astronomy.swin.edu.au). It has also been my experience that the Wikipedia web pages, at least with respect to astronomy data, are well written, up-to-date, and fully worth consulting for general astronomy questions and even for specific astrophysics inquiries. I also present an extensive list of references and suggested readings in Chapter 6.

And finally, astronomers are notoriously bad for mixing rather than matching the units that they use, interchanging meters with light years, and astronomical units with parsecs. This text will be no different to the norm, and according to context, different units will be applied to the quantities that appear. In general, however, and wherever reasonable, the *System International* (SI) units of meters, kilograms, and seconds will be used. A short selection of the more important constants to appear in the following chapters is given below:

Length:

- 1 Astronomical unit (AU) $= 1.4959 \times 10^{11}$ m
- 1 Parsec (pc) $= 206, 264.8$ AU $= 3.0857 \times 10^{16}$ m
- 1 Light year $= 9.460536 \times 10^{15}$ m $= 0.307$ pc

The Sun:

- Mass $M_\odot = 1.9891 \times 10^{30}$ kg
- Radius $R_\odot = 6.9627 \times 10^{8}$ m
- Temperature $= 5780$ K
- Luminosity $L_\odot = 3.85 \times 10^{26}$ Watts
- Apparent magnitude $= -26.75$
- Absolute magnitude $= +4.82$

Physical constants:

- Speed of light $c = 2.99792458 \times 10^{8}$ m/s
- Gravitational constant $G = 6.672 \times 10^{-11}$ Nm^2/kg^2
- Planck's constant $h = 6.6261 \times 10^{-34}$ Js
- Boltzmann constant $k = 1.381 \times 10^{-23}$ J/K
- Stefan-Boltzmann constant $\sigma = 5.67 \times 10^{-8}$ W/m^2/K^4
- Mass of the electron $= 9.1004 \times 10^{-31}$ kg
- Mass of the proton $= 1.6726 \times 10^{-27}$ kg
- Elementary charge $e = 1.6022 \times 10^{-19}$ C
- Ideal gas constant $\mathfrak{R} = 8314.46$ J/K/kg
- Radiation density constant $a = 7.5675 \times 10^{-16}$ J/m^3/K^4

Contents

Chapter 1
Knowing the Stars

On a human scale, the stars are both ancient and modern. They are ancient in the sense that the value of the stars for gauging time and the seasons has been recognized since the earliest beginnings of human civilization. The constellations of the zodiac, which straddle the projection of the Earth's orbit onto the celestial sphere (the ecliptic), were named over 3500 years ago by astronomer-priests in ancient Babylon. Yet despite the relentless weight of past history, the stars remain modern objects in the sense that they continue to challenge our intellect and because there are still many aspects of their structure and evolution that astronomers do not fully understand.

The nearest star is the Sun, of course, and its physical properties are described in the introductory constants list. The time at which the Sun demonstrably became a star and the stars became suns is a concept of recent genesis, dating back to about 150 years ago. Prior to the nineteenth century, all that could be said about the Sun was that it is very much brighter for terrestrial observers than any nighttime star. The Sun is relatively large to the human eye, about 0.5 degrees across, while in contrast, the stars are faint, point-like sources of light. Until one knows more about the physical arrangement of the stars, however, they remain indeterminate objects. They could in principle be objects a million times larger than the Sun located at inconceivable large distances from us, or they could be many times smaller than the Sun and hovering just beyond the boundary of the outermost planet in the Solar System.

To know that the Sun is a star and that the stars are suns requires that very specific measurements be made. The concept required a knowledge of how far away these objects were from the Earth, along with information concerning what they are made of and how much energy they radiate into space. Not only does one need to have good and accurate observational data on the intrinsic properties of the stars (see Sect. 1.1), but one also needs a theoretical framework within which to interpret the observations. The two disciplines—theory and observations—go hand in hand, one leading the other according to the moment in history.

History tells us that towards the close of the nineteenth century, both the theoretical picture and the observational outlook were sufficiently developed that the first

© Springer Nature Switzerland AG 2019
M. Beech, *Introducing the Stars*, Undergraduate Lecture Notes in Physics,
https://doi.org/10.1007/978-3-030-11704-7_1

scientific statements about the stars and the Sun could be articulated. In the first half of the nineteenth century, accurate distance measures to at least the nearby stars were made, and it was found that even they were located at vast distances away from the Solar System. The next nearest star to the Sun is situated some 29 million solar diameters away. In the last half of the nineteenth century, astronomers began to apply the new art of spectroscopy to the heavens. German optician Joseph von Fraunhofer had first noted that the Sun's spectrum was interrupted by dark lines in 1812, but it was not until the early 1860s that Robert Bunsen and Gustav Kirchhoff were able to outline the possibility of matching the spectral features of elements observed in the laboratory with the Fraunhofer lines observed in the Sun and in this manner identify what the Sun was physically made of. The first astronomer to observe and accurately record dark lines crossing the spectra of several bright stars, just like the Sun's Fraunhofer lines, was William Huggins in 1864. Only from this time onwards has it been clear that the Sun is a star and that stars are at least similar[1] to the Sun.

1.1   What Is a Star?

A short search on the internet will reveal a dictionary definition for the word 'star' that typically runs along the lines: 'a large, self-luminous ball of gas in space seen as a point of light in the night sky'. Under the editorial requirement that dictionary definitions be short, sharp and accurate, then all appears well with the star description provided by the internet sources. The problem with such definitions, however, is that they don't actually tell us anything particularly useful in a scientific sense. Yes, stars are large—but how large? Yes, stars are self-luminous—but by what mechanism? Yes, stars are spherical and are located in space—but how do we known they are spherical, as opposed to looking like a stellated icosahedron. And yes, they are located in space, but at what sort of distance and how are such distances measured? The dictionary definition also fails to explain to us how stars might form and age; what kind of gas they are made of; and what variations of temperature, density and pressure might take place within their interiors.

The dictionary definition certainly provides us with the essence of the concept, but to know and understand what such objects are is an entirely different story. The dictionary definition provided above is perhaps appropriate to the state of affairs that prevailed towards the close of the nineteenth century. Our modern twenty-first century view of what a star is, however, is much more detailed and much more complicated, and the question we need to address now is: How did this modern understanding come about? The modern theory concerning stellar structure takes us back to the mid-nineteenth century—a time when the Brontë sisters were in the

---

[1]The term similar is used here since, as we shall see, the stars show a wide range in their associated masses, sizes, temperatures and luminosity.

process of writing *Jane Eyre* and *Wuthering Heights*, the potato famine was deci-
mating the Irish poor, the gold rush was about to start in California, the Mormon
migration to Utah was in progress, Karl Marx and Friedrich Engels were working on
The Communist Manifesto, and the Great auk had just been driven into extinction
through human overexploitation. With all this and of course much more human
drama as a backdrop, German physician Julius Mayer asked himself an obvious
(in hindsight) question: What physical process powers the Sun?

Sometimes, in spite of everything, there are moments in life when one has to stick
to one's guns and simply accept that the critics are wrong. Such was the situation for
Mayer in 1848. In that year he published at his own expense a text with the
somewhat uninspiring title *Beitrage zur Dynamik des Himmels in popularer
Darstellung* (*A contribution on the dynamics of the heavens in popular terms*). We
shall get to the details of Mayer's text shortly. In the mean time, however, we must
backtrack 6 years in Mayer's life, to 1842. Indeed, it was in this latter year that
Mayer proposed something that was then entirely unacceptable to the established
scientific community. What he suggested was that all bodies, whether living or
inanimate, contain a specific and completely indestructible quantity called 'energy',
and that this energy was conserved under conversion. The energy could change and
take on different forms—that is, it could be converted from one kind of energy into
another, but the key point was that the total amount of energy at all times was some
fixed and constant value.

This idea—the conservation of energy—was a radical idea at the time, but it is
now accepted as one of the most fundamental of all principles in science. It took
several decades for Mayer's idea to be brought into mainstream thinking, largely
through the efforts of such famed practitioners as James Prescott Joules, William
Thomson (later Lord Kelvin), Herman von Helmholtz and Rudolph Clausius, and it
now sits at the heart of the theory of thermodynamics—the theory of heat, mechanical energy and energy transfer. We will consider the thermodynamic properties of
stars in a later section and for the present return to Mayer's 1848 book. In this book,
Mayer proposed the first model that attempted to account for the Sun's power
supply. This in and of itself was a radical idea, since most practitioners of the time
simply assumed that the Sun, and by assumption the stars, didn't need a power
supply—rather, they were objects that somehow formed as hot balls of gas and then
simply cooled off, growing fainter and fainter as they aged.

Mayer's argument that the Sun needed a source of energy was built around the
then-developing concept of a very old Earth, as implied by the fossil and geological
record and the fact that there was no evidence to indicate that the Sun was actually
getting cooler. To solve this age and non-cooling-off problem, Mayer invoked the
existence of an external energy source, arguing that the Sun was heated by the
continual infall of material onto its surface, in the form of extensive meteor showers.
As long as material continued to fall onto the Sun's surface, Mayer argued, it could
maintain a constant temperature and energy output. To see how this idea works we
must first use the conservation of energy to estimate the speed V with which a
quantity of mass m, falling in from an initial distance D, will impact upon the Sun's
surface. Take the Sun to have a mass M and radius R and the initial velocity of the

infalling mass to be V_0. In this infall situation the energy will be distributed between the gravitational potential energy U and the kinetic energy of motion K. Accordingly, the total energy at any instant is the sum $E = U + K$. The conservation of energy now requires that

$$[U + K]_{\text{end}} = [U + K]_{\text{start}} \tag{1.1}$$

and upon substitution for the gravitational potential energy and the kinetic energy terms, Eq. (1.1) becomes:

$$-G\frac{Mm}{R} + \frac{1}{2}mV^2 = -G\frac{Mm}{D} + \frac{1}{2}mV_0^2 \tag{1.2}$$

At this stage, we make the assumption that the initial distance at which the mass m begins its motion towards the Sun is much larger than R, and indeed, so large that the first term on the right-hand side of (1.2) is effectively zero. Likewise, we can also assume that the initial infall speed is effectively zero, so the second term on the right-hand side of (1.2) is also zero. With these starting conditions in place, the impact velocity V of mass m upon the surface of the Sun is

$$V = \sqrt{\frac{2GM}{R}} \tag{1.3}$$

and importantly, we know all of the numerical values appropriate to the right-hand side of (1.3). Substituting numbers gives: $V = 6.2 \times 10^5$ m/s. The result presented in Eq. (1.3) can also be worked the other way around and provides us with the so-called 'escape' velocity—the velocity that an object would have to have at the surface of the Sun if it was to eventually escape from the Sun's gravitational influence and never fall back again. At this stage, we have an impact velocity for each of our in-falling masses, but recall that what we want is a measure of how much matter must fall onto the Sun's surface so that it can maintain a constant energy output L. Here, L is the luminosity of the Sun (expressed in units of energy per second). Mayer's argument builds upon the notion that the energy that the Sun loses into space per second must be exactly compensated for by the kinetic energy of the in-falling meteoric matter. If the total amount of matter falling onto the Sun's surface per second is dm/dt and if all of the kinetic energy of the infalling matter goes into powering the Sun, so

$$L = dE_m/dt = \frac{1}{2}(dm/dt)V^2 \tag{1.4}$$

Eq. (1.4) builds upon the fact that all of the material being accreted by the Sun strikes its surface with the same speed, irrespective of initial mass m. Additionally, what Eq. (1.4) also allows us to do is calculate how much matter the Sun will have to accrete in order to maintain its observed luminosity $L = 1\,L_\odot$. The accretion rate comes out to be $dm/dt = 2L_\odot/V^2 = 2.0 \times 10^{15}$kg/s. Given that the Sun has a mass $M_\odot = 1.9891 \times 10^{30}$ kg, we can define an accretion timescale by the ratio: $T_{acc} = M_\odot/(dm/dt) = 10^{15}$ seconds $= 31.7$ million years. This timescale gives us the characteristic interval over which the Sun's mass will increase by an amount equal to 1 $M_\odot$. In other words, after 31.7 million years the initial Sun (of presumed mass 1 $M_\odot$) will have a mass equal to 2 $M_\odot$, and after another 31.7 million years it will have a mass of 3 $M_\odot$, and so on. Clearly, something has gone wrong with our calculation—or more correctly, what we really have to look at more closely is the way in which the kinetic energy of the accreted material is converted into internal energy that is usable to the Sun. To do this, we would need to know much more about the temperature of the Sun's outer layers and also describe the exact process by which the kinetic energy of the in-falling material is converted into thermal energy within the Sun's outer layers. Indeed, Mayer realized that this was the point at which his model of the Sun had to stop, and he simply asserted that the accreted matter was *somehow* converted into radiant energy, and accordingly the Sun's mass remained constant over the age of the Solar System.

(☺) **Exercise 1.1** Determine the mass flux of the in-falling material at the Sun's surface in units of kg/s/m². **Hint**: Divide the accretion rate by the Sun's surface area.

(☺) **Exercise 1.2** Determine how much matter the Earth would accrete per second on its night side hemisphere if Mayer's accretion model were actually correct. Take the Earth's orbital radius to be 1 AU $= 1.495 \times 10^{11}$ m, and the Earth's radius to be 6.371×10^6 m. **Comment**: The Earth actually accretes about 4×10^4 kg of meteoritic matter per day—compare this influx with that deduced from the Mayer model.

(⊛) **Exercise 1.3** If the mass of the Sun did actually increase by 1 $M_\odot$ every T_{acc} years, how long would it be before the Earth's orbital radius changed by 10% and will its orbit become smaller or larger? **Hint**: Assume that the Earth conserves its orbital angular momentum at all times t such that $m_{Earth} V r(t) = $ constant, where V is the Earth's orbital velocity and $r(t)$ is its orbital radius. Also assume that Kepler's 3rd law applies at all times: $M(t) + m_{Earth} = (4\pi^2/G)r^3/P^2$, where $M(t)$ is the mass of the Sun at time t, and P is the Earth's orbital period.

Mayer's solar accretion model never proved particularly popular, but he found solace and historical fame with respect to his ideas on the conservation of energy. The solar accretion model did not survive for very long at all, although it was championed in the late 1890s by British astronomer Norman Lockyer (who, in fact, was the first person to ever hold a university position with the title Professor of Astronomy). The idea of an accretion-powered Sun was essentially abandoned within a decade of its first appearance, being replaced by a new model developed by Alexander von Helmholtz and William Thomson (Lord Kelvin) in the 1850s and 60s. This new solar model grew out of a more refined analysis of Mayer's ideas on the conservation of energy, but rather than assuming the Sun had some external energy supply, it thought that the Sun generated energy internally through physical contraction. To see how this alternative Sun model works, let us first introduce a

Fig. 1.1 Consider the Sun to be composed of many spherical shells, each of thickness dr, arranged one inside of the other out to some outer radius r. The gravitational binding energy relates to the total energy required to disperse each shell to infinity. Credit: Author

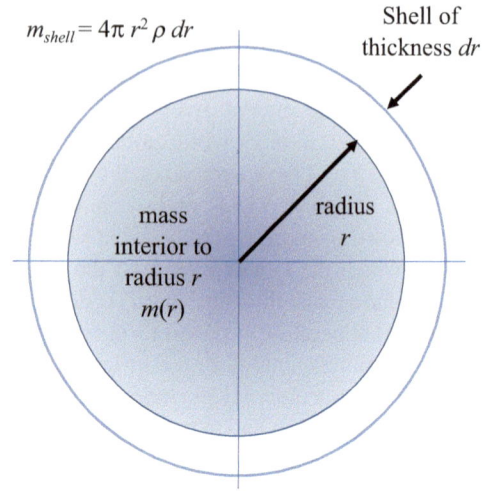

$m_{shell} = 4\pi r^2 \rho \, dr$

Shell of thickness dr

mass interior to radius r
$m(r)$

radius r

quantity called the gravitational binding energy U. To begin, imagine that a sphere (the Sun or a star in general) of radius R is being pulled apart by successively removing thin spherical shells (Fig. 1.1).

The gravitational binding energy is defined as the total amount of energy needed to disperse the star, shell by shell, to infinity. The mass of each shell, at a radial distance r from the center of the star, can be taken as $m_{shell} = 4\pi r^2 \rho dr$, where ρ is the (assumed at this stage) constant density of the material within the star and dr is the shell thickness. The mass of material inside the star out to radius r is $m(r) = 4\pi r^3 \rho/3$. The gravitational potential energy of each shell can accordingly be written as

$$dU = -G \frac{m(r) \, m_{shell}}{r} = -\frac{16}{3} G\pi^2 \rho^2 r^4 dr \tag{1.5}$$

If we now integrate Eq. (1.5) over the entire radius of the star, from its center $r = 0$ to its surface at $r = R$, we obtain the gravitational binding energy of the star. Accordingly

$$U = -\frac{16}{3} G\pi^2 \rho^2 \int_0^R r^4 dr = -\frac{16}{15} G\pi^2 \rho^2 R^5 \tag{1.6}$$

If we now take the density to be the average density of the star: $\rho = M/(4\pi R^3/3)$, then Eq. (1.6) reduces to the expression

$$U = -\frac{3}{5} G \frac{M^2}{R} \tag{1.7}$$

where U is the gravitational binding energy of a uniform density spherical star. This expression, Eq. (1.7), can be used to determine the change in the Sun's gravitational binding energy as it contracts from some initially large radius R_0 to its currently observed radius $R < R_0$. This change will be $\Delta U = U_0 - U$, with

$$\Delta U = -\frac{3GM^2}{5}\left[\frac{1}{R_0} - \frac{1}{R}\right] \approx \frac{3GM^2}{5R} \tag{1.8}$$

the second term on the right-hand side of Eq. (1.8) is derived under the condition that R_0 is effectively infinite or at least it is sufficiently large that it is safe to take the approximation that $G M^2 / R_0 \approx 0.0$. Equation (1.8) tells us how much energy the Sun generates in contracting from some initial radius R_0 to a final radius R. The question now is where does this energy go? Does it go into heating the Sun, or is it radiated into space, or both, or does something else happen?

(☺) **Exercise 1.4** (1) Assuming the Earth is a sphere and has a constant density, how much energy would be required to disperse it, shell by shell, to infinity? (2) How long will it take the present-day Sun to radiate the same amount of energy into space?

Equation (1.7) tells us that as the Sun shrinks, so its radius R becomes smaller and its gravitational binding energy U becomes larger and more negative. By shrinking, the Sun has effectively gained energy, and this energy can be transformed into the thermal heat energy of its constitutional gas as well as radiant energy that can leak out at the Sun's surface. To determine just how much of the energy gained by the Sun by shrinking is converted into radiant energy, we must first make reference to an important result derived by Rudolph Clausius in the late 1860s. The result we need is called the Virial theorem,[2] which relates the total thermal energy K of a stable, self-gravitating, spherical distribution of particles (a hot gas in the case of the Sun and the stars) to the total gravitational potential energy U. What Clausius specifically found is that the negative gravitational binding energy of a star is equal to twice its thermal energy: $U = -2 K$. Since, however, the total energy of a star E is the sum of its thermal and gravitational binding energy terms, $E = K + U$, the Virial theorem indicates that $E = U / 2 = -K$—that is, the total energy of a star is negative and is equal to half the gravitational binding energy.

Bringing this result back to Eq. (1.8), we determine that by shrinking, the amount of energy E_{rad} that the Sun can radiate into space is $E_{rad} = \frac{1}{2}\Delta U = (3/10)G M^2/R \approx 10^{41}$ Joules. The other half of the energy goes into internal heating. This is a remarkable result: by shrinking, the Sun (and by analogy any similar such star) becomes hotter and at the same time finds a source of energy that can be radiated into space at its surface. It is important to note, however, that in the analysis just presented we have essentially put the cart before the horse. The reason for this lies in the workings of the second law of thermodynamics. This law requires that hot objects must radiate their energy into colder surroundings—the Sun is hot, and

---

[2]The word 'virial' is derived from the Latin word *virias*, meaning forces.

therefore it must radiate energy into the coldness of space. Accordingly, it is the fact that stars are hot and thereby have a surface luminosity that (without any other internal energy supply to draw upon) drives them to contract. If we imagine placing the Sun in a very large oven such that it absorbed as much energy from the oven walls as it radiated into space, then it would not shrink. Indeed, if the Sun absorbed more energy from its surroundings than it radiated into space, then it would actually expand and get cooler.

The fact that a gaseous star by shrinking can draw upon its internal gravitational binding energy to maintain its luminosity is sometimes called Lane's Law, after the American physicist John Homer Lane, who first discussed the behavior of self-gravitating gas spheres in an 1870 article published in the *American Journal of Science*. Lane's analysis was revolutionary for the time, since prior to its appearance it was generally assumed that stars would simply cool off as they aged.[3]

(☺) **Exercise 1.5** By how much would the Sun have to shrink per year in order to maintain its present luminosity. *Hint*: Differentiate the expression for E_{rad} (assuming that the mass of the Sun remains constant), set $dE_{rad}/dt = L_\odot$ and show that $dR/dt = -2.4 \times 10^{-6}$ m/s.

(☺) **Exercise 1.6: Term Paper Topic** Can you think of a way in which any changes in the Sun's diameter (as well as changes in the Earth's spin period) on timescales of hundreds to even thousands of years might be tested? *Hint*: Think about ancient eclipse observations.

The characteristic time over which the Sun might maintain its present luminosity as a result of gradually shrinking is $T_{KH} = E_{rad}/L_\odot = 3 \times 10^{14}$ seconds = 9.3 million years: this is the so-called Kelvin-Helmholtz timescale, which is usually written (ignoring the 3/10 factor) as

$$T_{KH} = \frac{GM^2}{RL} \tag{1.9}$$

The Kelvin-Helmholtz timescale of the Sun being of order many millions of years was an important result when it was first derived in the 1860s, since earlier estimates for the Sun's age made it just a few thousand years old. This latter result would have been highly problematic since by the late nineteenth century it had become abundantly clear that the Earth must be at least several millions of years old—to have a Sun younger than the Earth would be a troubling contradiction. While the Kelvin-Helmholtz contraction model initially provided for a Sun older than the Earth by several millions of years, it was soon found that even this extended lifetime was nowhere near long enough. The problem of course with the contraction model is that after a time T_{KH} has elapsed, there is no more solar body that can undergo contraction—indeed, if there were no physical mechanisms to halt the contraction, the Sun

---

[3]It is interesting to note that Lane's paper was published under the journal section heading of 'Extraterrestrial Geology'; in 1870, there was no such topic as the physics of the stars.

would simply collapse into a black hole after about 10 million years. In reality, there are processes that do stop stars like the Sun from collapsing to become a black hole, but we shall discuss these later. As the geologists, paleontologists and evolutionary biologists pushed the age of the Earth every higher, the physicists in the early twentieth century were left wanting in the identification of a long-enough-lived energy source for the Sun. Mayer's accretion model could have been resurrected to save the day, since in principle this mechanism can power the Sun for as long as there is interstellar material to fall onto its surface, but unfortunately, there is no such source of material. Once again, it was time and circumstance that came to the rescue, and the key discovery was made by French physicist Henri Becquerel in 1896. With Becquerel's discovery of radioactive decay, scientists were forced to take on the idea that the Sun might have somehow found a way to extract energy from the very atoms distributed within its interior (Chap. 4). While it is now known that the Sun and the stars in general do not generate their internal energy through either gradual collapse and/or accretion, there are times when both of these mechanisms are important. Indeed, it is core contraction on the Kelvin-Helmholtz timescale that carries a star from one nuclear fusion regime to another, and it is accretion of material from a binary companion that makes stellar-mass black holes detectable—the classic case being Cygnus X-1.

While Lane made little play of his discovery that stars must heat up as they contract, structural engineer August Ritter in Germany realized that this result implied a very definite theory and a very specific direction for stellar evolution. Ritter argued that stars in general must form by gravitational collapse, starting out as large, low-density clouds of cool gas in the interstellar medium (see Chap. 2), which then, while gradually shrinking, heat up through the process described in Lane's law. Ritter then assumed that at some point during the contraction, a maximum temperature would be achieved and the interior density of the star would become so high that it would liquefy.

Accordingly, by the turn of the twentieth century, a two-phase theory of stellar evolution had been developed. The first phase envisioned the gravitational collapse and subsequent heating of an initially large, low-density, low-temperature cloud of gas. The second stage envisioned the simple cooling off of a liquefied, constant-radius body. What is historically interesting about this theoretical picture is that it was in harmony with the then new results being obtained by observational astronomers. These new observations published between 1911 and 1915 indicated the existence of two distinct types of stars: the giants, which were very luminous, very large and had low surface temperatures, and the dwarfs, which were small, high-temperature, low-luminosity objects. The observational results were eventually arranged in what is known today as the Hertzsprung-Russell (HR) diagram, and it is this very same diagram (to be described later) that constitutes the great testing ground between the theoretical models, as derived from stellar evolution theory, and the truths indicated by the observations and the annotation of intrinsic stellar parameters.

(☺) **Exercise 1.7** Determine the average density of the Sun in kg/m^3. Compare the Sun's average density with that of air at sea level (1.2 kg/m^3), that of liquid honey (1433 kg/m^3), and that of lead (11,342 kg/m^3). (☺) What important physical factors are being ignored in this (purely numerical) comparison of densities?

At this stage in our narrative we have only considered timescales and potential modes of solar energy generation (accretion and/or contraction). Both Lane and Ritter, however, developed their ideas to describe the run of the pressure, density and temperature within the Sun, working upon the assumption that at any one instant, the Sun could be regarded as being in a state of quasi-static (or hydrostatic) equilibrium. We shall leave the discussion of these developments until Chap. 3 and turn for the remainder of this chapter to the question of how stellar parameters are determined observationally.

By 1915, dramatic developments were beginning to take place within the worlds of both physics and astronomy. The physicists were just starting to develop a better understanding of the laws of quantum mechanics, special and general relativity and the properties of the atom, while the astronomers were beginning to ponder the implications of new and detailed tabulations of intrinsic stellar properties. Indeed, it was the realization that the intrinsic parameters of the stars are closely correlated (mass and luminosity, for example, as we shall see below) that inspired theoretical physicists to take a much more detailed look at exactly what processes are at play inside of a star.

1.2   What Can Be Measured?

Stars radiate energy from their surface (technically their photosphere) into space in the form of electromagnetic radiation. We know this from the simple fact that we can see stars with our eyes. Specifically, what our eyes are responding to is the energy flux received from a particular star. Let us introduce two definitions at this stage:

Definition 1.1 The luminosity (L) is the total amount of energy radiated by a star over all wavelengths of electromagnetic radiation per second into space. The units of luminosity are Joules per second, or Watts.

Definition 1.2 The flux (F) is the amount of energy received by a specific detector per square meter per second. The units of energy flux are Joules per square meter per second, or Watts / m^2.

The luminosity is the intrinsic parameter that we wish to know, since it is a measure of the total energy output of a star. Unfortunately, we cannot measure the luminosity directly—astronomers must rely upon a measurement of the energy flux and then convert this to luminosity. This latter step can only be made once the distance to a star is known. Accordingly, for a star of luminosity L, a distance d away and with a measured energy flux F, the relationship of interest is

$$F = \frac{L}{4\pi d^2} \tag{1.10}$$

(☺) **Exercise 1.8** Why can we not measure the luminosity of a star directly? What is the important assumption about how stars radiate their energy into space that enters into Eq. (1.10)? **Hints**: Think about the definition for luminosity and what it would technically require to measure it directly. Think about what the $4\pi\,d^2$ term corresponds to.

(☺) **Exercise 1.9** What is the energy flux from the Sun at the Earth's orbit, and how much energy does the sunward-facing side of the Earth intercept per second? Compare the latter number to the energy consumption by human activity, which over the entire year of 2013 amounted to some 5.67×10^{20} Joules.

The task of the observational astronomer is to determine an accurate measure of the energy fluxes to as many stars as possible, and then for these same stars determine how far away they are from the Sun. Since this is not a text concerning photometric techniques, at this stage we will simply assert that astronomers can measure fluxes, and measure them in the modern era to very high accuracy and very small values. For example, the solar flux (also called the solar constant[4]) is determined as $F_\odot = 1362$ W/m^2, while the flux from Sirius, the brightest star in the sky, is $F_{Sirius} = 1.1 \times 10^{-7}$ W/m^2; that for Vega, the 5th brightest star in the sky, is $F = 2.7 \times 10^{-8}$ W/m^2.

(☺) **Exercise 1.10** At what distance would an observer have to view the Sun from in order for its measured flux to be the same as that derived for Sirius?

Having introduced the idea of energy flux measurements, most astronomy texts would at this stage go on to describe the idea of stellar magnitudes. This is a strange, anachronistic process that purely because of past history has been used by observational astronomers. It is an odd method since it essentially takes the fundamental number and physical quantity measured—that is, the flux—and then converts it into an entirely artificial number, the magnitude m, through the relationship: $m = -2.5$ Log (measured flux) + a constant. We will not consider why this is done here, but simply note that in spite of it being an anachronism, there are sound enough practical reasons for working with the magnitude scale. There are many very good texts that cover the topic of photometry in great detail, and the student is directed towards a repeat of (⊗) student exercise number 3 from the Introduction to find them. As far as we need be concerned in this text, a measure of the energy flux from a specific star can be obtained, and provided that a distance to the same star is also available, then Eq. (1.10) provides a means of determining the star's total energy output and luminosity L.

---

[4]While called the solar constant, the energy flux actually varies according to the Earth's distance from the Sun, being 1412 W/m^2 in early January and 1321 W/m^2 is early July.

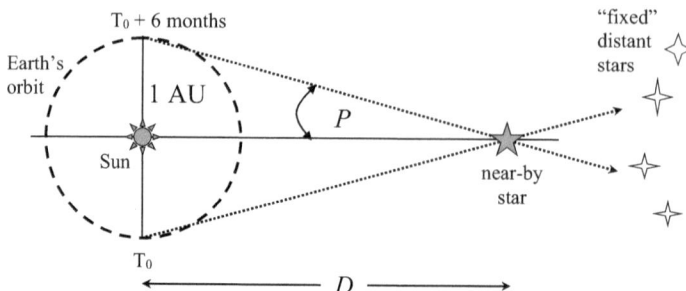

Fig. 1.2 The essential geometry behind the measure of the angle of stellar parallax P. (Credit: Author)

To determine the distance to a star, astronomers use the idea of parallax. They use the motion of the Earth about the Sun over several months to induce an apparent change in the position of a nearby star compared to the backdrop of very distant stars. The idea of stellar parallax is illustrated in Fig. 1.2. The basic concept behind the parallax distance d measure is to use the known baseline length across the Earth's orbit and a measure of the half-angular shift P across the sky by a nearby star, relative to the much more distant background stars, over a six-month interval. The result of such measurements is the evaluation of the expression $\tan(P) = 1$ AU $/ d$, where 1 AU is the radius of Earth's orbit.

As with all diagrams that paint simple pictures, such as Fig. 1.2, the observational realties behind parallax measurements are much more challenging than might at first be imagined. Indeed, the first believable measures of stellar parallax were not to be published until the late 1830s, although the geometry of the method had been known since antiquity. With the first measurements, it was apparent that even the closest stars must be at great distances away from the Sun. The parallax of the closest star, Proxima Centauri, is just 0.769 seconds of arc (that is just 0.00021 of a degree), placing it at a distance of $d = 1$ AU $/ \tan(0.00021) = 272{,}837.0$ AU. At this stage we introduce a distance measure change, in which the distance to an object is defined to be exactly 1 parsec (pc) if it has an angle of parallax $P = 1$ second of arc (or equivalently $1/3600^{\text{th}}$ of a degree). In this manner, $d = 1\text{pc} = 1$ AU $/ \tan(1/3600) = 206{,}264.8$. Accordingly, Proxima Centauri is located at a distance of 1.32 pc from the Sun.

(☺) **Exercise 1.11** There is a useful mathematical dodge that can be associated with the stellar parallax distance formula. The simplification relates to the series expansion of the $\tan(P)$ term, when P is expressed in radians and when P is very small. (i) Use the Web or a mathematical handbook to find the series expansion for $\tan(\varphi)$, and convince yourself that for very small angles, $\tan(\varphi) = \varphi$, where φ is in radians. By definition, 2π radians is equivalent to 360 degrees, and accordingly, $P(\text{radians}) = (2\pi / 360) P(\text{degrees})$. (ii) Combine the small angle approximation formula for $\tan(P)$ with P expressed in radians and show that $d(\text{pc}) = 1 / P''$, where$''$ indicates that the angle is expressed in seconds of arc.

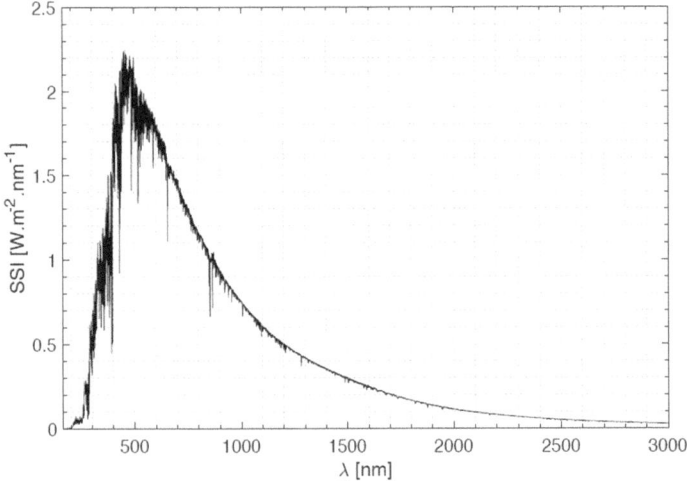

Fig. 1.3 The spectrum of the Sun. The wavelength region covered on the x-axis is from the ultraviolet to the near infrared. (Image courtesy of CNRS/ESA)

Once a measure of the total flux and the distance to a specific star has been obtained, its luminosity can be determined. The next intrinsic parameter to be estimated is that of a star's surface temperature. This particular characteristic is found by looking at a star's spectrum—the diagram that plots the variation of the energy flux (W/m^2) against wavelength. The Sun's spectrum is shown in Fig. 1.3, and while certain key features do vary according to the temperature of a star (specifically the absorption line features—see below), the basic spectral profile for all stars is that of a blackbody radiator. Given the specific focus of this text, little will be said about stellar spectra, and as with stellar photometry earlier, the student is directed towards a (third) repeat of (⊗) student exercise number 3 from the Introduction to go and find an appropriate text that covers this extensive topic.

The characteristics of blackbody radiators were first explained by German physicist Max Planck at the very beginning of the twentieth century, and indeed, it was Planck's analysis of these objects that eventually ushered in the new era of quantum mechanics. The important point about blackbody radiators is that energy flux at any given wavelength is entirely determined by the temperature. Planck's formula provides an annotation of basic stellar spectra, and specifically that the flux at very short wavelengths (from the ultraviolet and smaller) and at very long wavelengths (the far infrared and longer) is small, with there being a reasonably well-defined maximum flux in the optical part of the electromagnetic spectrum. The specific form of Planck's formula for the spectral radiance[5] $B(\lambda, T)$ at a specified wavelength λ and blackbody temperature T is

---

[5]The *SI* units for the spectral radiance are watts per steradian per square meter per nanometer.

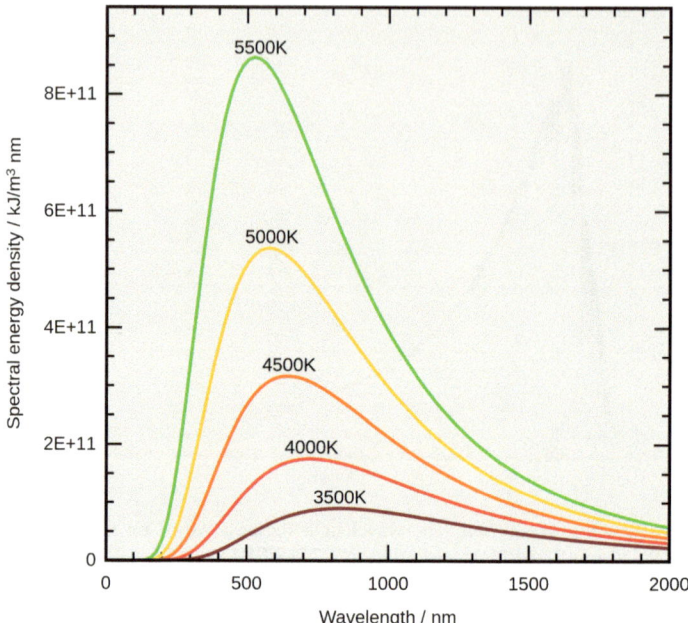

Fig. 1.4 Blackbody radiation curves for a range of temperatures between 3500 and 5500 K. (Image from Wikimedia commons (http://commons.wikimedia.org/wiki/File:wiens-law.svg))

$$B(\lambda, T) = \frac{2hc^2}{\lambda^5} \frac{1}{[\exp{(hc/\lambda kT)} - 1]} \tag{1.11}$$

The new fundamental constant introduced in Eq. (1.11) is that of Planck's constant h, and it is accompanied by the additional fundamental constants c, the speed of light and the Boltzmann constant k. The spectral radiance versus wavelength profiles for several blackbody radiators having different temperatures are shown in Fig. 1.4. The profiles show a systematic displacement of the wavelength at which the maximum energy flux is emitted, and this displacement is expressed according to Wien's law, named after the German physicist Wilhelm Wien, who discovered the result experimentally in 1893. Specifically, Wien's law indicates that

$$\lambda_{\max} T = 2.89777 \times 10^{-3} \tag{1.12}$$

where $\lambda_{\max}$ is the wavelength, measured in meters, at which the maximum energy flux is recorded, and T is the temperature of the blackbody radiator expressed in Kelvin. Equation (1.12) provides a crude means by which stellar temperature can be determined: $\lambda_{\max}$ is in principle provided in the observed spectrum of a star. This method, however, is not especially precise, since the maximum is typically broad and not well defined, and astronomers' invariably use measured characteristics of specific absorption lines (see later) to determine stellar temperatures.

The area under the blackbody curve for a specified temperature will correspond to the total amount of energy radiated over all wavelengths per square meter per second, which is the total energy flux F. This result is associated with another important law, the so-called Stefan-Boltzmann law, which indicates that

$$F = \sigma T^4 \tag{1.13}$$

where σ is the Stefan-Boltzmann constant. From Eq. (1.13) we obtain an immediately useful astronomical result. Given that stars are reasonably good approximations to blackbody radiators, and that stars are also to a good approximation spherical in shape, if we multiply the flux F (in units, recall, of energy per meter squared per second) by the star's surface area $4\pi R^2$, where R is the star's radius, we recover the luminosity L, and accordingly

$$L = 4\pi R^2 F = 4\pi R^2 \sigma T^4 \tag{1.14}$$

Equation (1.14) provides a means of finding the size of a star—that is, the radius once the luminosity and temperature have been deduced from its spectrum and parallax measure.

(⊛) **Exercise 1.12** Technically, Eq. (1.12) is obtained by setting the differential of Eq. (1.11), with respect to the wavelength, to zero. Likewise, Eq. (1.13) is the result of integrating Eq. (1.11), with respect to the wavelength, from 0 to infinity. See if you can verify these results. **Suggestion**: There are actually a number of mathematical tricks required to obtain the final formula, so find a step-by-step derivation (on the web or in a physics text) and work through the details.

(☺) **Exercise 1.13** The luminosity and radius of the Sun are measurable from the Earth. Given that the Sun has an observed angular diameter of 0.53 degrees, and, as indicated earlier, a measured energy flux at the Earth's orbit (the solar constant) of $F_\odot = 1362$ W/m^2, deduce the Sun's temperature.

At this stage, we have discussed methods by which the luminosity, temperature and size of a star can be deduced through observations. The next quantity of interest, and indeed, the quantity of greatest importance, is mass. The Sun's mass can be determined directly from Kepler's 3rd law, which gives:

$$\frac{P^2}{a^3} = \frac{4\pi^2}{G(m_1 + m_2)} \tag{1.15}$$

where P and a are the orbital period and semi-major axis of some specified planet (or asteroid or comet), and where (say) $m_1 = 1$ M$_\odot$ is the mass of the Sun, and m_2 is the mass of the planet (or asteroid or comet). In using Eq. (1.15) (such as in exercise 1.14), we make use of the fact that the Sun is very much more massive than any other object within the Solar System, and this gives us the approximation that $m_2 / m_1 \approx 0$. For stars within a binary system, however, this approximation no longer applies, and

Fig. 1.5 A schematic
binary system composed of
two stars of mass m_1 and m_2,
having orbital radii a_1 and a_2
and orbital velocities V_1 and
V_2 respectively. (Credit:
Author)

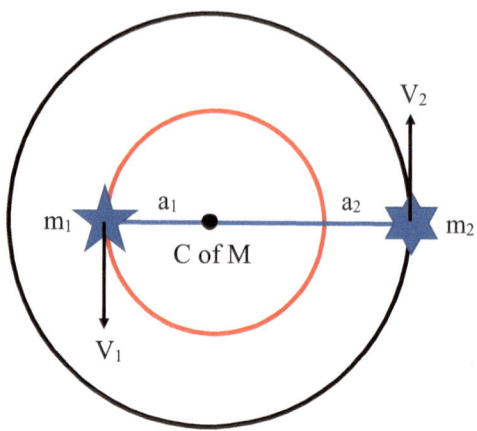

accordingly we need to find one more equation linking the specific mass terms to observable quantities. This can fortunately be done through the equality of the first moment of each star about the system's center of mass. For ease of argument, let us assume that we have a binary star system in which the stars move along circular orbits of radius a_1 and a_2 as shown in Fig. 1.5.

(☺) **Exercise 1.14** Halley's Comet has an orbital period of $P = 75.316$ years, an orbital semi-major axis $a = 17.83$ AU and a mass of 2.2×10^{14} kg. Use Eq. (1.15) to determine the mass of the Sun. **Hints**: Watch your units—recall that for Eq. (1.15) it is *SI* units that apply.

The equations that will enable us to find the individual masses of the stars within a binary system are Eq. (1.15)—Kepler's 3rd law—and the condition of first moment equality (or balance about the center of mass), which gives

$$a_1 m_1 = a_2 m_2 \tag{1.16}$$

To use Eq. (1.15), we need to specify the a term appropriately, and it transpires that $a = a_1 + a_2$ (see exercise 1.15). With Eqs. (1.15) and (1.16) in place, we have two equations that provide the sum and the ratio of the masses in terms of observable quantities—specifically a_1, a_2 and P. In order to determine stellar masses in kilograms, however, the radial terms a_1, a_2 must be expressed in units of meters. Likewise, the period of the system must also be expressed in seconds.

(☺) **Exercise 1.15** Since the two stars in a binary system have the same angular velocity, $\omega = 2\pi / P$, about the center of mass (that is, they have the same orbital period P), we have equality between their centrifugal accelerations: $m_1 \omega^2 a_1 = m_2 \omega^2 a_2$. The centrifugal acceleration of the stars can in turn be equated to the gravitational force acting between them: $F_{\text{grav}} = G m_1 m_2 / a^2$, where $a = a_1 + a_2$. Using Eq. (1.16) we also have $a_1 = a / (1 + m_1/m_2)$. So, setting $F_{\text{grav}} = m_1 \omega^2 a_1$, and substituting for a_1 and ω, show that Kepler's 3rd law—i.e., Eq. (1.15)—is the result.

There are many different kinds of observed binary systems. In some, the visual binaries, the stars are sufficiently far apart that both stars can be resolved in the telescopic eyepiece. The star α Centauri, although a single star to the human eye, is actually a binary system, readily resolved into two components α Cen A and α Cen B by even a modest-sized telescope. Observations of the system over many decades has revealed an orbital period of 79.91 years, along with a semi-major axis of 10.78 AU for α Cen A and 12.76 AU for α Cen B. This is enough information to determine the component masses. Firstly, from Eq. (1.16) we have

$$\frac{m_A}{m_B} = \frac{a_B}{a_A} = \frac{12.76}{10.78} = 1.18 \tag{1.17}$$

where the A and B subscripts refer to α Cen A and α Cen B respectively. Additionally, from Eq. (1.15) we have

$$m_A + m_B = \left(\frac{4\pi^2}{G}\right)\frac{(a_A + a_B)^3}{P^2} = 4.07 \times 10^{30} \tag{1.18}$$

Equations (1.17) and (1.18) allow for the determination of the individual masses, and we find that α Cen A has a mass of 2.20×10^{30} kg $= 1.11$ M$_\odot$ while α Cen B has a mass of 1.87×10^{30} kg $= 0.94$ M$_\odot$. Accordingly, we deduce that the masses of the stars in the α Centauri binary straddle that of the Sun by factors of about $\pm 0.1\%$—they are indeed solar analogs.

(☺) **Exercise 1.16**   Verify the mass determinations given for α Cen A and α Cen B.

The mass range deduced for the stars ranges from about 0.08 solar masses at the lower end of the spectrum to about 150 solar masses at the top. The closest star to the Sun at the present epoch is Proxima Centauri, and it is one of the lowest mass stars known, at 0.123 M$_\odot$. At the other end of the mass spectrum, the star η Carina, located some 2300 pc away, has an estimated mass of between 100 and 200 times that of the Sun. As we will see in Chap. 5, nature prefers to make low-mass stars, and indeed, the low-mass stars outnumber the high-mass stars in our galaxy by many thousands to one. The lower-mass limit for an object to be called a star is set according to the definition that a star is an object capable of initiating hydrogen fusion within its core (see Chap. 4), while the upper-mass limit is not well understood. In principle there are no specific limits to how massive a star might be when it first forms, although we shall look at some limiting mechanisms in Chap. 2. Additionally, as we will see in later chapters, the lifetime of a star is strongly influenced by its mass, with the most massive stars taking just a few million years to go from formation to supernova disruption. Massive stars are rare and short-lived and are therefore largely mysterious beasts.

In order to employ Eqs. (1.17) and (1.18) in the evaluation of stellar masses, some absolute measure of the orbital radii must first be made. In relatively well-spaced and well-observed systems, these quantities can be obtained from a measure of the

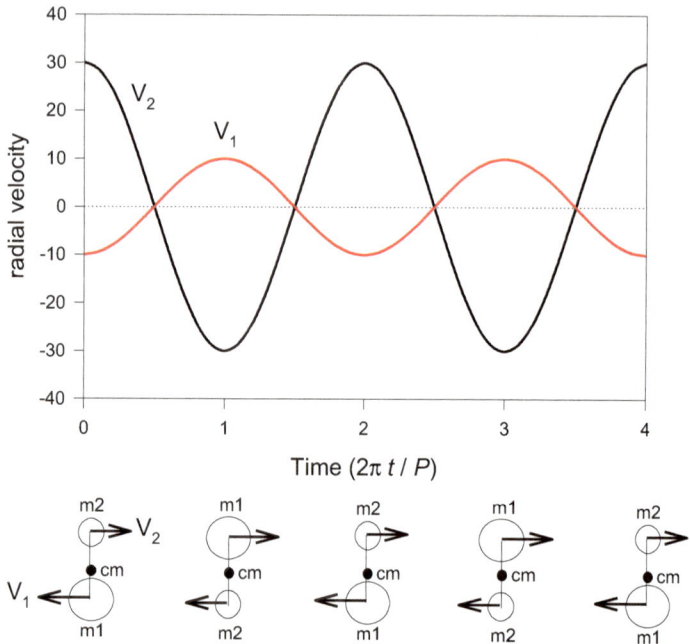

Fig. 1.6 The radial velocity variation for two stars within a binary system, with $m_1 > m_2$, $V_2 > V_1$. The orientation and direction towards or away from the observer (located at the far left) are illustrated below the x-axis. (Credit: Author)

system distance (via a parallax study) and the measure of the angular separation of the two components. For compact systems, which may not be resolvable as two stars in a telescope, the orbital radii can be deduced by monitoring changes in the orbital velocity. In this latter case, Doppler shift observations of selected absorption lines are used to determine the speed with which each star is moving about the system center of mass as a function of time. With reference to Fig. 1.6, so-called radial velocity measurements will give: $V_1 = 2\pi a_1 / P$ and $V_2 = 2\pi a_2 / P$, where P is the orbital period.

The Doppler shift velocity is determined by looking at the displacement of specific absorption lines away from their rest wavelength λ. Indeed, if a specific absorption line is observed at some wavelength λ_{obs}, then the line-of-sight (or radial) velocity V between the observer and the star is given as:

$$\frac{\lambda_{obs} - \lambda}{\lambda} = \frac{V}{c} \qquad (1.19)$$

where c is the speed of light. If the star is moving away from the observer, then $\lambda_{obs} > \lambda$ and the velocity will be positive. If the star is moving towards the observer, then $\lambda_{obs} < \lambda$ and the velocity will have a negative sign.

Fig. 1.7 Eclipse configuration and contact times. T_1 corresponds to first contact when the smaller star just begins to move behind the larger star in the observer's line of sight. Contact times T_2 and T_3 correspond to the beginning and end of the total eclipse phase, while the fourth (or last) contact T_4 indicates the time at which the eclipse ends, and the light from smaller star is once again fully revealed to the observer

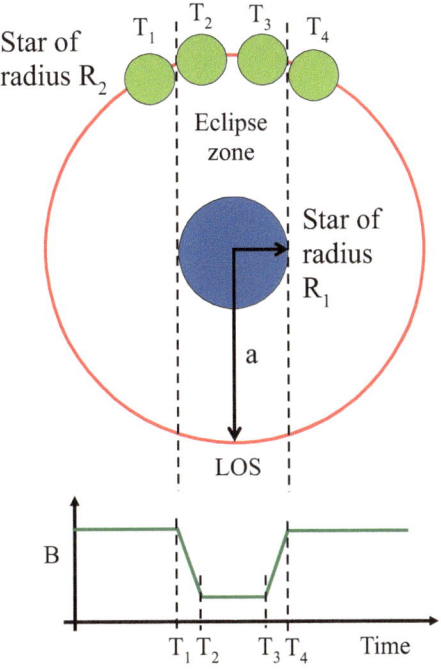

While the orientation of a binary system's orbital plane can in principle be at any angle to an observer's line of sight, some purely by chance will be aligned so that mutual eclipses can be seen. Such eclipses will occur when one star passes in front of the other (at half orbital period intervals) in the observer's line of sight. While rare, such eclipsing binary systems afford a direct means by which the radius of a star can be measured. To determine the radii, special attention needs to be focused on the timing of the brightness variations. Specifically, the time interval between the so-called first and fourth contact (T_4-T_1) as well the second and third contact (T_3-T_2) need to be determined. The contact conditions are illustrated in Fig. 1.7, where it is assumed that the smaller, less massive star is passing behind the larger star in the system and the observer's line of sight.

With reference to Fig. 1.7, in the time interval from T_1 to T_4 the smaller star (of radius R_2) moves a distance along its orbit corresponding to $D_{14} = 2R_1 + 2R_2$, where R_1 is the radius of the larger star. In the time interval T_2 to T_3, the smaller star moves a distance $D_{23} = 2R_1-2R_2$ along its orbit. We also know that the velocity with which the smaller star moves relative to the larger star is constant and given by $V = 2\pi a /P$, where $a = a_1 + a_2$ and P is the orbital period. Since the velocity is constant, we can now construct the following relationships:

$$V = \frac{D_{14}}{(T_4 - T_1)} = \frac{2(R_1 + R_2)}{(T_4 - T_1)} = \frac{2\pi a}{P} \tag{1.20}$$

and

$$V = \frac{D_{23}}{(T_3 - T_2)} = \frac{2(R_1 - R_2)}{(T_3 - T_2)} = \frac{2\pi a}{P} \tag{1.21}$$

From Eqs. (1.20) and (1.21), we obtain expressions for the sum and the difference of the radii in terms of measurable quantities. Accordingly, from the measured eclipse times T_1 through to T_4 and a measure of the system separation a, we have:

$$R_1 + R_2 = \frac{\pi a}{P}(T_4 - T_1) \tag{1.22}$$

$$R_1 - R_2 = \frac{\pi a}{P}(T_3 - T_2) \tag{1.23}$$

Once again, the system separation term can be determined from the known (parallax determined) distance d to the system and the maximum angular separation φ of the two stars, giving $a = 2\,d \tan(\varphi / 2)$. Alternatively, the separation a can be determined via radial velocity observations with: $a = a_1 + a_2 = (P / 2\pi)(V_1 + V_2)$.

An example of an eclipsing binary system that is seen almost exactly edge-on is that of CM Draconis. This system, located some 14.5 pc away in the direction of the constellation of Draco, is composed of two nearly identical M-dwarf stars (this spectral type designation will be discussed shortly) and has recently been studied in great detail by Juan Carlos Morales (*Institut d'Estudis Especials de Catalunya*, Spain) and colleagues. The observed system information gives us:

CM Draconis: Period: $P = 1.268$ days
                    $V_1 = 72.23$ km/s
                    $V_2 = 77.95$ km/s
                    First to fourth contact time: $(T_4 {-} T_1) = 4571$ s
                    Second to third contact time: $(T_3 {-} T_2) = 128$ s

From the radial velocity data we have $a_1 = PV_1/2\pi = 1,259,935.5$ km, and $a_2 = PV_2/2\pi = 1,359,711.6$ km, and accordingly, $a = a_1 + a_2 = 2,619,647.1$ km. Turning to Eqs. (1.22) and (1.23), the sum and difference of the component radii are: $R_1 + R_2 = 343{,}236.4$ km, and $R_1{-}R_2 = 9611.5$ km. Unraveling these two terms provides the individual radii as: $R_1 = 176{,}424.0$ km $= 0.253\ R_\odot$, and $R_2 = 166{,}812.5$ km $= 0.239\ R_\odot$. Both stars are indeed small compared to the Sun.

(☺) **Exercise 1.17**   Verify the radii determinations given for CM Draconis.

(☺) **Exercise 1.18**   Determine the mass of the two stars in CM Draconis.

(☺) **Exercise 1.19**   Given that the stars move along circular orbits, and working to a scale of 4 cm to 1 $R_\odot$, draw a scale diagram of the CM Draconis system. Also draw the Sun to scale (centered on the center of mass) in your diagram.

Figure 1.8 summarizes our present ability to gauge and annotate the fundamental (intrinsic) properties of the stars. Accordingly, we now have observational methods that can deliver to the theoretician actual numbers for stellar distances, luminosity,

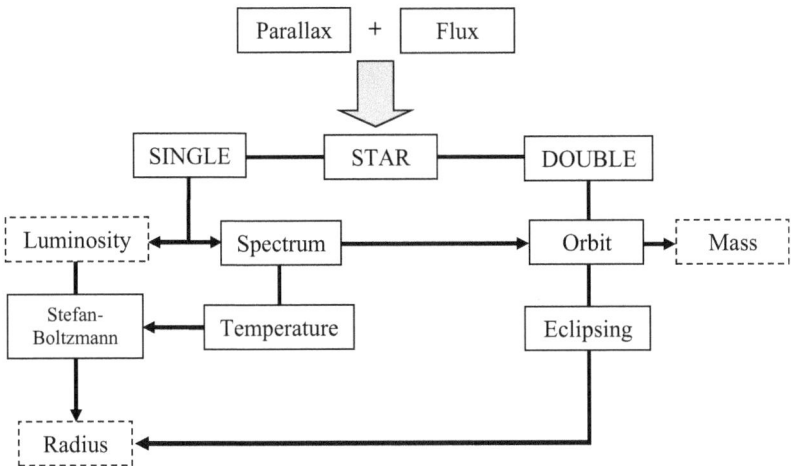

Fig. 1.8 Flowchart of relationships between what can be observed and quantified with respect to intrinsic stellar parameters (the latter being shown in boxes having dashed side lines). (Credit: Author)

approximate temperature, radii and mass. Our next step is to analyze and tabulate the data. First, a few words will be said on the stellar classification scheme, and then a brief outline will be given on how the composition of a star is determined.

1.3 The HR Diagram

Having developed the tools to determine the intrinsic properties of the stars—the mass, radius, luminosity, and temperature—the next key step is to arrange, collate and display the numbers for these terms in a meaningful fashion. This step enables the development of calibration schemes, which place stars with similar characteristics into consistent and (hopefully) well-defined groups. These data tables enable not just calibration but also the search for correlations. To the theoretician, correlations are of vital importance and inspiration. If two or more quantities of a specific subgroup of stars are correlated, say by some power law relationship, then the immediate question arises as to why are they correlated and what does the correlation tells us about the internal structure of the stars. The HR diagram was one of the first diagrams to indicate the existence of distinct correlations between a star's mass, radius, luminosity and temperature, and it also revealed the existence of various distinct groups of stars.

The HR diagram has its origins set in the second decade of the twentieth century. The diagram was a natural outcome of the astronomy of its time, and sooner or later somebody would have arranged the data display. History tells us that the diagram, which is a plot of a star's luminosity against its surface temperature, was first

Fig. 1.9 HR diagram from the GAIA spacecraft data release of 2018. Various axis scales are displayed. The luminosity is given on the right-hand axis, while on the left-hand side the scale is an equivalent absolute magnitude scale. The lower x-axis shows a stellar temperature measure based upon flux measurements through two specified wavelength region transmission filters. The equivalent temperature scale is shown at the top of the diagram along with the corresponding spectral type. (Credit: European Space Agency)

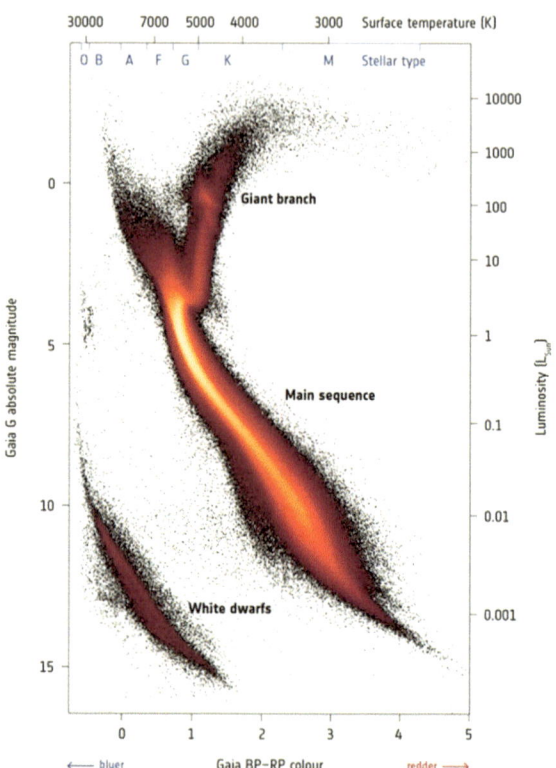

→ **GAIA'S HERTZSPRUNG-RUSSELL DIAGRAM**

constructed by Eijnar Hertzsprung (circa 1910) and independently by Henry Norris Russell (in 1914)—their respective surnames giving us the HR moniker. The HR diagram is full of treasures, but it took most of the twentieth century to unravel the various deeper messages that it encodes.

Figure 1.9 shows the HR diagram based upon data derived by the GAIA spacecraft. One of the first features to note about the HR diagram is that it is not a scatter plot, with the stars falling in three main regions. The vast majority of stars (about 92%) plot along a diagonal that runs from high-temperature, high-luminosity stars (the upper left-hand corner) to low-luminosity, low-temperature stars (lower right-hand corner) in the diagram. The diagonal feature is called the main sequence (MS). About 7% of the stars lie below the main sequence in the lower left-hand corner of the diagram, in what is known as the white dwarf region. These stars are small, hot and of low luminosity. About 1% of the stars in the diagram plot toward the upper right-hand corner. These are the red giants that are large, low-temperature, high-luminosity stars. Many questions immediately result from these descriptive results—why are stars grouped into the specific regions identified; why are the vast majority of stars found on the main sequence; what, if any, are the linkages between the various stellar domains; and what are the specific correlations between stellar characteristics?

The main sequence can be reasonably well fit by two simple power laws in the luminosity and temperature, with $L \sim T^{5.5}$ for $M < 1.5$ $M_\odot$ and $L \sim T^{8.5}$ for $M > 1.5$ $M_\odot$. This variation in the slope of the main sequence relates to a changeover in the fusion sequence by which hydrogen is converted into helium within stellar cores (see Chap. 4). Since the HR diagram is a plot of luminosity against temperature, the Stefan-Boltzmann law (Eq. 1.14) dictates that the loci of constant radii will plot as diagonal lines in the diagram, with

$$\mathrm{Log}\,L = 2\,\mathrm{Log}\,R + 4\log T + \text{constant}$$

The smallest sized stars will fall in the lower left-hand corner of the diagram, in the white dwarf region. The largest stars are located towards the upper right-hand corner of the diagram, in the red giant region. Since white dwarfs do not change in size as they evolve, the luminosity-temperature relationship for these stars follows directly from the Stefan-Boltzmann law, with $L \sim T^4$.

While the dwarf and giant labels are relatively self explanatory and separate out stars according to their physical size, the red and white labels separate out stars according to their temperature. This color distinction is a direct consequence of Wien's law (Eq. 1.12), which places λ_{max} towards the red and infrared end of the electromagnetic spectrum in the case of low-temperature stars, and λ_{max} towards the blue and ultraviolet end of the electromagnetic spectrum in the case of high-temperature stars. Accordingly, red giants and red dwarfs are distinguished according to size but are similar in temperature. Red dwarfs and white dwarfs are distinguished according to their temperature but have similar sizes.

The variation of other star characteristics are not so easily derived and/or seen in the HR diagram, but on the main sequence the red dwarfs (low temperature and small sized, located towards the lower right-hand corner) are the lowest mass stars, while the most luminous and hottest stars (in the upper left-hand corner) are the most massive. As will be discussed below, this latter observation is expressed through the luminosity-radius-mass relationship. The red giants and the white dwarfs have different mass, radius and temperature relationships than those found for main sequence stars. The reason for these relationships and why they are different for main sequence, red giant and white dwarf stars will be explored in later chapters— suffice to say here that they relate to age, mass and mode of energy generation.

In terms of basic numbers, we have already stated that if we pick a star at random from any location in the Milky Way galaxy, the odds are over 90% in favor of it being a main sequence star. In terms of spectral type across the main sequence, the vast majority of stars are located in the red dwarf region, corresponding to stars of low mass, low temperature, small size and spectral type M. Little will be said about the spectral classification scheme, other than that it is a temperature-based sequence that is centered upon the presence and measured characteristics of various absorption lines in the spectra of stars. Table 1.1 provides a brief overview of the spectral classification scheme. The final column in Table 1.1 shows the number per cubic

Table 1.1. Elementary properties of the spectral classification scheme for main sequence stars. Each of the spectral types (column 1) is divided into a set of subgroups running from 0 to 9. The spectral type changes according to temperature (column 2) and is determined according to the strengths of specific absorption lines (column 3). Column 4 gives a characteristic main sequence mass for each of the spectral types, and column 5 indicates the typical number of the specified spectral type stars (over all subgroups) per unit volume of space

Sp Type	T(K)	Spectrum	$M/M_\odot$	N / pc^3
O5	38,000	He II lines	40	2×10^{-8}
B0	30,000	He I + weak H	20	
B5	16,500	$S_i(\lambda = 4128) > He(\lambda = 4121)$	7	1×10^{-4}
A0	10,800	Balmer lines at maximum	3	
A5	8620	K line > Hδ	2	5×10^{-4}
F0	7240	Fe I lines	1.7	
F5	6540	Ca II ~ H	1.3	3×10^{-3}
G0	5920	Fe II > Fe I	1.1	
G2	5780	Sun like spectrum	1.0	
G5	5610	Strongest Fe II line	0.9	6×10^{-3}
K0	5240	H & K lines at maximum	0.8	
K5	4410	Strongest Ca II	0.7	1×10^{-2}
M0	3920	Fe I ~ Ca I	0.5	
M5	3120	Strong TiO lines	0.2	5×10^{-2}

parsec of main sequence stars arranged according to spectral type. This last column clearly indicates that nature prefers to make small, low-mass M spectral type stars, and the statistics indicate that for every massive, high-temperature, high-luminosity O spectral type star, there are something like one million M spectral type red dwarf stars.

In addition to relating a (temperature-specific) spectral type to a star according to measured features within its spectrum, astronomers also ascribe a spectrum-based measure relating to the star's luminosity. The luminosity class is set according to a Roman numeral between I and V. Rather than being based upon line strength ratios, as in the case of spectral types, the luminosity class it is based upon the line widths. The width characteristic being measured relates to what is called pressure broadening: the more compact and the higher the atmospheric density, the broader a spectral line will appear for a given temperature. In contrast, the more extended and the lower the density of a star's atmosphere, the sharper its spectral lines will appear. The luminosity class is constructed so as to pick out higher luminosity, larger radii stars—the giants. The spectral type and luminosity class uniquely defines the location of a star in the HR diagram, with the Sun for example being a G2 V star. The loci of constant luminosity class in the HR diagram are shown schematically in Fig. 1.10.

The three distinct regions identified in the HR diagram (the main sequence, the red giant region and the white dwarf region) are delineated by stars in different evolutionary stages (or ages), with the white dwarfs being the most evolved objects. After the end of the star formation phases (see Chap. 2), a star settles into its main sequence phase, generating energy via hydrogen fusion reactions (Chap. 4). The red

Fig. 1.10 Schematic loci of constant luminosity class for stars in the HR diagram. The main sequence corresponds to luminosity class V, while the most massive, most luminous, hypergiant stars correspond to luminosity class Ia and Ib. The dashed vertical line indicates the location of the G2V (Sun-like stars), G2IV, G2III, etc. stars in the HR diagram. This sequence of stars has the same surface temperature but increasing larger luminosity and radius. (Credit: Author)

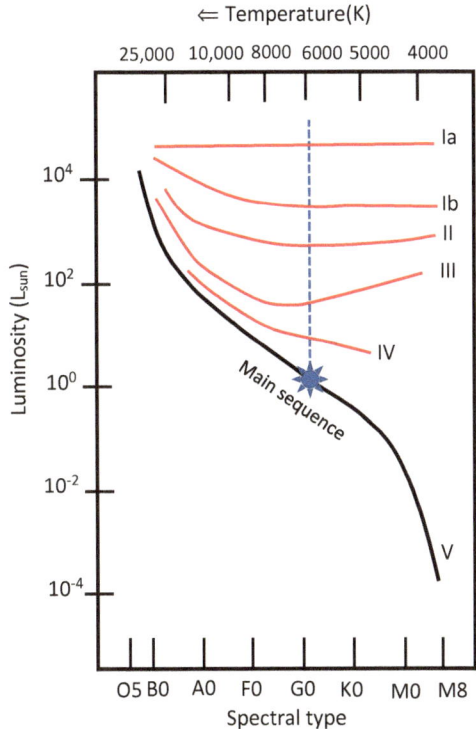

giant region is delineated by those stars that have exhausted hydrogen in their inner cores and have commenced steady fusion reactions converting helium into carbon. Once stars like the Sun (see Chap. 5 for a more general discussion) have exhausted helium within their cores, their stellar race is run, and they evolve into the white dwarf region. Technically, white dwarfs no longer qualify as being stars since they are no longer generating energy within their interiors through fusion reactions (nor by gravitational collapse either) and are on a cooling sequence, ultimately fated to become, after many billions of years, black dwarfs of zero luminosity and near absolute zero temperature. The relative number of stars within the various regions of the HR diagram is explained in terms of evolutionary timescales. In Chap. 4 the nuclear timescale will be defined, and it will be shown that the longest lived nuclear fusion phase is that corresponding to hydrogen fusion, accounting for why most stars are located on the main sequence. Stars converting helium into carbon reside in the red giant region of the HR diagram, but the helium fusion timescale is relatively short, which explains why few stars are captured in this phase and why relatively few stars are located in the red giant region. Once a star evolves into the white dwarf region, it has entered a very slow, long-lived cooling-off phase, and accordingly this stellar graveyard region acts as a kind of bottle neck, with new white dwarfs continually entering the high-temperature, high-luminosity end of the zone, with very old low-temperature, low-luminosity white dwarfs slowly dropping-out of the HR diagram as black dwarfs at the other.

1.4   The Luminosity-Radius-Mass Relationship

Like the HR diagram, the luminosity-mass and radius-mass diagrams reveal many important insights. The mass-luminosity (ML) relationship was first constructed circa 1915, at about the same time as the HR diagram first appeared. Again, it was a product of the astronomy of that time—a time when good parallax measurements were becoming available and the data relating to the mass, luminosity and radii of stars located within eclipsing binary systems was first appearing. The diagrams once plotted out were as revolutionary to the understanding of the stars as the HR diagram. The ML and MR diagrams for main sequence stars are shown in Figs. 1.11a and 1.11b. With respect to the ML diagram, the remarkable feature is how well defined the relationship is, and while it is not a line of constant slope, it is a line of near-constant slope in specific mass domains. The ML relationship over various mass domains is as follows:

$$L \sim M^{2.3}, \qquad \text{for } M < 0.5\,M_{\odot} \tag{1.24a}$$

$$L \sim M^{4.0}, \qquad \text{for } 0.5 < M/M_{\odot} < 2 \tag{1.24b}$$

$$L \sim M^{3.5}, \qquad \text{for } 2 < M/M_{\odot} < 50 \tag{1.24c}$$

$$L \sim M, \qquad \text{for } M > 50\ M_{\odot} \tag{1.24d}$$

The reasons for the slope changes in the ML diagram will be explored in Chap. 4, but suffice to say here that they relate to the mode of energy transport within a star (convective versus radiative), the importance of radiation pressure and the manner in which electromagnetic radiation interacts with the stellar material (the opacity).

Fig. 1.11a The mass-luminosity relationship for main sequence stars. The power-law relationship is described by Eqs. (1.24a, 1.24b, 1.24c and 1.24d). (Credit: Author)

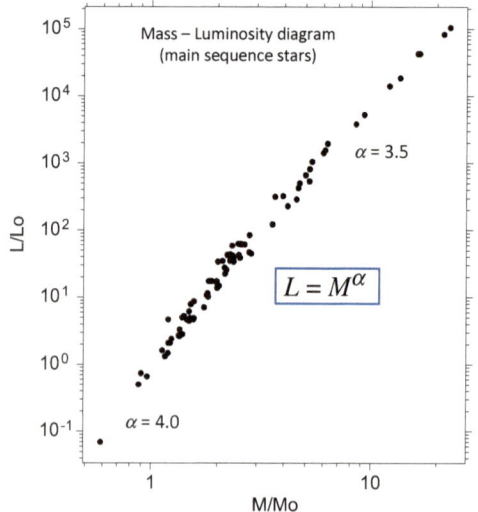

Fig. 1.11b The mass-radius relationship for man sequence stars. The vertical spread in radii for a given mass is a consequence of increasing main sequence age. The power-law relationship is described by Eqs. (1.25a and 1.25b). (Credit: Author)

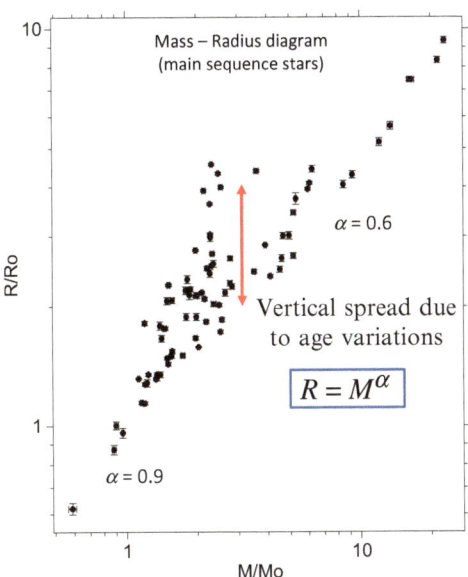

The radius-mass relationship shows a relatively large spread in radius for a given mass; this, as we shall see later, is an age effect that sees the radius essentially double in size over a star's main sequence lifetime. The locus that follows the minimum radius with respect to mass is composed of two essentially linear relationships, with the slope of the lines changing at about a mass of 1.5 $M_\odot$. The relationships are

$$R \sim M^{0.9}, \qquad \text{for } M < 1.5 M_\odot \tag{1.25a}$$

$$R \sim M^{0.6}, \qquad \text{for } M > 1.5 M_\odot \tag{1.25b}$$

The change in the slope of the MR locus is a result of the changeover in the mode by which hydrogen is converted into helium (via CNO cycle at higher masses and PP chain at low masses; see Chap. 4). Also, stars with masses much smaller than about 1.5 Msun begin to develop extensive convective envelopes, and since energy transport by convection is very efficient, such stars tend to be smaller in size.

1.5   Star Clusters and Stellar Populations

Two kinds of star clusters are generally recognized, and they are distinguished primarily according to their dynamics, composition and age. The galactic clusters, also called open clusters, are typically composed of relatively young, newly formed stars. Such clusters are found within the disk and spiral arms of the Milky Way galaxy and have near-circular orbits about the galactic center.

In contrast to the galactic clusters are the globular clusters. These structures are composed of the oldest, first-formed stars within the galaxy. Galactic clusters contain perhaps a few hundred stars spread over a region perhaps a few parsecs in scale. Globular clusters contain perhaps a few hundred thousand stars distributed within a spherical volume of space perhaps a few tens of parsecs across. Additionally, globular clusters are composed of low-mass, low-metallicity stars and have elongated, elliptical orbits that can carry them well away from the galactic plane. 'Metallicity' is the term used to describe the abundance of elements other than hydrogen and helium within a star. The very first stars to form in the universe would have been entirely made of hydrogen and helium, but as successive generations of stars were born, evolved and proceeded through to their end phases, so the metallicity of the interstellar medium increased in an incremental manner. Metallicity therefore is a measure of the time at which a star formed in galactic history. Stars forming now in the galaxy will be the most metal rich, and these will be the massive O and B spectral type stars that delineate the spiral arms.

Astronomers divide the stars found within the galaxy into so-called Populations, with the Populations being distinguished according to dynamics, composition and galactic location. Population I are the youngest, more recently formed, highest metallicity stars. They are located within the galactic disk and move along circular orbits within the galactic plane and about the galactic center. Population II objects correspond to the oldest stars with low metallicity, and orbits that can carry them to regions well above and below the galactic plane. Galactic clusters, being relatively new structures within the galaxy, are composed of Population I stars. Globular clusters in contrast are composed of the oldest and least chemically enhanced Population II stars. The Population I and II division is really a sliding scale that maps out the history of star formation and star death within the galaxy, and it describes the gradual increase in the metallicity of the interstellar medium over time. Table 1.2 and Fig.1.12 illustrate the basic features that are inherent to the Population I and II stars and their galactic domains.

Table 1.2. Characteristics of various stellar Populations. The metallicity is expressed through the Z term in column 3. If X and Y correspond to the mass fractions of hydrogen and helium within a star (see Chap. 3), then the heavy element (metallicty) mass fraction is $Z = 1—(X + Y)$. For the Sun, $X = 0.71$, $Y = 0.271$ and $Z = 0.019$. The last column indicates the characteristic height, above and below the galactic plane, over which the various Populations are found. The extreme Pop I stars have a patchy distribution and are found within very young galactic clusters and delineate the Milky Way's spiral arms. Older Population I, thin disk stars also have a patch distribution, while the (thick) disk Population II stars have a relatively smooth, space-filling, and uniform distribution. The characteristic ages associated with the various Population types is given in columns 4

Population	Orbit	Metallicity	Age (Gyr)	<h(pc)>
Extreme Pop I	Spiral arms / circular	$Z > Z_\odot$	< 0.1	120
Older Pop I	Thin disk / circular	$Z \approx Z_\odot$	0.1—10	160
Disk Pop II	Thick disk / circular	$Z \approx 0.1 Z_\odot$	3—10	400
Halo Pop II	Eccentric / inclined	$Z < 0.01 Z_\odot$	> 10	2000

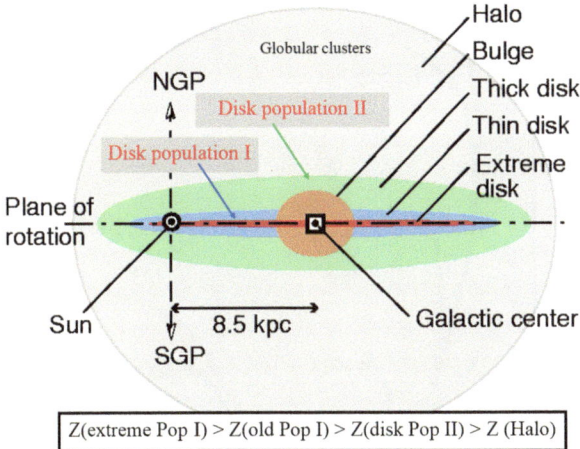

Fig. 1.12 Schematic edge-on view of the Milky Way galaxy showing its basic disk and bulge structures. The Sun is located approximately 8500 parsecs away from the galactic center. The globular clusters move about the galactic center along eccentric and inclined orbits that can carry them deep into the galactic halo region. The spherical galactic bulge, which surrounds the galactic center, is composed of a mixture of Population I and II stars. The distance to the galactic center is not well constrained, and various recent estimates place the Solar System at between 7.5 and 8.5 kpc from the galactic center. (Credit: Author)

(☺) **Exercise 1.20** Assuming a galaxy to be disk-shaped with radius of 15 kpc and height h (pc): (1) Determine the mass of stars in the extreme Pop I disk, assuming that this region is populated by O and B spectral type stars with characteristic mass of 10 $M_\odot$. (2) Calculate the mass of the Pop II thick disk, assuming it is populated by M spectral type stars with a typical mass of 0.5 $M_\odot$. (3) Given that the characteristic luminosity of O and B spectral type stars is 2500 $L_\odot$, and that of M spectral type stars is 0.01 $L_\odot$, in which disk structure does most of the galaxy mass reside, and which disk structure produces the most luminosity? **Hints**: The volume V of a disk of radius r and height H is $V = \pi r^2 H$. Take the disk height from Table 1.2 (note $H = 2 < h >$) and use Table 1.1 to find the approximate number of stars of the given spectral type per cubic parsecs.

The evolutionary stellar models are constrained and tested by their ability to explain cluster HR diagrams. The important characteristic of any cluster as far as stellar evolutionary models are concerned is that all the constituent stars must have formed from the same interstellar cloud, at essentially the same time and with the same composition. In this respect, it is stellar mass alone (and how stars of different mass evolve at different rates) that must account for the various observed features. One important feature in the HR diagram of a cluster is the so-called turnoff point, since it can be used to find the cluster formation age (Fig. 1.13). The turnoff point is delineated by those stars that are about to transition from the main sequence to the red giant region. This phase is marked by a changeover in the central fusion reactions that power a star (from hydrogen fusion on the main sequence to helium fusion

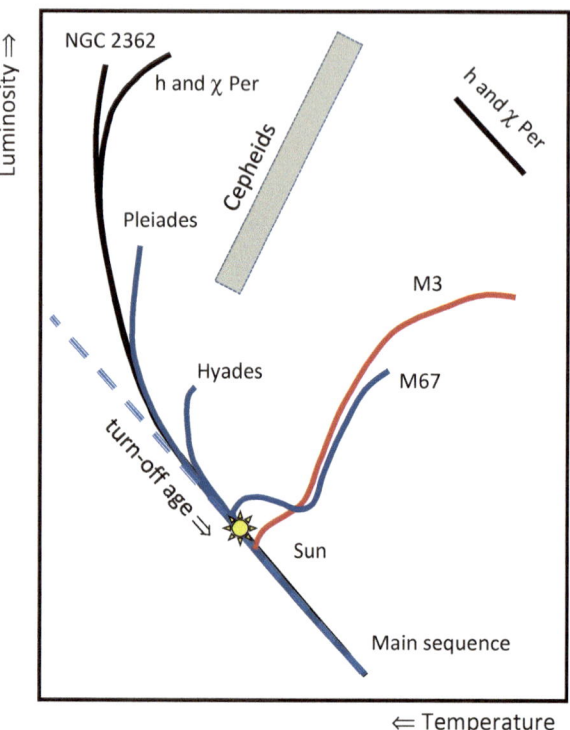

Fig. 1.13 Schematic HR diagram for several galactic and one globular cluster (M3). The estimated ages for the various clusters are: NGC 2362 ~ 5 million years; h and χ Perseus (the Double cluster) ~ 13 million years; the Pleiades ~ 75–150 million years; the Hyades ~ 625 million years; M 67 ~ five billion year; (globular cluster) Messier 3 ~ 11.4 billion years. The dashed line shows the extended main sequence, and the cluster age is gauged by where the turnoff point occurs. The location of the Cepheid variable (see Sect. 1.6) strip is shown in the upper center of the diagram. The Sun, having an age of 4.5 billion years, is shown in its main sequence position

reactions in the red giant region; see Chap. 5). Not only do the properties of a star (its radius, luminosity and temperature) change dramatically during this transition phase, but they do so rapidly on a Kelvin-Helmholtz timescale rather than that of the much slower nuclear timescale. Specifically, the key point about fitting the turnoff location is that the more massive a star is, the more rapidly it evolves away from the main sequence (as will be shown in Chap. 4). As stellar models give a mass dependent main sequence lifetime (again see Chap. 4), so the stars at the turnoff point correspond to the oldest main sequence stars for the age of the cluster.

Table 1.3 provides a set of approximate ages for clusters having various turnoff point luminosity (mass and spectral type) characteristics. Given that the universe is estimated to be some 13.7 billion years old, it is clear that no cluster is going to be found with a turnoff point characterized by a star of spectral G5 or cooler.

Table 1.3 Characteristic cluster ages based upon the spectral type (luminosity) of the turnoff point. The mass data is taken from Table 1.1

Sp type @ turnoff	Mass ($M_\odot$)	Luminosity ($L_\odot$)	Cluster age
O5	40	500,000	10^6
B0	18	20,000	6×10^6
B5	7	800	8×10^7
A0	3	80	6×10^8
A5	2	20	2×10^9
F0	1.7	6	3×10^9
F5	1.3	2.5	5×10^9
G0	1.1	1.3	8×10^9
G5	0.9	0.8	13×10^9
K5	0.7	0.2	24×10^9
M5	0.2	0.01	559×10^9

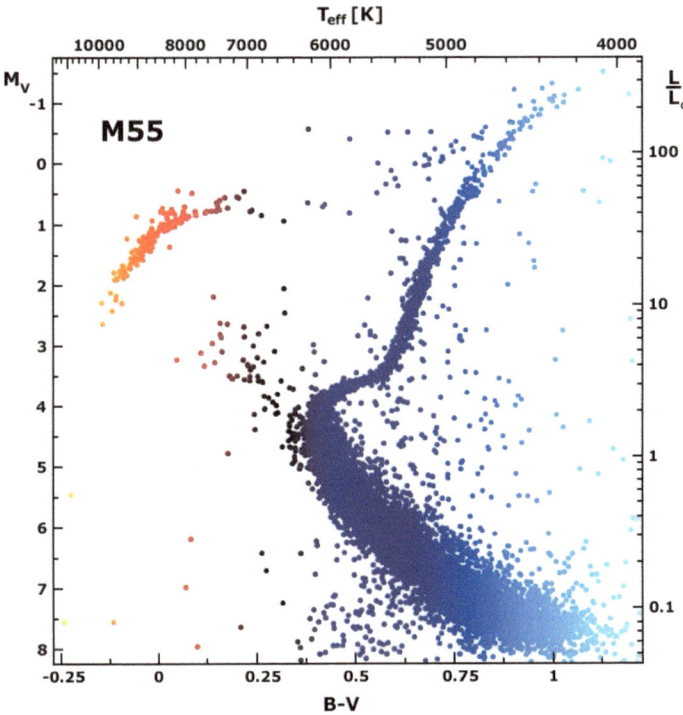

Fig. 1.14 The HR diagram for the globular cluster M 55. Image courtesy of APOD/NASA

(☺) **Exercise 1.21** The HR diagram for the globular cluster M55 is shown in Fig. 1.14. Estimate the temperature and luminosity of those stars at the turnoff point. Use the data in Table 1.1 to estimate the spectral type corresponding to the turnoff temperature, and then use the data in Table 1.3 to estimate the age of the cluster.

1.6   The Period-Luminosity Law

That stars can vary in brightness has been known since at least 1638, when Johannes Holwards first recorded variations in the brightness of the red giant star Mira (Omicron Ceti). This particular star shows an 11-month cyclical change in brightness. The variations in the brightness of stars in general can be periodic, aperiodic and one-off, as in the case of supernovae (see the stellar taxonomy below). And, the variations can be large or barely detectable—at some level, all stars vary in their brightness. The total solar irradiance (that is, the amount of solar energy incident upon the Earth's upper atmosphere) for example shows about a 0.1% variation with an approximately 11-year periodicity—a period that coincides and is in synchronization with the sunspot cycle, which in turn is driven by a 22-year magnetic field polarity cycle. The Sun also undergoes small amplitude oscillations with a period of about 5 min. These oscillations are a consequence of sound waves trapped within the Sun's outer envelope.

Indeed, there are a whole host of variable star prototypes, and several are identified in Fig. 1.15. One of the key features revealed in Fig. 1.15 is the instability strip that runs from the lower left in the HR diagram to the upper right (see also Fig. 1.13). The stars that fall within the boundaries of this zone have just the right conditions that allow for the development of a pulsation instability. Described according to the kappa mechanism (kappa signifying opacity; to be discussed in

Fig. 1.15 A schematic HR diagram showing the location of the instability strip and the various domains of variable star prototypes (see also Fig. 1.13). (Credit: Author)

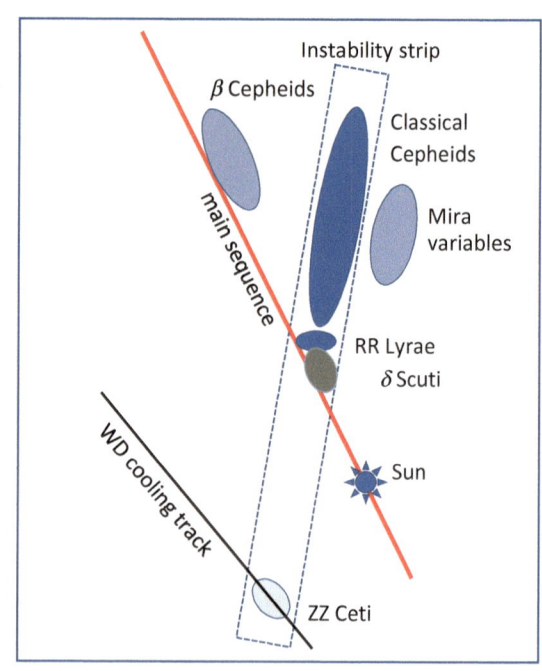

Chap. 4) the Cepheid's are pulsation variables, with the brightness variations being accompanied by a periodic radial expansion and contraction. The pulsation of Mira variables is driven by the existence of a partial ionization zone of hydrogen situated in the near-surface stellar envelope; the pulsation of δ Cepheid variables is due to a partial helium ionization zone, while the pulsation of β Cepheid variables is driven by a partial iron atom ionization zone. Key to explaining the limited width of the instability strip is the physical location of the partial ionization zones. If the zones are too deep, any pulsations tend to be damped out by convective motion; if they are too close to the surface, there is not enough envelope mass to drive any significant motion.

American astronomer Harlow Shapley first drew attention to the idea that the Cepheid variables might be undergoing radial pulsation in 1914, and Arthur Eddington argued shortly thereafter that the pulsation variation could be explained in terms of a heat-engine analogy with the star rhythmically breathing in and then out again. Eddington's model envisions the expansion and contracting being caused by the generation of resonant sound waves within the interior of a star, and accordingly the pulsation period will be of order $P = 2\,R\,/\,V_s$, where R is the average stellar radius and V_s is the speed of sound in the interior. We shall see later in Chap. 3 that the period of oscillation relates to the inverse square root of a stars characteristic density—what this means for us here, however, is that the larger the characteristic density of a star so the more rapidly can it pulsate. The pulsation period P is expressed according to the pulsation constant Q and the average density, where

$$P = Q \sqrt{\overline{\rho}_{sun}/\overline{\rho}} \qquad (1.26)$$

Detailed numerical models indicate that $Q \sim 0.06$. The two important ionization zones that drive the radial pulsations within a Cepheid variable are those associated with hydrogen and helium. In the temperature range 1—2×10^4 K, both hydrogen and helium will be partially ionized, with

$$H\,I \Leftrightarrow H\,II + e^-$$

$$He\,I \Leftrightarrow He\,II + e^-$$

Here the ionization state is expressed via a Roman numeral, with I indicating a neutral state, II indicating a singly ionized state, III indicating doubly ionized state and so on. Deeper within a star at a temperature $\sim 4 \times 10^4$ K, helium will become fully (that is doubly) ionized, with

$$He\,II \Leftrightarrow He\,III + e^-$$

The development of radial pulsations depends critically on where these partial ionization zones are located. For stars with surface temperatures greater than about 7500 K, the zones are located too close to the surface to drive oscillations. For surface temperatures less than about 5500 K, the partial ionization zones are located too deep within the star, and these two surface temperature limits effectively define the width of the instability strip on the main sequence—that is, it crosses the main sequence between spectral types B and F. Partial ionization zones are relevant in driving radial pulsations in that they rob energy from the envelope gas during a compression cycle by undergoing further ionization, and this changes the ability of radiation to move through the zone (a topic to be further discussed in Chap. 4). This is the opacity effect. Importantly for the pulsation cycle, the opacity within the partial ionization zones must increase with increasing temperature; by slowing the rate of energy flow through the compressed zone, that region becomes hotter. The pulsation process begins when a region in the outer layers of a star looses pressure support and starts, under gravity, to fall inwards. The following sequence of events then takes place:

1. A layer in the outer envelope of the star loses its pressure support and begins to fall inward.
2. The inward motion of the falling layer compresses its underlying layers, causing those layers to become more opaque to radiation, allowing them to heat up and become more ionized. The greater the ionization in these layers, the higher the opacity and the harder it is for radiation to flow through the region.
3. Eventually, the rising temperature in the lower, more compressed layer causes the outward pressure to increase, pushing it outward. As this layer moves outward it cools, with a concomitant reduction in the ionization and opacity, thereby allowing radiation to move more freely through the region.
4. The expanded region, now more transparent to radiation, cools down, looses its pressure support and begins to fall inwards again.

From (4), the cycle returns to (1), and the breathing, expanding and contracting cycle begins to repeat itself. The star has become a periodic variable. There are many additional details to the pulsation cycle not described here, but it transpires that a Cepheid variable is at its brightest and hottest after passing through its minimum radius phase. The subtleties lie in the fact that during the pulsation cycle, the locations of the partial ionization zones change with respect to both the linear radius of the star and the mass fraction of material above and below them. The Cepheid is at its brightest when the least amount of mass exists above the hydrogen ionization zone. Usefully, however, the observations reveal that Cepheid variables follow a cyclical loop in the luminosity versus temperature diagram (Fig. 1.16), and this can be used to determine the characteristic size variations of such stars. To extract this latter information about a Cepheid, it is necessary to determine the radius of the star during the pulsation cycle $R(t)$. This can be done by looking at the rate of change of the radius and by making a series of Doppler shift observations to determine the expansion velocity (recall Eq. 1.19). Accordingly, we have

$$\frac{dR}{dt} = V(t) \tag{1.27}$$

Fig. 1.16 The luminosity-temperature loop for a Cepheid variable during one pulsation cycle. The star will attain its smallest size when the luminosity and temperature attain their greatest values. (Credit: Author)

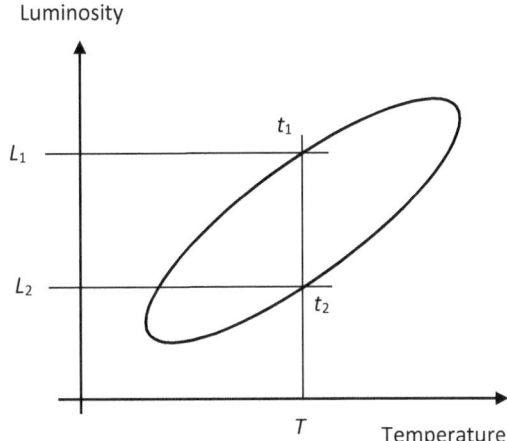

where $V(t)$ is the star's expansion velocity at time t (as deduced from Doppler measurements). Also,

$$R(t_1) - R(t_2) = \int_{t_1}^{t_2} V(t)\,dt \tag{1.28}$$

Now, if the times t_1 and t_2 are chosen such that the star has some specific temperature T (see Fig. 1.16), then from the Stefan-Boltzmann law (Eq. 1.14) we have

$$\frac{L(t_1)}{L(t_2)} = \frac{R^2(t_1)}{R^2(t_2)} \tag{1.29}$$

With Eqs. (1.28) and (1.29), we have enough information to determine the radius of the Cepheid at times t_1 and t_2. Changing the specified reference temperature T will give the star's radius at another set of times, and accordingly the full variation of $R(t)$ can be mapped out.

(☺) **Exercise 1.22** A Cepheid variable is observed at two times t_1 and t_2 in its cycle, where its effective temperature is the same. The ratio of the Cepheid's luminosity at these two times is $L(t_1)/L(t_2) = 1.166$. Between the two observation times the radial velocity is found to be approximately constant with a value of 15 km/s towards the observer. Given that the time interval between observations is $(t_2 - t_1) = 1.073$ days, determine the radius of the Cepheid at times t_1 and t_2.

One of the pivotal events in the history of astronomy was the 1908 discovery by Henrietta Swan Leavitt of the period-luminosity relationship associated Cepheid variables—the first such star of this group being δ Cephei, which was noted as early

Fig. 1.17 The period-luminosity relationships for classical Cepheids (Type I), W Virginis (Type II) variables and RR Lyrae stars. (Credit: Australia Telescope National Facility)

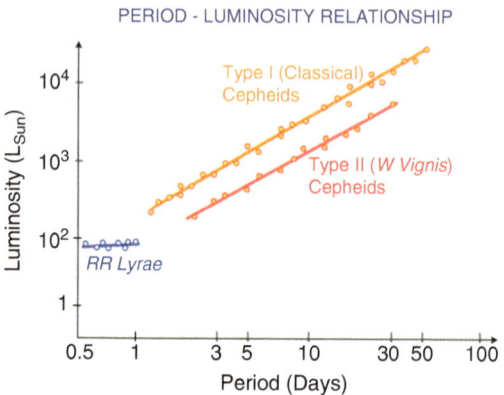

as the mid-1780s as showing a near-5.4-day variation in its brightness. Leavitt's discovery has been of immense importance with respect to the determination of the scale of the universe. Importantly, Cepheids have a highly distinctive light curve profile: a rapid rise to maximum brightness followed by a relatively slow decay to minimum. Additionally, Cepheids are inherently bright, which means they can be detected at great distances, and even more importantly, they have a pulsation period that correlates with their luminosity—the longer the period, the more luminous the star. In this manner, the identification of a Cepheid variable in a distant galaxy can be used to determine the distance to the galaxy.

Observations of a Cepheid's light curve are used to determine the pulsation period; the calibrated period-luminosity relationship then provides a value for the Cepheid's luminosity. This luminosity combined with an average of the measured energy flux can then be used to determine the distance (recall Eq. 1.10) to the Cepheid and its host galaxy. Figure 1.17 shows a series of calibrated period-luminosity curves for the classics Cepheids, the W Virginis variables and the RR Lyrae stars. Classical Cepheids (also called Type I Cepheids) are relatively young (Population I objects) with masses between 5–20 times that of the Sun and pulsation periods that can vary from a few days to of order 100 days. The characteristic relationship between the luminosity L (expressed in solar units) and the pulsation period P (expressed in days) for classical Cepheid variables is:

$$\text{Log}\,(L) = 2.47 + 1.15\,\text{Log}\,P(\text{days}) \tag{1.30}$$

The W Virginis (or Type II Cepheids) are older Population II stars with pulsation periods typically falling between 1 and 50 days. The Type II Cepheids have a lower metallicity than their Type I companions, and they characteristically have much lower masses (being no more massive than the Sun) and lower luminosities for a given pulsational period—hence why the two lines shown in Fig. 1.17 are essentially parallel, but that for the W Virginis (Type II Cepheids) is displaced downward to lower luminosity. The RR lyrae stars are commonly found in globular clusters and are accordingly old, low (sub-solar) mass stars. The characteristic feature of RR

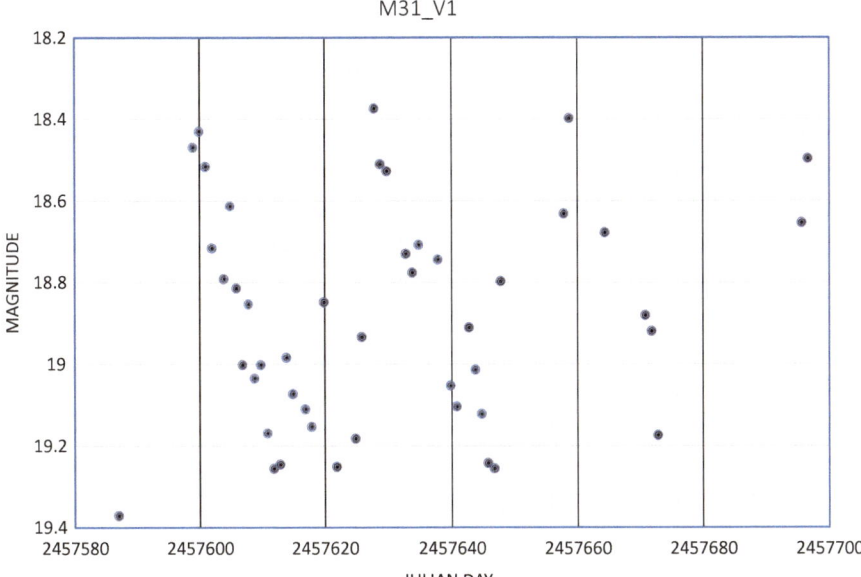

Fig. 1.18 AAVSO light curve for M31_V1. The x-axis is in Julian days, and the thin vertical lines correspond to time intervals of 20 days. (Data courtesy of The American Association of Variable Star Observers at https://www.aavso.org/)

Lyrae stars is that they have a near-constant average luminosity (about 80 times that of the Sun's luminosity) irrespective of the pulsation period (which can vary from a few hours to a day).

(☺) **Exercise 1.23** Figure 1.18 shows the light curve for the classical Cepheid variable star HV1 located in the Andromeda galaxy (M31). The HV1 label indicates that it is the first Cepheid variable found by Edwin Hubble in M31 (in 1924). Determine the star's pulsation period from the time interval between successive maxima. Use Fig. 1.17 to estimate the star's luminosity. Given an average flux measurement of $F = 6.3 \times 10^{-16}$ W/m^2 (corresponding to an average visual magnitude of 18.8), determine the distance to HV1.

(☺) **Exercise 1.24: Term Paper Project** Investigate the history of Cepheid variables as standard candles.

1.7 A Stellar Taxonomy

A complete listing of all observed stellar phenomenon and the organizational principles underlying star classification and groupings is beyond the scope of this book, but Fig. 1.19 provides a compacted taxonomy of features, forms and commonalities for a selected series of objects. The scheme developed in the figure

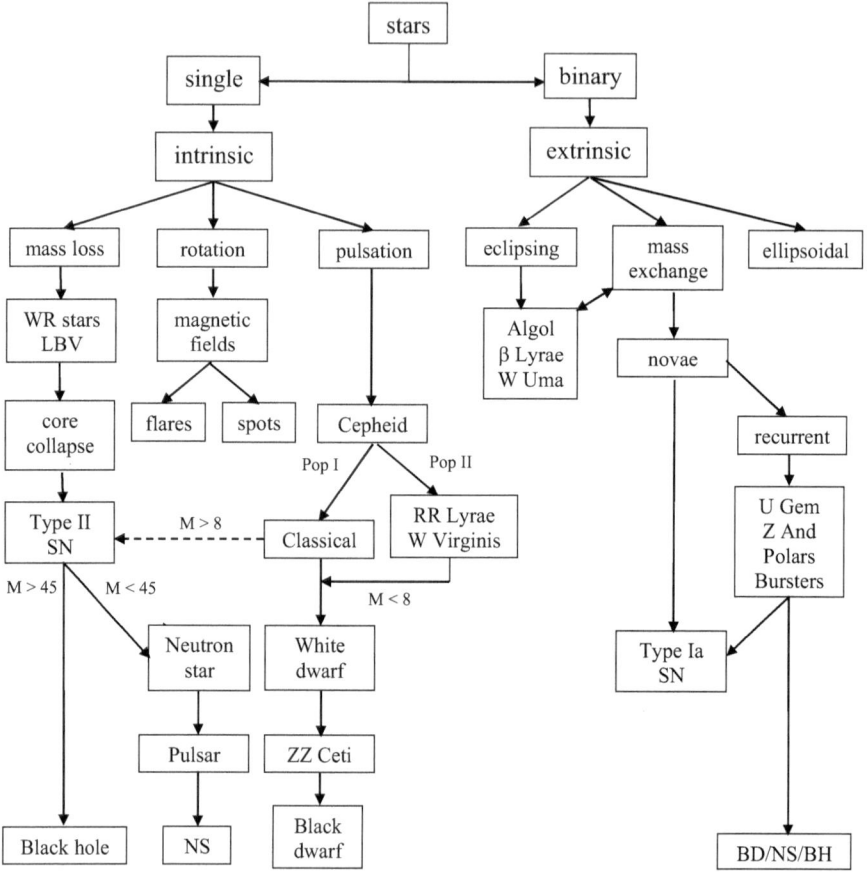

Fig. 1.19 A basic stellar taxonomy. Here the key mechanisms that modulate the appearance and structure of stars are indicated by various pathways. Although shown as distinct channels in the diagram, at some level, all stars are subject to the effects of mass loss, rotation, magnetic fields and pulsation. Time effectively runs from the top of the diagram to the bottom, where stellar end states of evolution are highlighted. **Key:** BH = black hole; BD = black dwarf; NS = neutron star

follows that of variability and separates out specific mechanism for the underlying modulations. Working from the top to the bottom of the diagram, the flow chart splits initially according to whether the star is a *bona fide* single object or a member of a close binary system. This division separates out close binary systems for scrutiny, since stars in wide binary systems effectively evolve as isolated objects. In contrast, a star located in a close binary systems can undergo interactions with its companion via mass exchange (as in the case of an Algol system), and it might show brightness variations due to gravitational distortion effects (e.g., as an ellipsoidal variable or as a W Ursae Majoris system); it may also show eclipse effects when one star passes in front or behind its companion in the observer's line of sight.

These two domains—single versus close binary—lead to observable variations in the brightness of a star, such variations being either intrinsic or extrinsic, or possibly both. Moving from the top to the bottom of Fig. 1.19, an aging sequence is also (partially) portrayed. Following the intrinsic, single star branch first, the most massive stars are most affected by mass loss. This steady ejection of material results in a significant reduction in their mass and produces the phenomenon associated with Wolf-Rayet (WR) stars, which have shed perhaps half their initial mass to reveal nuclear processed material at their surface.

The luminous blue variables (LBV) show extreme variations in temperature and also undergo dramatic changes in brightness during great outbursts of mass ejection. Indeed, the LBVs hover on the boundary of the Humphreys-Davidson limit, which sets an upper boundary in the HR diagram (in the extreme upper right-hand corner) to the luminosity, temperature and size to the most massive red giants. The quintessential LBV is the η Carina binary star system, which contains one of the moist massive stars known and which underwent an outburst in 1843, when it briefly became the second brightest star in the sky. Eta Carina is believed to be a binary system containing a primary star with a mass greater than 100 $M_\odot$ (the secondary star has a mass some 50 times that of the Sun). During the 1843 outburst, at least 10 solar masses of material was ejected from the envelope of the primary star (this material being observed in a feature called the Homunculus). The condensation of carbon grains within the expanding envelope of gas about the *star* caused it to dim, the *star* falling to well below naked-eye visibility by 1890.

The ultimate fate of WR and LBVs is to undergo supernova disruption (see Sects. 5.8 and 6.2). Such stars form so-called Type II supernovae as a result of core-collapse. The end result of core-collapse is the production of a neutron star, a black hole or possibly no remnant. If a neutron star forms, then it may be observed as a pulsar. The formation of a pulsar is predicated on the presence of a very strong magnetic field and rapid rotation. Ultimately, the pulsar will spin down but the underlying neutron star will remain. If a black hole forms as a consequence of the core-collapse of a massive star within a close binary system, then an X-ray variable system such as Cygnus X-1 might be produced via mass exchange and accretion of material from its companion. Detailed model calculations indicate that stars with an initial mass more than eight times that of the Sun will typically produce a neutron star. Only once the initial mass is greater than about 40 times that of the Sun will a black hole be produced during core-collapse.

Mass loss can be produced and modulated by a number of physical mechanisms, of which one is rotation. Certainly stars spin, although the rate of spin varies dramatically according to star mass, spectral type and age. Importantly, however, rotation can couple with convective motion within the outer layers of a star to produce a magnetic field. The Sun rotates at its equator once every 25 days or so, and it shows (at the present epoch) a 22-year magnetic cycle. This magnetic cycle is usually described according to the 11-year sunspot cycle, in which the number of sunspots varies from near-zero to some maximum number that changes from one cycle to the next. Many stars show brightness variations due to the intermittent formation of massive starspots (e.g., the RS Canum Venaticorum variables), and other stars show irregular and variable magnetic field related variations through energetic flare activity. Some stars show anomalous abundances of some elements at their surface due to magnetic levitation effects (e.g., the Ap and Bp, chemically peculiar stars). In addition to mass loss, rotation and magnetic fields, stars can also (as discussed earlier) undergo radial pulsation. These latter stars, if less massive than eight times that of the Sun, will eventually evolve via a mass-losing planetary nebula phase into white dwarfs. These latter objects will gradually cool off to become, after many billions of years, black dwarfs—such objects having about half the mass of the Sun, half the the size of the Earth, but zero luminosity and a temperature just above absolute zero. Before becoming a black dwarf, a white dwarf will pass through the instability strip, where the kappa mechanism can drive rapid—a few tens of seconds—radial pulsations (such objects being recognized as ZZ Ceti variables).

Stars within close binary systems (the extrinsic variable group in Fig. 1.19) can certainly undergo mass loss, will certainly be spinning and may have strong surface magnetic fields. Because of the close proximity of a companion star, they can also show additional variable effects. The most obvious additional effect is a geometrical one and comes about if the orbital plane of the two stars is coincident with our line of sight. As one star passes in front of the other, a decrease in system brightness will be observed. If the stars are very close to each other, then their outer layers will follow so-called Roche lobe profiles (these are equi-gravitational force surfaces), making them appear decidedly non-spherical in shape. This can result in sinusoidal-like variations in brightness, such as those seen in ellipsoidal variables. In the extreme case of closeness, the two stars may develop a common envelope (the system adopting a profile somewhat like an unshelled peanut) and be observed as a W Ursae Majoris star. Not only can a star lose mass, but it can also gain mass when located within a close binary system. Such mass exchange can produce so-called Algol and β Lyrae systems, and, if the receiving star is a white dwarf, it can also result in the production of Type Ia supernovae (see Sect. 5.8). These latter objects, as calibrated standard candles, are of vital importance with respect to gauging the size of the universe.

Algol, the so-called winking demon star, is visible to the naked eye and is the second brightest star in the constellation of Perseus. Its variability was noted in antiquity, and it is in fact an eclipsing binary system with a period of 2.86 days. When the system was first studied in detail, it was found that the more massive star in the binary was on the main sequence, while the less massive star was a subgiant. This is the exact opposite of what was expected from stellar evolution theory, which would predict that the more massive star should evolve more rapidly than its lower mass companion. The solution to this apparent paradox is mass transfer, where mass has been transferred from the originally more massive component (now the subgiant) onto its initially lower mass companion (still on the main sequence). Astrophysicist Fred Hoyle colorfully described the mass exchange solution to the Algol paradox as a 'strange dog-eats-dog evolution'. Beta lyrae variables are also eclipsing variables, but now they are sufficiently close that the stars have become distorted through mutual gravitational interactions. Accordingly, the light curve of a β Lyrae system shows a smooth and continuous variation rather than the angular dips observed in wider separation eclipsing binaries (recall Fig. 1.7). The stars within a β Lyrae system will also experience the effects of mass transfer.

If one of the companions within a close binary system is a white dwarf, then mass transfer can drive dwarf nova and/or supernova outbursts. Dwarf novae show sudden and sometimes repeated outbursts in brightness, whereas supernovae represent a terminal, fully destructive end phase. U Geminorum stars are composed of a red dwarf–white dwarf pairing, and mass falls onto the white dwarf via an encircling accretion disk. In the case of Z Andromedae stars (also called symbiotic stars), the binary is composed of a red giant–white dwarf pairing. The outburst mechanism is related to the buildup of hydrogen on the surface layers of the white dwarf. The hydrogen gas, upon becoming compressed and heated, undergoes a brief but violently energetic set of runaway fusion reactions. In some situations where the white dwarf supports a particularly strong magnetic field, matter is funneled directly towards its polar regions, rather than falling onto its equator via an accretion disk and a so-called Polar is observed.

If rather than a white dwarf the accretion is onto a neutron star—again a processed controlled by strong magnetic fields—a so-called X-ray burster is observed. Here the X-ray emission is modulated by the rotation of the stars within the system. If the accreting object is a black hole, then the X-ray emission is produced at the inner edge of an accretion disk formed around the black hole. Prior to the gas in the disk being lost to sight, upon crossing the black hole's event horizon, its temperature may exceed several hundred thousand degrees, and accordingly it will emit strongly at X-ray wavelengths. This appears to be the case with respect to the Cygnus X-1 system, where the heated accretion disk surrounds a 15 solar mass black hole; the mass donor (identified as HDE 226868) has a mass of about 30 times that of the Sun and is an O9.7 Iab spectral type supergiant. The two objects, black hole and supergiant, are sufficiently close to each other, just 0.2 astronomical units apart, that the supergiant has been distorted into an ellipsoidal shape—this being observed and deduced from a 5.6 day light curve modulation in brightness. To produce a 15 solar mass black hole, the initial progenitor star mass must have been of order 45 solar masses.

As will be described in later chapters, both white dwarfs and neutron stars have upper mass limits around 1.44 and 3 times the mass of the Sun, respectively. If pushed towards these upper mass limits by matter transfer and/or mergers, a supernova can result. This dramatic disruption phase results in the formation of a Type Ia supernova in the case of white dwarf disruption. The properties and physics of supernova disruption will be considered in more detail in Chap. 5, Sect. 5.8.

(☺) **Exercise 1.25**   The WD + A8 spectral type binary system IK Pegasi is presently located some 46 pc away from the Solar System. The white dwarf component currently has a deduced mass of 1.15 $M_\odot$. Given the matter transfer rate from the A star companion is 10^{-7} $M_\odot/yr$. Estimate when IK Pegasi will undergo Typa Ia supernova disruption. Given that the relative velocity between the Sun and IK Pegasi is about 21.3 km/s, is the supernova disruption of IK Pegasi likely to be a threat to the Earth in the case of a critical detonation stand-off distance of 10 pc? *Hints*: Type Ia disruption will take place in approximately three million years from the present. No threat is posed to the Earth since at the time of detonation IK Pegasi will have moved at least 63 pc away from its current location. Even if heading straight for the Solar System (which it is not) it would be located 17 pc away at the time of detonation.

Not shown directly in Fig. 1.19 is another remarkable pathway that some stars might chance to follow when formed within a close binary system, being the result of the two stars physically merging into one object. For example, this happens for a Type Ia supernova if the two coalescing objects chance to be white dwarfs, or for a kilonova if the two objects are neutron stars. If the two stars are main sequence stars, then a single, rejuvenated main sequence star can form, such objects being observed as blue stragglers in cluster HR diagrams (recall exercise 1.21). These stars stand out in the cluster diagram in the sense that while they are definitely cluster members, they have a higher luminosity and temperature to that of the cluster turn off point.

Our foray into stellar taxonomy is now complete, even though it is far from being fully or even partially described. Indeed, there is almost an endless, even bewildering list of types, categories, interactions and evolutionary channels through which stars can move. To the theoretician, all these stellar types and observations are grist to the mill, and the aim is to work from the general model to the circumstances relevant to an individual object. The list of observed phenomenon and the delineation of their associated physical mechanisms is also a theoretician's nightmare. Clearly, rotation, magnetic fields, mass loss, mass gain, shape distortion, coalescence, convection, thermonuclear reactions, pulsation and many more physical effects are involved in the understanding of stars and star related phenomena. Fortunately for us, it transpires that many of the complex physical phenomena are only dominant in the outer stellar layers. The deep interior of a star, as we shall see over the next several chapters, can be well explained under the conditions of spherical symmetry, and the effects due to rotation and magnetic fields can largely be ignored—at least for a general understanding. The devil of course is in the details. And while, for example, we shall describe later on an elementary theory for the transport of energy via convection, we must also remember that there is in fact no complete theory to describe convection and turbulence under stellar conditions. On this point, Canadian astrophysicists Jean-Louis and Monique Tassoul have dramatically (and correctly) noted in their text, *A concise history of solar and stellar physics* (Princeton University Press, 2004), that 'until such time as an adequate hydrodynamical theory of the phenomenon of turbulence and convection has been formulated, there is. . . no hope of describing the sun and stars from first principles alone'. For all this, however, the physical description of stellar interiors and of star phenomenon has attained a remarkably advanced state. The remainder of this text will discuss, trace out and explore how some of this remarkable understanding of the stars has come about.

1.8   Answers to Exercises

Following are the answers to all exercises in this chapter. The exercises for which no answer is provided are those in which the student is asked to check a result or complete the algebra steps between two equations given in the text.

Exercise 1.1 The mass flux is 3.3×10^{-4} kg /m^2/s.

Exercise 1.2 The mass flux at the Earth's orbit is 7.11×10^{-9} kg/m^2/s. The surface area of the Earth's night-side hemisphere is 2.55×10^{14} m^2, and the mass accreted on the night-side hemisphere is 1.81×10^6 kg/s.

Exercise 1.3 The orbit becomes smaller, and the time required to accumulate a 10% change in the orbital radius is 3.5 million years.

Exercise 1.4 For part (1) $U_{Earth} = 2.2 \times 10^{32}$ Joules. For part (2) evaluate $U_{Earth} = L_\odot \times time$, which yields a time interval of 6.7 days.

Exercise 1.5 About 76 meters per year.

Exercise 1.6 The timing of ancient eclipse events indicates that the Earth's spin rate is decreasing by about 2 milliseconds per century. The available data on eclipse durations and type (annular or total) at a specific location, however, show no evidence for the Sun's diameter having changed during the interval of recorded history.

Exercise 1.7 The Sun's bulk density is 1384.4 kg/m^3.

Exercise 1.8 It is assumed that the star radiates its energy in an isotropic manner (equally in all directions), and this is what the $4 \pi d^2$ term implicitly assumes. To measure the luminosity directly would require the detector to wrap around the entire star—this is not a practical, or indeed possible, exercise.

Exercise 1.9 The solar flux is $F_\odot = 1371$ W/m^2 and $E = (\pi R^2) F_\odot = 1.75 \times 10^{23}$ J/s (taking the Earth's radius to be $R = 6.371 \times 10^6$ m). Human energy consumption per second is ten billion times smaller than the energy received from the Sun by Earth's sunward pointing hemisphere per second.

Exercise 1.10 The distance is: $d = 1.76$ light years.

Exercise 1.13 The temperature is: $T = 5790$ K.

Exercise 1.14 The mass is: $m_1 \equiv M_\odot = 1.984 \times 10^{30}$ kg.

Exercise 1.18 The masses are $m_1 = 4.6 \times 10^{29}$ kg $= 0.23$ M$_\odot$ and $m_2 = 4.26 \times 10^{29}$ kg $= 0.21$ M$_\odot$.

Exercise 1.20 The Mass of the extreme Pop. I disk is ~ 170 million M$_\odot$; the mass of the Pop. II thick disk is ~ 14 billion M$_\odot$. The Luminosity of thin disk stars is ~ 42 billion L$_\odot$, and the luminosity of the Pop. II thick disk is ~ 280 million L$_\odot$.

Most of the mass resides in the Pop. II thick disk, but the extreme Pop. I think disk provides most of the galaxy's luminosity.

Exercise 1.21 The turnoff point temperature is located between 5800 and 6400 K (with a corresponding turnoff point luminosity of about 1–2 $L_\odot$). These temperature and luminosity characteristics correspond to stars with a spectral type between F5 and G0, giving a cluster age of between 5–8 billion years. Note that the cluster diagram shows some considerable scatter—this is largely due to foreground contamination by non-cluster member stars. Also note that there are a number of stars located at higher temperatures and luminosities than the turnoff point (with temperatures of about 7000 K and luminosities of order 5–10 $L_\odot$). These stars are called blue stragglers, and they have come about due to the physical merger of two stars that were initially located in a close binary (or triple star) system.

Exercise 1.22 Equation (1.28) reveals that $R(t_1) - R(t_2) \approx 1.39 \times 10^6$ km $\approx 2\, R_\odot$. And, the final result is that $R(t_1) \approx 27\, R_\odot$ and that $R(t_2) \approx 25\, R_\odot$.

Exercise 1.23 The pulsation period is $P \approx 30$ days. From Eq. (1.30) $L = 1.5 \times 10^4$ $L_\odot$, and the distance (via Eq. 1.10) is accordingly 2.7×10^{22} meters ≈ 875 kpc.

Chapter 2
Star Formation

Where do stars come from? Or, perhaps more pertinent to our discussion, when does the process of star formation begin and end, and when does that of stellar structure and evolution begin? To say that star formation is simply the result of happenstance seems inappropriately non-scientific. But similar to how the chaos theorists may tell us (entirely incorrectly in terms of physical causality) that the beating of some butterfly's wings in the Australian outback can cause a snowstorm in central Saskatchewan, so star formation begins through the chance concourse of atoms in the interstellar medium. It is a fine balance of forces, past history and present circumstance, which dictates whether a star will form or not. Enough raw material must be brought together at the right temperature, at the right density and in the right place. It is easier to not make stars than it is to make them, but as with all difficult things, nature has found a way.

From a physical perspective, the very first moments of the star formation process are ill-defined, yet the end of this first step in the process is instantly recognizable. Serendipity having brought enough raw material together in a small enough volume of space, the generation or not of a *bona fide* star initially hangs on a knife-edge. The would-be star-producing cloud hovers, as if uncertain about the future, continuously testing its environment and comparing the opposing forces that infuse its mass. As if holding its breath, the natal cloud waits for the right set of conditions to occur, before—if the timing is right—rushing headlong into its star-building phase. Sometimes, however, the right conditions do not evolve, and the would-be natal cloud is dispersed. The scattered remnants of our still-born cloud may regroup elsewhere, co-mingled with the material from other stars unborn or long dead. Star formation is chancy, chaotic and complex, but in spite of all the apparent adversity, stars do form. Indeed, stars have been forming in the universe for at least the past 13 billion years, and in our galaxy alone they will continue to form for many trillions of years yet.

© Springer Nature Switzerland AG 2019
M. Beech, *Introducing the Stars*, Undergraduate Lecture Notes in Physics,
https://doi.org/10.1007/978-3-030-11704-7_2

2.1   The Interstellar Medium

Let us imagine that we can teleport ourselves to any location in the Milky Way galaxy. If the teleportation places us at an entirely random spot, what is the likelihood that we will find ourselves inside of a star? To answer this question, let us take the galaxy to be a disk that is 15,000 pc in radius and 1000 pc in height, and within in this disk let us place 400 billion stars. Let us further assume that each star has a size of one-half that of the Sun, characteristic of the most common K and M spectral type stars. The volume of the galaxy is $V_{MW} = 7 \times 10^{11}$ pc$^3 = 2 \times 10^{61}$ m^3. The volume of each star will be $V* = 2 \times 10^{26}$ m^3, and the total volume of the galaxy occupied by stars is $V_{stars} = 7 \times 10^{37}$ m^3. From these numbers the volume of the galaxy actually taken up by stars is $V_{stars} / V_{MW} = 3.5 \times 10^{-24}$, and accordingly the odds are astronomically small against a random teleportation landing placing us within the body of a star. If you were able to visit a new random location within the galaxy every second, then it would be of order 9×10^{15} years before you might land in an actual star—this is some 650,000 times longer than the current age of the universe. While the stars make up the visually obvious component of our galaxy, they most certainly do not occupy the greater part of its volume.

(☺) **Exercise 2.1** If each of the 400 billion stars within our Milky Way galaxy has a characteristic mass of 10^{30} kg (half that of our Sun), and the matter within these stars is uniformly spread throughout the galaxy, what then would the density of star-matter be? Assume that the stars are entirely composed of protons (ionized hydrogen atoms).

The density of matter within the interstellar medium is about one proton per cubic centimeter, over one hundred times smaller than that of the dispersed proton density from the stars (see answer to Exercise 2.1). In this manner we deduce that most of the material mass of the galaxy is held within stars, but it is the interstellar medium that occupies most of the galaxy volume. Taking an even larger perspective view, the typical density of matter within the intergalactic medium is found to be about one atom per cubic meter. From these characteristic numbers we deduce that stars are very rare, extremely small, very high-density oddities within the context of the observable universe. And yet, as we shall see, the radiation that stars produce within their interiors and then radiate into space is fundamental to both the structure of the universe and our very existence.

(☺) **Exercise 2.2** There are something like 10^{22} atoms in each cubic centimeter of air at Earth's surface. The lung capacity of a typical adult human is about Six litres (6000 cm^3) of air. Taking air to be composed of nitrogen only (with the nucleus of each nitrogen atom containing 7 protons), what spherical volume of interstellar space will need to be sampled in order to take one lungful of air?

(☺) **Exercise 2.3** The Hubble Space Telescope Deep Field image, obtained in 1996, reveals that there are some 1600 galaxies visible in a 2×10^{-3} square degree region of the sky. Assuming each galaxy contains some 300 billion stars, and given that the entire sky encompasses at total of 41,253 square degrees, approximately how many galaxies N_{gal} and how many stars $N*$ are there in the observable universe?

Assuming that each star in the universe has a characteristic mass equal to half that of the Sun, we deduce from Exercise 2.3 that there are of order $10^{30} \times 10^{22} / 10^{-27} = 10^{79}$ protons incorporated within stars in the observable universe. Remarkably, however, this 10^{52} kg worth of matter constitutes only about 0.5% of the total estimated mass of the universe. Indeed, estimates for the matter content of the universe indicate that dark matter and dark energy significantly overshadow that contained within stars and gas. Dark matter we shall discuss later in the text, and it accounts for the anomalously high rotation velocities observed within individual galaxies and galaxy clusters. Dark energy is really the topic of cosmology, but suffice to say it accounts for the observation that the universe appears to be expanding at an accelerating rate. Given the deduced breakdown of the matter content within the universe, we once again find that stars really are the rare and special gems of the cosmos.

(☺) **Exercise 2.4** If the number of stars derived in Exercise 2.3 were spread uniformly throughout the universe, making one meta-galaxy, what would the density n_U of stars within the meta-galaxy be? Express your answer in stars per cubic parsec. Take the present radius of the observable universe to be 14 Gpc. Compare the value of n_U with the 0.1 stars per cubic parsec observed within the solar neighborhood.

Given that the density of matter in the interstellar medium is of order one proton per cm^3, then the total mass of the interstellar medium is of order $1 \times V_{MW} \times m_P \times 10^6 = 3.3 \times 10^{40}$ kg ≈ 17 billion $M_{\odot}$ (where m_P is the mass of the proton and the 10^6 term accounts for the number of cubic centimeters within one meter cubed). A more detailed accounting would indicate that our estimate for the amount of matter in the interstellar medium is too large by a factor of about two, but it gives us the correct order of magnitude for the amount of material within the Milky Way galaxy that might eventually be involved in the star formation process. The deduced rate at which interstellar material is converted into stars within the Milky Way galaxy is just one solar mass per year. On this basis, given the deduced reserves of interstellar gas, star formation might be expected to continue for at least another ten billion years, that is, for about another Hubble time.[1] However, at the end of their lifetimes stars return most of their constitutional mass back into the interstellar medium, so star formation will in fact run for perhaps another trillion years into the future.

Just as British poet John Keats once bemoaned that science had reduced the Sun from a glowing angel to a 'ball with spots', so our description of the interstellar medium will be reduced to the consideration of one table and figure. The data in the table has been hard won by astronomers over many decades, and the picture that has

---

[1]The Hubble time is a measure of the age of the universe (currently set at about 13.8 billion years). It is expressed in terms of the inverse value of Hubble's constant. The Hubble constant in turn expresses the expansion rate of the universe at the present epoch.

Table 2.1 The various phases of the interstellar medium in descending order of associated temperature

Phase	Temperature (K)	Number density (m^{-3})	Characteristic size (pc)	Observations
Coronal gas	10^6	10^4–10^2	Galactic (50%)	UV & X-ray
Warm neutral	10^4	5×10^5	Galactic (15%)	21-cm radio
Warm ionized	10^4	5×10^5	Galactic (25%)	21-cm radio
H II	10^4	10^8–10^{10}	1–10 (< 1%)	Optical
Cold neutral	100	5×10^7	Galactic (5%)	21-cm radio
GMC	15	10^8	45 (< 1%)	IR & radio
MC	10	5×10^6	10	IR & radio
MC clump	10	10^9	5	IR & radio
MC core	10	10^{11}	0.25	IR & radio

In column 4, the term 'Galactic' implies that these phases are widespread across the galactic disk, and the bracketed terms indicate a characteristic volume-filling factor. **Key**: GMC giant molecular cloud, MC molecular cloud

emerged is that the interstellar medium (ISM) is made-up of several substructures all in near-pressure equilibrium. The various components have characteristic temperatures, densities, composition and galactic volume-filling factors. Table 2.1 provides the key numbers.

Table 2.1 indicates that the very low-density, but high-temperature coronal gas fills most of the volume of interstellar space. This material is far too hot and far too diffuse to form stars. Most of the mass of the interstellar medium, however, is held within the low-temperature, relatively high-density molecular cloud complexes— regions of many parsecs in scale that can contain as much as a million solar masses of molecular hydrogen. These complexes are characteristically located within the spiral arm regions of the galaxy, and they hold the raw material out of which stars will eventually form. Active star formation complexes (such as the iconic Orion nebula) are identified at optical wavelengths with glowing (ionized hydrogen) HII regions, and these regions are invariably embedded within the dense cores of larger molecular cloud structures (only observed at radio and infrared wavelengths). Indeed, it is now recognized that the locations where stars are going to form can be identified with the cool, relatively small and high-density MC and MC core regions. MC cores are the highest densest regions within the interstellar medium, but these structures (unlike the coronal gas through to the cold neutral gas clouds within the ISM) are not in approximate pressure equilibrium (Fig. 2.1). Indeed, the highly irregular shape and long filamentary structures adopted by molecular clouds indicate that they are dynamic and capable of relatively rapid change. And indeed, if the conditions are right, stars will form within giant molecular clouds, producing star clusters containing many thousands of stars, or smaller substructures containing perhaps a handful of stars (Fig. 2.2). The question now is: What are the mechanisms that control star formation?

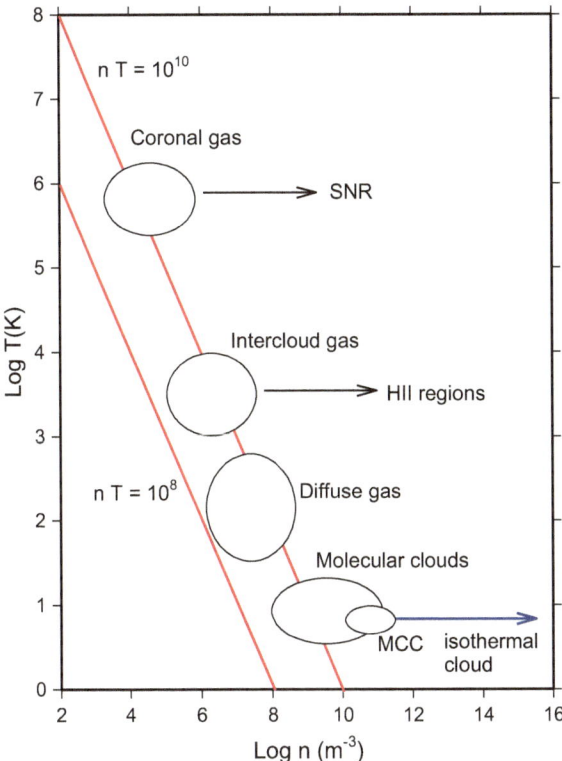

Fig. 2.1 A schematic plot in the temperature number density domain showing the locations of the various phases of the interstellar medium. The two diagonal lines correspond to constant values of n $T = 10^{10}$ and 10^8 k/m^{-3}, and these indicate constant pressure domains. Associations between the coronal gas and supernova remnants, intercloud gas and HII regions are indicated by arrows, and the star formation pathway from molecular clouds through molecular cloud cores and isothermal clouds is also indicated at the bottom right of the diagram. (Credit: Author)

2.2 Jean's Criterion

Gravity is the midwife of star formation. To make stars, this all-pervasive, long-range and attractive force applies its influence within the low-temperature, relatively high-density molecular cloud cores. There, having set its grip upon the core material, it works to transform the extensive cloud into a compacted star. Yet while gravity works to gather the material together, it does not rule unopposed; other forces act to disrupt the process. Most of the time, the penchant for gravity to make stars is held in check by the internal thermal pressure within a cloud, turbulence and the effects of cloud rotation, and the presence of galactic magnetic fields. Indeed, if this were not the case, there would be no interstellar medium. The cloud collapse condition, in which gravity wins against its rivals, can be expressed in terms of energies. Accordingly, we write:

Fig. 2.2 Star formation in the Monoceros R2 molecular cloud complex. (Image courtesy of the Two Micron all Sky Survey)

$$\left|E_{\text{grav}}\right| > E_{\text{thermal}} + E_{\text{rotation}} + E_{\text{turbulence}} + E_{\text{magnetic}} \tag{2.1}$$

where $|E_{\text{grav}}| = \varepsilon\, GM^2\,/\,R$ is the absolute value of the gravitational binding energy of a cloud of mass M and radius R. The ε term accounts for the distribution of matter within the collapsing cloud, with $\varepsilon = 3/5$ for a uniform density distribution. The total thermal energy of the cloud, assuming an ideal isothermal gas of temperature T is

$$E_{thermal} = \frac{3}{2}\Re\left(\frac{T}{\mu}\right)M \tag{2.2}$$

where $\Re = k/m_{\text{p}}$ is the gas constant (with k being the Boltzmann constant and m_{p} being the proton mass), and μ is the mean molecular weight (to be discussed more fully at the beginning of the Chap. 3). The rotational energy is:

$$E_{rotation} = \frac{1}{5}M(R\Omega)^2 \tag{2.3}$$

where $\Omega = 10^{-13} - 10^{-14}\,\text{s}^{-1}$ is the angular velocity of the cloud, and the 1/5th term corresponds (again) to the assumption of uniform density. The turbulent kinetic energy is

$$E_{turbulence} = \frac{1}{2}M\sigma_{turb}^2 \qquad (2.4)$$

where σ_{turb} is the characteristic turbulent velocity. Typically, σ_{turb} is found to be much larger than the sound speed within and the rotational velocity of a molecular cloud, making $E_{turbulence}$ an important quantity in controlling cloud collapse. The magnetic energy term is given by a volume integral

$$E_{magnetic} = \frac{1}{8\pi}\int B^2 dV \approx \left(\frac{B^2}{8\pi}\right)\left(\frac{4\pi}{3}R^3\right) \qquad (2.5)$$

where B is the assumed uniform magnetic field. Gathering together Eqs. (2.1), (2.2), (2.3), (2.4) and (2.5) yields a complex interplay of quantities and no straightforward analytic solutions. Historically, however, it has been assumed that one term or another dominates on the right-hand side of (2.1), and indeed, pioneering astrophysicist James Jeans made the assumption that the thermal energy of the cloud dominates over all other gravity-opposing effects. In this manner, he developed an expression for the so-called Jeans length R_{Jeans}, with (from Eqs. 2.1 and 2.2)

$$\frac{3}{5}\frac{GM^2}{R} = \frac{3}{2}\mathfrak{R}\left(\frac{M}{\mu}\right)T$$

which re-arranges to the Jeans length:

$$R_{Jeans} = \frac{2}{5}\left(\frac{G\mu}{\mathfrak{R}}\right)\frac{M}{T} \qquad (2.6)$$

Collapse of the cloud (of mass M) will proceed once its radius $R < R_{Jeans}$. A solar-mass molecular cloud core, with a temperature of 50 K, will therefore undergo collapse if its radius is smaller than about 2.6×10^{14} meters $\approx 370,000\ R_\odot$. Assuming that the collapsing cloud has a uniform density ρ, then Eq. (2.6) can be transformed to define the so-called Jeans mass M_{Jeans}, where

$$M_{Jeans} = \left[\left(\frac{5\mathfrak{R}}{2\,G}\right)^3\left(\frac{3}{4\pi}\right)\right]^{1/2}\left(\frac{T}{\mu}\right)^{3/2}\frac{1}{\sqrt{\rho}} \qquad (2.7)$$

With Eq. (2.7), the collapse of a molecular cloud core of temperature T and density ρ will proceed once its mass $M > M_{Jeans}$. Given the characteristic temperature of a molecular cloud core is $T \approx 50$ K and the density is of order 10^{-17} kg/m^3, (taking $\mu = 2$) $M_{Jeans}\sim 10^{32}$ kg$\sim$50 M$_\odot$. The implication of this latter result is that those molecular core regions that do undergo collapse to form stars must undergo fragmentation as the collapse proceeds, since, as seen in Chap. 1, most stars in the galaxy have a mass of just a few tenths of a solar mass.

(☺) **Exercise 2.5** Complete the algebra that takes Eq. (2.6) to Eq. (2.7). This requires the straightforward substitution $M = \rho \, (4\pi R^3 / 3)$. Show that the constant in the square brackets in Eq. (2.7) evaluates to 2.7×10^{21} (*SI* units).

(☺) **Exercise 2.6** (1) Derive a formula for the Jeans length R_{Jrot} when it is the cloud rotation term that dominates on the right-hand side of Eq. (2.1). (2) Compare R_{Jeans} and R_{Jrot} for a solar mass molecular cloud. (3) Show that $\beta = |E_{rot} / E_{grav}|$ is typically small for a solar mass molecular cloud confined within a Jeans radius. (4) Show that there is a critical rotation velocity for the cloud such that gravitational collapse can only occur if $V_{rot} < R_{Jrot} (4\pi \, G \, \rho)^{1/2} \approx 220$ m/s (taking $\rho = 10^{-17}$ kg/m^3). We will see in the next chapter that the characteristic collapse time for an unstable gas cloud is of order $(G \, \rho)^{-1/2}$, and accordingly the result from (4) indicates that collapse is possible only if V_{rot} is smaller than the characteristic velocity involved in radial (that is non-rotating cloud) collapse.

In the case of a magnetic field supporting a molecular cloud against gravitational collapse, the analog to the Jeans mass is obtained from the relation $|E_{grav}| > E_{magentic}$, which gives

$$M_{Jmag} > \left(\frac{5}{18\,\pi^2\,G} \right)^{1/2} \left(\pi R^2 \right) B$$

For our standard solar mass cloud, the magnetic support condition requires a magnetic flux density of $(\pi \, R^2)B = \phi > 10^{26}$ Tesla. For a characteristic galactic magnetic field strength of a few nano-Tesla, this implies a critical radius of $R \sim 10^{17}$ meters, or about 3 parsecs, which is the characteristic size of a giant molecular cloud (recall Table 2.1). There are additional complex caveats (not considered here) with respect to considering the effects of magnetic fields on cloud support. Suffice to say that the current consensus among astronomers attributes a more important role to magnetic field and internal random motion (turbulence) in molecular cloud support than cloud rotation and/or internal thermal pressure support. Thus, star formation is largely dependent upon those mechanisms that are responsible for the dissipation of turbulence and magnetic fields within giant molecular cloud structures.

2.3   Collapse to Complexity

Once the collapse of a cloud has started, the gas molecules (mostly in the form of molecular hydrogen H_2) will begin to migrate towards the cloud center, gaining kinetic energy as they do so. Collisions between the molecules will result in the heating of the gas and the temperature of the central cloud region will begin to increase. If nothing opposes the cloud collapse, then the central core region would become infinitely hot and infinitely compressed—but, of course, no such situation comes about. Importantly, as the collapse continues, the cloud temperature eventually reaches a level where collisions between neighboring H_2 molecules result in their dissociation: $H_2 \Rightarrow H + H$, and later collisions between H atoms results in their ionization: $H \Rightarrow P + e^-$. The energy E_{DI} needed to drive the dissociation and ionization of the hydrogen in the production of a protostar of mass M is

$$E_{DI} = \varepsilon_D \frac{M}{2m_H} + \varepsilon_I \frac{M}{m_H} \tag{2.8}$$

where $\varepsilon_D = 7.2 \times 10^{-19}$ Joules (or 4.5 eV)[2] is the energy required to dissociate all the molecular hydrogen in the cloud and $\varepsilon_I = 2.2 \times 10^{-18}$ Joules (or 13.6 eV) is the energy required to ionize all the hydrogen atoms. This energy, of course, has to come from somewhere, and indeed, the energy is supplied by the release of gravitational binding energy (recall Eq. (1.7)) as the cloud collapses from its initial radius R_1 to a final (protostar core) radius R_2. Accordingly:

$$\frac{3}{5}\frac{GM^2}{R_2} - \frac{3}{5}\frac{GM^2}{R_1} = \varepsilon_D \frac{M}{2m_H} + \varepsilon_I \frac{M}{m_H} \tag{2.9}$$

(☺) **Exercise 2.7**   Show that the energy required to dissociate and then ionize a solar mass molecular cloud composed of molecular hydrogen is $E_{DI} \approx 3 \times 10^{39}$ Joules.

(☺) **Exercise 2.8**   Given that a solar-mass molecular cloud began its collapse with a size equivalent to the Jeans length $R_{Jeans} = 370,000\ R_\odot$, what is the cloud radius once ionization is essentially complete? *Hint:* Use Eq. (2.9) with $R_1 = R_{Jeans}$ and take E_{DI} from exercise 2.7.

When most of the hydrogen has been fully ionized and the protostar becomes increasingly opaque to its own radiation, the continued release of gravitational binding energy is translated into the random thermal energy of the electrons and ions. This increase in the thermal energy establishes an internal pressure gradient, and the collapse of the protostellar core is slowed until it eventually reaches hydrostatic equilibrium (we shall have much more to say about this condition in Chap. 3). Accordingly, once the internal temperature reaches some 30,000 K, a stable protostellar core is formed at the center of the collapsing cloud. Detailed hydrodynamic calculations indicate that, (almost) irrespective of the starting conditions, the initial stable protostellar core has a mass of about 0.01 $M_\odot$ and a radius of about 3 $R_\odot$. This core once formed continues to accrete material, growing in mass and radiating energy into space from a surface accretion shock front (Fig. 2.3). The accretion luminosity L_{acc} is described according to Eq. (1.4)—see also exercise 2.9 below—and if the in-fall rate of material onto the accretion front is $\dot{m}$, then:

---

[2]The energy unit of the electron volt is commonly used in many physics texts, with 1 eV $= 1.602 \times 10^{-19}$ Joules

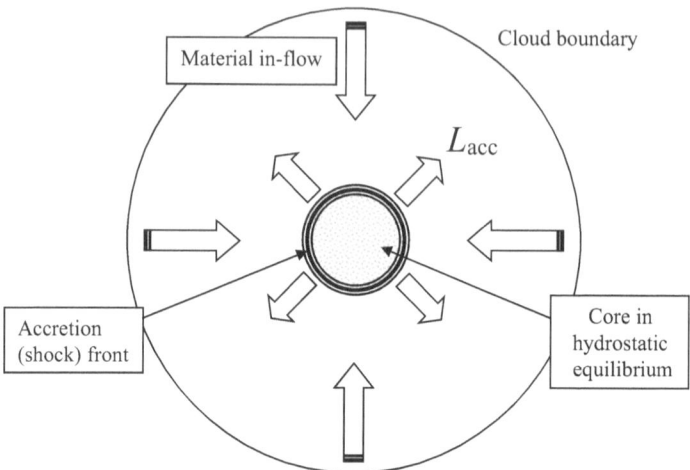

Fig. 2.3 A schematic diagram of an accreting protostellar core. Once a stable core has formed at the center of the collapsing cloud, it develops a surface accretion shock. (Credit: Author)

$$L_{acc} = G\frac{M_{core}}{R_{core}}\dot{m} \tag{2.10}$$

The observed (and deduced from detailed computational models) accretion rate for protostellar cores is typically found to be about 10^{-6} $M_\odot$/year, and it is the accretion process that is now recognized as being the controlling mechanism behind star production. In general, the accretion of material by the protostellar core will not be spherically symmetric, that is, the infalling material will not settle evenly upon its entire surface. Indeed, rather than moving along straight radial paths onto to the surface of a protostar, the infalling material forms an accretion disk (to be discussed shortly).

(☺) **Exercise 2.9** In this extended exercise you will determine the approximate evolutionary path of a very young protostellar core through the HR diagram. Let the protostar have an initial core mass of $M_0 = 0.01$ $M_\odot$ and radius of $R_0 = 3$ $R_\odot$. Take the accretion rate to be $\dot{m} = \dot{m}_0(1 - t/\tau)$, where $\dot{m}_0 = 2 \times 10^{-6} M_\odot$/yr is the initial accretion rate, and where t is the time and $\tau = 1$ My is the time at which accretion stops. (1) Show that the mass of the protostar core increases as

$$M(t) = M_0 + \dot{m}_0 t\left(1 - t/2\tau\right)$$

where $M_0 = 0.01$ $M_\odot$ is the initial protostar core mass, and determine the final mass (*Answer:* 1 $M_\odot$). (2) Furthermore, assuming that the size of the protostar remains constant as it grows in mass, show that the accretion luminosity varies as

$$L(t) = \frac{1}{2}\frac{GM(t)}{R_0}\dot{m}_0(1 - t/\tau)$$

Hint: In this step the conservation of energy (recall Chap. 1) dictates that the infall kinetic energy per unit mass of material onto the protostar core will be $\frac{1}{2}u^2$, and the energy inflow rate will be $\left(\frac{1}{2}u^2\dot{m}\right)$. Half of this energy, by virtue of the Virial theorem, will go into heating the core and the remainder will be radiated into space. Accordingly: $L_{acc} = \frac{1}{2}\left(\frac{1}{2}u^2\dot{m}\right) = \frac{1}{2}(GM/R)\dot{m}$, as given earlier in Eq. (2.10). (3) Produce a table (by hand calculation, or write a short computer program) showing the values of $L(t)$ and $M(t)$ for a range of time values from $t = 0.05\tau$ to $t = 0.95\tau$. (4) Determine the temperature $T(t)$ of the protostar (again) assuming that its radius remains constant and equal to R_0; add these numbers to your table from part (3). **Hint**: Use Eq. (1.14). Also add another column to your table showing the maximum emission wavelength λ_{max} determined upon the basis that the protostar behaves like a blackbody radiator. **Hint**: Use Eq. (1.12) and show that λ_{max} is typically of order 1 micron—the protostar radiates strongly in the near-infrared part of the spectrum. (5) Now, plot an HR diagram showing the evolution of the temperature and luminosity. (Fig. 2.4) illustrates what you should find.

An accretion disk forms about a protostar during cloud collapse in response to the conservation of angular momentum. This fundamental physical law requires that for a small blob of material of mass m at a distance h from the spin axis (see Fig. 2.5), the quantity $L = m\,V_\varphi\,h = m\,V_\varphi\,R\,\sin(\varphi)$ must be conserved at all times. It is observed that molecular clouds do have a small amount of rotation, with velocities of order a few hundred meters per second, and while this rotational velocity is small, it does result in a big problem with respect to describing disk structure and formation. In the extreme case, we can imagine material located in the equatorial plane of the initial collapsing cloud falling inward to accrete onto the surface of a solar mass protostar. Taking the initial radius to be the Jeans length R_{Jeans}, and the initial rotational velocity to be $V_{\varphi = 90} = 0.15$ km/s, the angular momentum per unit mass will be $L = 4 \times 10^{16}$. Since angular momentum is conserved, at the surface of the protostar at, say, 3 $R_\odot$, the rotation velocity V_* will be: $V_* = L / (3 \times 6.9 \times 10^8) \approx 2 \times 10^7$ m/s. The problem with this number is that the implied rotation velocity at the surface of the protostar is some 6% of the speed of light. Clearly, something must happen within an accretion disk to stop such high velocities from coming about, since stars do not spin at relativistic rates.

The simplest disk that we might imagine forming around a protostar is a disk in which the material is in centrifugal balance. That is, the infalling material settles into a disk at a radius r at which the gravitational and centrifugal forces balance, with a circular velocity given by $V_{circ} = \sqrt{GM/r}$. Accordingly, material in the collapsing cloud with some initial angular momentum L will settle into the disk at a radius given by $L = r\,V_{circ}$, or, $r = L^2 / G M$. For a solar mass protostar, the maximum disk radius (corresponding to $L \sim 4 \times 10^{16}$ as given above) will be $r \approx 1.2 \times 10^{13}$ meters $\approx$ 80 AU. This outer disk radius (for a 1 $M_\odot$ protostar) is about the size expected for our Solar System, since it is within the Sun's remnant accretion disk that the asteroids, comets, planets and Kuiper belt objects (KBOs) will eventually form.

So far, we have skirted around the key problem of accretion disks. The centrifugal disk is all well and good, but it does not allow for material to move through the disk

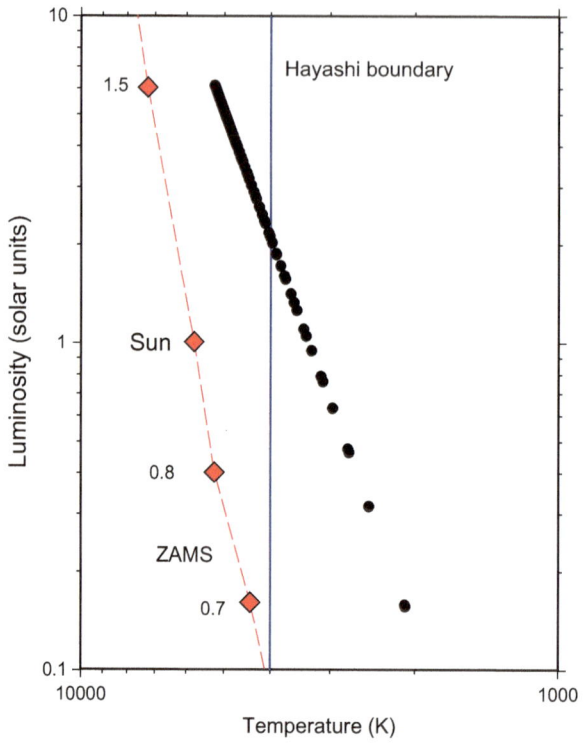

Fig. 2.4 HR diagram showing the luminosity and temperature variation for an accreting protostar. As a consequence of its construction, the protostar completes a loop up and then down, the 3 $R_\odot$ constant radius diagonal in the diagram. The present location of the Sun is shown in the diagram along with the zero age main sequence (shown in red (see Chap. 4) with the masses of several additional stars being indicated), and the vertical line at $T = 4000$ K corresponds to the so-called Hayashi boundary. The Hayashi boundary is a stability delineator (described later in Chap. 5), with objects at the boundary becoming unstable to convective energy transport within their interiors. At this point, our model assumptions essentially fail, and the evolution is modified from that shown. (Credit: Author)

Fig. 2.5 The angular momentum L of the small mass m is determined by its distance from the spin-axis h, with $L = m\, h\, V_\varphi$. (Credit: Author)

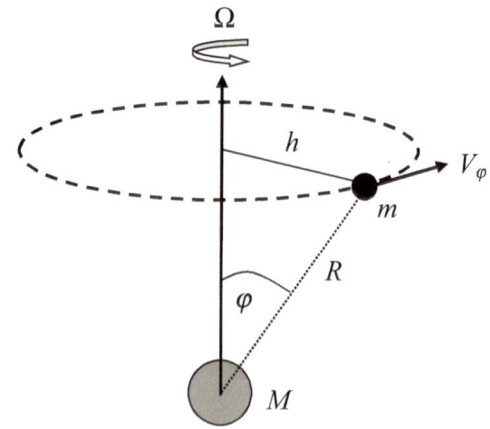

to accrete upon the central protostar. The issue, remember, is what happens to the angular momentum. For matter to move inwards within the disk, it must effectively leave its angular momentum behind. How nature achieves this trick is somewhat unclear, but the answer lies between the physical interaction of the matter within the disk and all those other factors, such as turbulences and magnetic fields, that we have so far not considered. To include such effects takes us well beyond the scope of this text, and we must accordingly let the topic lie unresolved. Suffice to say that detailed numerical models of disk structure can allow for shear interactions between the gas particles at each radial zone within a disk, with the result that material can spiral inwards and successfully leave its excess angular momentum within the disk. Collisions between the inward-moving disk particles and the infalling cloud material result in disk heating, and a characteristic temperature can be defined at each radius r of the disk, such that

$$T(r) = \left(\frac{L_{acc}}{2\pi \, r^2 \sigma}\right)^{1/4} = \left(\frac{G\,M\,\dot{m}}{2\,\pi\,r^3\sigma}\right)^{1/4} \tag{2.11}$$

In developing Eq. (2.11) we have adopted the Stefan-Boltzmann relationship between the luminosity (given by Eq. 2.10), area and temperature, and taken the disk surface area to be $2\pi \, r^2$. Figure 2.6 shows a plot of $T(r)$ against r for various disk radii and accretion rates. In this figure, we illustrate the point that accretion disks can be found in many different kinds of astrophysical systems, not just around protostars. Indeed, accretion disks can exist around ordinary stars within a binary system, as well as within binary systems that contain a white dwarf, a neutron star or a black hole. Also shown in Fig. 2.6 is the temperature variation expected for an accretion disk surrounding a super-massive black hole of mass 10^8 solar masses—typical of that found at the centers of many galaxies. As is evident from Eq. (2.11), the characteristic temperature will be highest at the inner edge of the disk where the radius is small. An accretion disk surrounding a small astrophysical object such as a solar mass black hole might attain an inner-edge temperature in excess of many tens of thousands of degrees, and accordingly emit copious amounts of X-rays (as indicated by Wien's law; Eq. (1.12)). In contrast, the portions of an accretion disk about a solar mass protostar with radii larger than the Earth's orbit will have temperatures of just a few hundred Kelvin and will radiate most strongly in the infrared part of the spectrum. Accretion disks around super-massive black holes, with masses many million times that of the Sun, will not only generate very high accretion luminosities but also attain very high temperatures, resulting in strong emission at X-ray and UV wavelengths.

(☺) **Exercise 2.10** (1) Construct a diagram similar to Fig. 2.6, but in which the accretion luminosity (in solar units) is plotted against radius. In this case, use Eq. (2.10) in the form

$$L_{acc} = \left(\frac{GM}{r}\right)\dot{m}$$

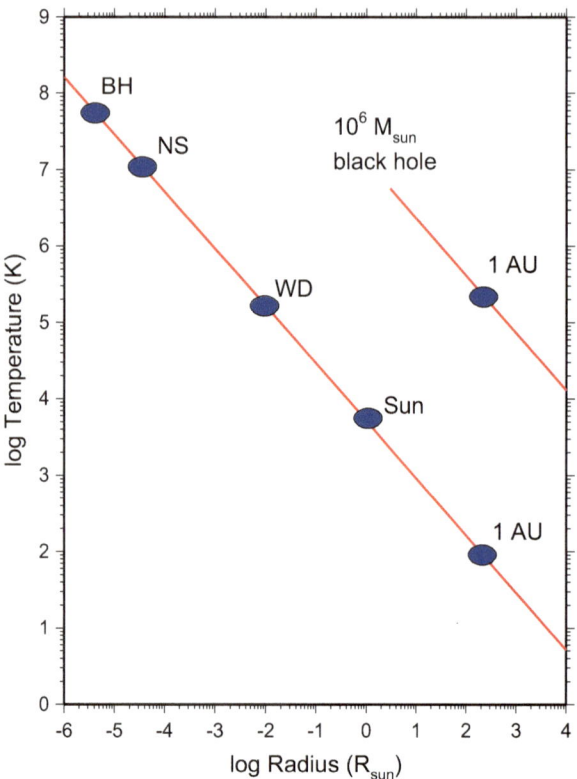

Fig. 2.6 Characteristic temperature variations for various accretion disk radii (in units of $R_\odot$) surrounding a solar mass object (radii are shown for the Earth's orbit (213 $R_\odot$), Sun (1 $R_\odot$), a white dwarf (0.01 $R_\odot$), a neutron star (3×10^{-5} $R_\odot$) and a black hole (4×10^{-6} $R_\odot$)) when the accretion rate is 10^{-8} solar masses per year. Also shown is the characteristic temperature variation for a disk surrounding a million solar mass (radius of 4 $R_\odot$) black hole (similar, in fact, to the super-massive black hole found at the galactic center) accreting matter at a rate of 0.1 solar masses per year. (Credit: Author)

Set $M = 1$ $M_\odot$ and allow for a mass accretion rate of 10^{-8} solar masses per year. (2) Compare the accretion luminosity associated with a Sun-like star to that of a solar mass white dwarf, neutron star, and black hole (see Fig. 2.6 for characteristic sizes). Your results should look something like Fig. 2.7.

While initial cloud rotation is the important characteristic property leading to the formation of an accretion disk, the effects of magnetic fields and turbulent motion cannot be ignored. Since the temperature of a disk increases inward, the ionization state of the gas must also increase inward, and accordingly the dynamical effects of any associated magnetic field will generally become stronger towards its inner edge. The combined effects of rotation and the winding up of the magnetic field lines result in the formation of bipolar jets along the spin-axis of the system, with material being carried away from the central object (Fig. 2.8).

Fig. 2.7 Characteristic luminosity values for various accretion disk radii (in units of $R_\odot$) surrounding a selection of solar mass objects; see Fig. 2.6 for characteristic sizes. (Credit: Author)

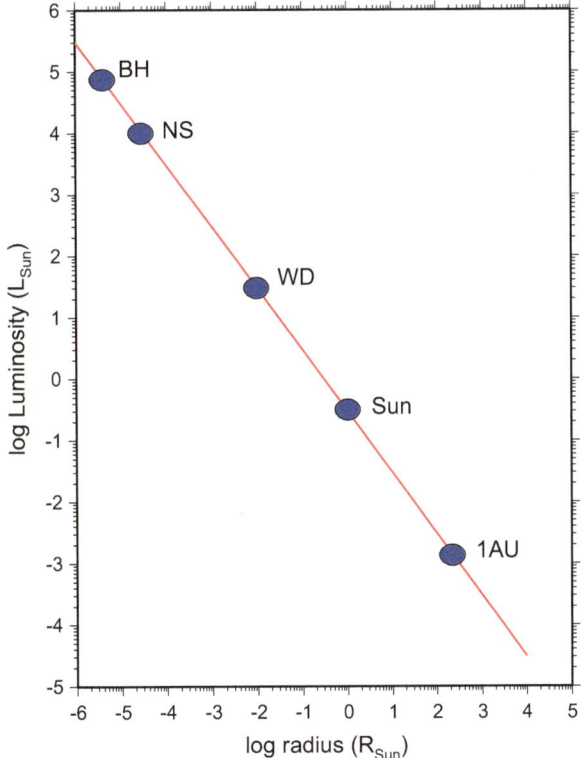

Accretion disks are ephemeral objects. In the ideal case of a newly forming star, the disk will last as long as there is material left within the natal cloud to accrete. In the case of matter transfer in a binary star system, the disk will last as long as material can be accreted from the companion star. In the case of a massive black hole at the center of a galaxy, a disk will likely form, disperse and re-form on numerous occasions according to the erratic supply of material (stars and entire gas clouds in this case) within the core region. Not only will the material supply limit the lifetime of an accretion disk, but so will the effects of the environment, with disruption coming about through tidal interactions with nearby stars as well as interactions with stellar winds and jets. Additionally, a disk might suffer strong wind and/or ultraviolet photoionization disruption from its parent star. In short, the study of accretion disk formation, structure and evolution is highly complicated and remains a topic of cutting-edge research in the modern era.

At this stage, we leave behind the topic of star formation and turn to the consideration of what kinds of stars might be formed. What is the mass range of the objects that we might call stars, and what processes are at play in determining the upper and lower limits to stardom? To do this, we return to a discussion of fundamental physical characteristics.

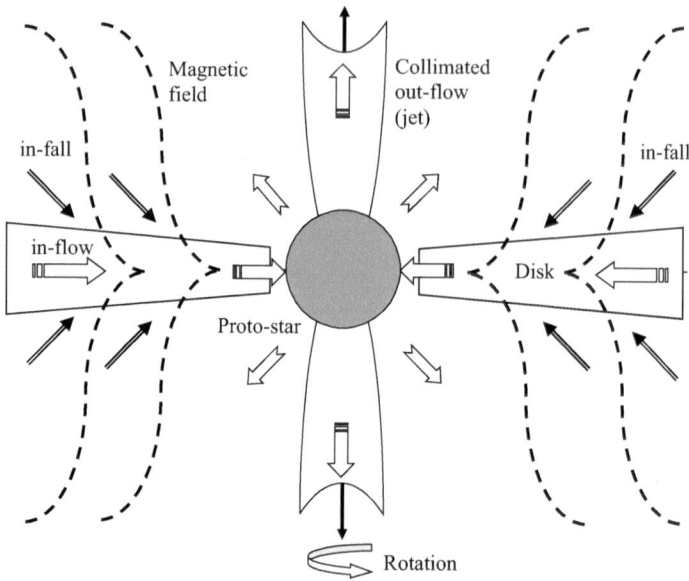

Fig. 2.8 Schematic cross section through an accretion disk located about a newly forming protostar. Material falls onto the disk surface from the surrounding (and collapsing) natal cloud, and with appropriate angular momentum sorting moves through the disk to be accreted by the central protostar. As a consequence of the high surface temperature of the protostar, however, material at the inner edge of the disk will become ionized and therefore begin to strongly interact with any disk-threaded magnetic field lines. As a consequence of disk rotation, the magnetic field lines wind up and become collimated about the system spin axis, forming mass-ejecting outflows (or jets). (Credit: Author)

2.4   Eddington's Blank Page

The philosophical reasoning of Arthur Eddington has fallen out of favor in recent decades, but it nonetheless contains many useful gems. One of Eddington's most useful and engaging philosophical ideas was the blank page principle. An empty page to Eddington was a challenge, and its blankness needed to be filled with equations and insight. Eddington's Fundamental Theory (published posthumously in 1946) was predicated upon the notion that the structure and working of the universe could be understood and divined entirely through thought, logic and pure mathematical reasoning. The popularity of Eddington's approach has ebbed and flowed, but his basic idea builds upon the concept of dimensional analysis and the existence of various dimensionless numbers—pure numbers that are either surprisingly very large or extremely small.

Eddington's principle of the blank page builds upon one of the most fundamental insights of physics: the fact that there are fundamental constants that underpin the very fabric and construction of all matter. As to whether these fundamental constants are *just what they are*, with no rhyme or reason, or whether they are constrained by some deep, currently unknown underlying physical theory is unclear (see Sect. 6.5). However, the present-day collection of physical theories that describe the atomic realm, the macroscopic world and the architecture of the universe requires the identification of some 32 fundamental constants. We need only consider a few of these constants here: c, h, G, e, m_p and m_e, corresponding to the speed of light, Planck's constant, the gravitational constant, the electron charge, the mass of the proton and the mass of the electron. It is truly astonishing that with this set of six constants, the basic nature and characteristic quantities that define a star can be explained to a good order of magnitude. The numerical toolset at our disposal (with the numbers in all their measured glory) is:

$$c = 2.99792458 \times 10^8 \text{ m/s}$$

$$h = 6.626070040 \times 10^{-34} \text{Js}$$

$$G = 6.67408 \times 10^{-11} \text{ m}^3/\text{kg/s}^2$$

$$e = 1.6021766208 \times 10^{-19}\text{C}$$

$$m_p = 1.672621898 \times 10^{-27}\text{kg}$$

$$m_e = 9.10938356 \times 10^{-31} \text{kg}$$

From these constants, a whole series of dimensional as well as pure number (dimensionless) constants can be constructed. The later reluctant founder of quantum mechanics, Max Planck, took these basic and fundamental constants and formulated a further set of characteristic units for size, mass and time. These are:

$$\text{Planck length}: R_{pl} = \left(\hbar G/c^3\right)^{\frac{1}{2}} = 1.6162 \times 10^{-36} \text{ m}$$

$$\text{Planck time}: \quad T_{pl} = \left(\hbar G/c^5\right)^{\frac{1}{2}} = 5.3912 \times 10^{-44} \text{ s}$$

$$\text{Planck mass}: M_{pl} = \left(\hbar c/G\right)^{\frac{1}{2}} = 2.1765 \times 10^{-8} \text{ kg}$$

Note that the so-called reduced Planck constant $\hbar = h / 2\pi$ (this notation was introduced by Paul Dirac) is used in the derivation of the Planck units. The units derived by Planck are often called 'natural units', as they are based entirely on universal constants rather than arbitrary human standards. For all this, however, there is no clear physical significance to be associated with them. It is generally assumed that the Planck length signifies the domain below which quantum gravity theories must apply. In string theory, for example, the Planck length is the approximate scale of the oscillating strings that form elementary particles. Again, the Planck time is taken as a measure of the shortest time over which quantum gravitational effects are likely to be revealed. The Planck mass is somewhat odd in the sense that it sits close to the realm of actual human experience. The Planck mass is large primarily because the gravitational force is so weak (that is G is small).

An additional set of dimensionless pure numbers can also be constructed from our toolset of six universal constants, revealing:

Fine structure constant: $\alpha = e^2/\hbar c = 7.3 \times 10^{-3} \sim 1/137$

Gravitational coupling constant: $\alpha_g = G m_p{}^2/\hbar c = (m_p/M_{p1})^2 = 5.9 \times 10^{-39}$

The fine structure constant characterizes the strength of the electromagnetic interaction between elementary charged particles. The gravitational coupling constant is a measure of the gravitational attraction between a given pair of elementary particles (note that the electron mass can also be used in the definition of α_g). With this diverse range of pure numbers and dimensional constants, we have a skeletal framework upon which the cloths of the stars can be hung. An immediate association between our constants and the Sun, a relatively middling mass star (see Sect. 6.1) of mass $M_\odot = 2 \times 10^{30}$ kg, can be established, where $M_\odot = N\, m_p$, with $N = \alpha_g{}^{-3/2}$. N is a measure of the typical number of nucleons within a star. Indeed, to within a numerical factor $0.1 < s < 10$, the mass range of the stars can be written as $M = s\,N\,m_p$.

In order to see why this relationship for stellar masses comes about, we need to backtrack to the Virial theorem introduced in Sect. 1.1, and consider the temperature evolution during gravitational collapse. The Virial theorem provides a relationship between the average kinetic energy E_K of the material components within a star and the time average of its gravitational potential energy E_P, giving $E_K = \frac{1}{2} E_P$. Since the kinetic energy of the electrons and protons within a star are the result of their thermal motion, the equipartition of energy dictates that the total internal kinetic energy of the star will be $3kTN$, where k is the Boltzmann constant and T is the temperature. This quantity can be equated to the gravitational potential energy $G\,(N\,m_p)^2\,/\,R$, where R is the radius of the collapsing cloud—here we have assumed that the typical separation between protons is of order the radius of the star (some will be just a few proton radii away; others will be the star's diameter away) and that there are N^2 pairs of protons. The equilibrium condition described by the Virial theorem now dictates that

$$E_K = 3kT\ N = \frac{1}{2} E_P = \frac{1}{2} G\big(N m_p\big)^2 / R \qquad (2.12)$$

which rearranges to provide an expression

$$kT = \frac{1}{6} \hbar c \left(\frac{N^{1/3}}{R} \right) \qquad (2.13)$$

where we have substituted[3] for $N = \alpha_g{}^{-3/2}$, and recall $\alpha_g = G\,m_p{}^2\,/\,\hbar\,c$. The bracketed term in (2.13) introduces a characteristic length d, which is the typical separation

---

[3] (☺) Exercise 2.11: Derive Eqs. (2.13) from (2.12).

between neighboring protons—it is the linear dimension of the volume that can be associated with each proton such that $d^3 = R^3 / N$. Upon substitution we obtain a useful result that expresses the typical spacing d between protons and the temperature T, such that

$$kT = \frac{1}{6}\hbar c \frac{1}{d} \tag{2.14}$$

Equation (2.14) underscores the earlier result that as the cloud of material out of which a star is going to form decreases in size, so the typical distance between protons decreases and the temperature of the cloud increases. The process of gravitational collapse will in principle continue to infinite compression, resulting in the formation of a black hole, unless one of two possible alternatives plays out. One alternative relates to the temperature evolution, while the other depends upon a quantum mechanical effect acting upon the cloud's constituent particles (namely the electrons; more on this later). We have seen that as the star-forming cloud collapses, so its internal temperature increases; at a temperature of about 10^4 K, the collisional energy between neighboring hydrogen atoms is sufficiently high that they become dissociated into a gas of protons and intermixed electrons, the electrons no longer occupying bound orbitals. Further contraction continues to drive the temperature upwards, such that once the temperature reaches some $T_{nuc} \geq 10^7$ K, proton pairs have sufficient energy to penetrate the Coulomb barrier active between them and accordingly come within range of the strong nuclear force. Under these circumstances, nuclear fusion can begin, and an important energy source has been unleashed. With $T \geq T_{nuc}$, the energy lost into space at the outer boundary of the collapsing cloud can be compensated for by the energy generated by fusion reactions, and accordingly the cloud need no longer collapse in order to generate energy through the release of gravitational potential energy. Indeed, at this stage a *bona fide* star has formed, leading us to a simple first attempt at a definition for the attainment of stardom:

Definition 2.1 A star is an object that is capable of maintaining a sustained period of dynamic equilibrium through the generation of energy by fusion reactions within its interior.

Like all definitions, the shorter they are, the more elastic they need to be, and there are certainly additional preconditions and qualifications that need to be added to definition 2.1; these will be explored at later points within the text.

(☺) **Exercise 2.12** Taking the temperature to be $T = T_{nuc} = 10^7$ K, (1) use Eq. (2.14) to determine the characteristic scale term d. (2) Express d in terms of the characteristic size of a proton (1×10^{-15} m), and (3) compare d to the Planck length R_{pl}. ☹ Think about what the latter two comparisons mean.

---

[3](☺) Exercise 2.11: Derive Eqs. (2.13) from (2.12).

Cloud collapse, as we have argued, will stop if the temperature reaches a level at which nuclear fusion reactions can begin (this is the topic of Chap. 4). The collapse can additionally be halted, however, if the electrons within a cloud begin to feel the presence of their neighbors. The reluctance of electrons to crowd together is described according to the Pauli Exclusion Principle. Introduced by Wolfgang Pauli in 1925, the exclusion principle is a quantum mechanical constraint that dictates that no two electrons in the same quantum state can occupy the same region of space.

The wave-particle duality that sits at the very core of quantum mechanics is expressed through two relationships: the de Broglie formula for the momentum, and Planck's formula for the energy. The de Broglie equation connects a particles momentum P to its associated wavelength λ, with $P = h / \lambda$. Planck's equation relates the energy E to a specific frequency f, with $E = hf$. When a particle such as an electron is confined to a finite region of space, the confined wavelength must be comparable to the linear dimensions of the confinement. Since the maximum wavelength via de Broglie's formula corresponds to a minimum momentum, a confined electron must have kinetic energy K_e at least as large as the one corresponding to the maximum wavelength, and accordingly: $K_e = \frac{1}{2} m_e V^2 = P^2 / 2m_e = 2\pi^2 \hbar / m_e \lambda^2$, where V is the velocity and m_e is the mass of the electron. To see where this becomes important for a collapsing cloud, we note that K_e increases with decreasing λ. Now, the kinetic energy of the electrons must come from the release of gravitational potential energy by the collapsing cloud, and as more and more of the lower energy states available to the electrons are filled, in accordance with the Pauli Exclusion Principle, the condition may arise that the gravitational energy per electron is comparable to the minimum kinetic energy level K_e. Under these conditions, the electrons are squeezed so tightly together that the contraction must stop. The contraction stopping condition can be written according to $d_{stop} \approx \lambda$, and by using Eq. (2.14) above, we have

$$kT = \frac{1}{6}\hbar c \frac{1}{d_{stop}} \approx \frac{2\pi^2 \hbar^2}{m_e d_{stop}^2}$$

which yields a value for d_{stop} of

$$\frac{1}{d_{stop}} = \frac{1}{12\pi^2}\left(\frac{m_e c}{\hbar}\right) \tag{2.15}$$

(☺) **Exercise 2.13** (1) Using Eq. (2.15), evaluate the characteristic stopping distance d_{stop}. (2) Compare d_{stop} to the typical separation between protons at the onset of nuclear fusion (recall exercise 2.12).

The d_{stop} condition given in Eq. (2.15) effectively sets a lower limit for the central temperature of a star and provides our second condition for the stopping of cloud collapse. In this second case, however, we do not have a star as defined in definition

2.1, when the contraction stops. Evaluating Eq. (2.15) with $d = d_{stop}$ yields a temperature of $T_{stop} \approx 8 \times 10^6$ K $< T_{nuc}$. This lower temperature condition can be further expressed in terms of a minimum stellar mass M_{min} (although the resultant equation for this mass is far from pretty). As argued earlier, the d_{stop} condition is established on the basis that $d_{stop} \approx \lambda = h / (m_e V)$, and we additionally have for the electrons that $\frac{1}{2} m_e V^2 = 3 k T / 2$, which enables us to write $d_{stop} \approx h (3m_e k T)^{-\frac{1}{2}}$. Now, the density of matter ρ under the d_{stop} condition will be $\rho = \mu m_p / [(4/3) \pi (d_{stop})^3]$, where μ is the mean molecular weight (to be more fully discussed in Chap. 3); for this calculation, we take $\mu = 1/2$. Using the Virial theorem one more time, the kinetic energy per proton can be written as

$$kT = \frac{1}{3} G \frac{M \mu m_p}{R}$$

where M and R are the mass and radius of the minimum mass star to be derived. We now substitute for the μm_p term, express the radius as $R = [3 M / (4 \pi \rho)]^{1/3}$ and set $M = M_{min}$. A good amount of algebra[4] eventually yields an expression (albeit a rather ugly one) for M_{min} as

$$M_{min} = \left(\frac{3}{4\pi}\right)^{1/2} \left[9 k T_{stop} \left(\frac{h}{G}\right)^2 \frac{1}{m_e (\mu m_p)^{8/3}}\right]^{3/4} \tag{2.16}$$

Equation (2.16) is our estimate for the minimum mass of a star—that is a stable (non-collapsing) body with a temperature $T \sim T_{stop}$. The right-hand side of Eq. (2.16) is largely expressed in terms of fundamental constants, and substituting for $T = T_{stop}$ and $\mu = \frac{1}{2}$, it is found that $M_{min} \sim 9 \times 10^{29}$ kg$\sim 0.4 M_\odot$. This result establishes, to an order of magnitude, our previously expressed lower mass limit for stardom: namely, $M_{min} \approx s N m_p$ with $s \sim 0.1$ and $N = \alpha_g^{-3/2}$.

To determine an upper mass limit M_{max} for stardom, we need to ask what happens within a star once the temperature becomes very high, and we have to anticipate a result to be derived later in Chap. 4. As will be revealed later, the temperature of a star increases linearly with its mass and inversely according to its radius. The higher the temperature of a star, the greater its radiation pressure P_{rad} will be in comparison to the gas pressure P_{gas}, and accordingly a maximum stellar mass condition is set according to the ratio $P_{rad} / P_{gas} > 1$. This condition provides the result

$$\frac{P_{rad}}{P_{gas}} = \frac{\frac{1}{3} a T^4}{n k T} > 1 \tag{2.17}$$

where a is the radiation constant and n is the number density of free particles: $n = 2\rho / m_p$, where it has been assumed that the star is composed entirely of

---

[4] (☺) Exercise 2.14. Take a large sheet of scrap paper and verify the derivation of equation (2.16).

hydrogen, which once ionized yields two particles, a proton and an electron, per initial hydrogen atom. Anticipating the result to be derived in Chap. 4, the temperature T of a star is characterized according to its mass M and radius R as

$$T \approx \frac{Gm_p}{2k}\left(\frac{M}{R}\right) \tag{2.18}$$

By combining Eqs. (2.17) and (2.18) with the substitution $\rho = M/(4\pi R^3/3)$, and setting $M = M_{max}$, we find that[5]

$$M_{max} = \left[\frac{36}{\pi}\frac{k^4}{a m_p^4 G^3}\right]^{1/2} \tag{2.19}$$

Upon substitution for the constants in Eq. (2.19), we find $M_{max} \approx 2 \times 10^{31}\text{kg} \sim 10 M_\odot$. This sets the order of magnitude estimate to our previously expressed upper mass limit for stardom: $M_{max} \approx s\, N\, m_p$ with $s \sim 10$ and $N = \alpha_g^{-3/2}$.

The order of magnitude results derived above carry us only so far with respect to understanding the stars—we have applied little more than dimensional analysis. Such arguments indicate characteristic values, sizes, masses and temperatures, but they do not reveal the whole story. To go further, we need to develop a set of differential equations that are capable of describing the full range of physical variables and their variation under the conditions pertaining to stardom—and that is exactly what the next three chapters will do.

(☺) **Exercise 2.16: Term Paper Project** Investigate the history, development and usage of Max Planck's natural units. A good introductory paper on this topic is given by John Barrow: Natural Unites before Planck, published in *The Quarterly Journal of the Royal Astronomical Society*, 24, 22–26 (1983).

2.5 Answers to Exercises

Following are the answers to all exercises in this chapter. The exercises for which no answer is provided are those in which the student is asked to check a result or complete the algebra steps between two equations given in the text.

---

[5](☺) Exercise 2.15. Once again, take a large sheet of scrap paper and verify the derivation of equation (2.19).

Exercise 2.1 About 116 protons per cubic centimeter.

Exercise 2.2 (1) This is some 4.2×10^{26}. (2) $R = 4646$ km. Accordingly, to account for the number of protons contained within just one lung full of air, a sphere having a radius ~73% that of the Earth of interstellar medium will need to be sampled.

Exercise 2.3 $N_{gal} = 33 \times 10^9$, and $N^* = 10^{22}$.

Exercise 2.4 $n_U = 9 \times 10^{-10}$ stars per cubic parsec of space, which is about 125 million times smaller than the density of stars in the solar neighborhood.

Exercise 2.6 (1) In this case, $R_{Jrot} = (3\,G\,M/\Omega^2)^{1/3}$. (2) For a 1 $M_{\odot}$ cloud, $R_{Jrot} = 2.4 \times 10^{15}$ meters (taking $\Omega = 10^{-13}\,\mathrm{s}^{-1}$), and accordingly $R_{Jrot} / R_{Jeans} = 10$. (3) $\beta = 0.001$ when $\Omega = 10^{-13}\,\mathrm{s}^{-1}$, which indicates that the rotational energy is very small. (4) Starting with the relationship derived in (1), substitute for $V_{rot} = \Omega\,R$ and $M = 4\,\pi\,\rho\,R^3 / 3$, and the rest follows.

Exercise 2.8 $R_2 \approx 76\,R_{\odot}$ (indicating that the cloud radius has decreased by a factor of about 24,000).

Exercise 2.12 (1) $d = 3.8 \times 10^{-11}$ meters. (2) $d / D(\text{proton}) \approx 38,000$. (3) $d / R_{pl} = 2.4 \times 10^{25}$. The latter two ratio tells us that the protons are typically well separated from each other and that quantum gravity effects can be safely ignored.

Exercise 2.13 (1) $d_{stop} = 4.6 \times 10^{-11}$ meters. (2) $d_{stop} / d = 1.2$. This latter ratio essentially sets a lower limit on the temperature for a star, since if d_{stop} is very much greater than d, then degeneracy sets in before nuclear reactions can occur.

Chapter 3
From the Outside In

A star's a star some matter in a ball
compelled to courses mathematical
amid the regimented, cold, inane,
where destined atoms are each moment slain

J. R. R. Tolkien, Mythopoeia

We tend to think of stars as being stable and ancient structures—the constants of the sky. Even in a long human lifetime of 100 years, the heavens hardly change: the constellations are the same, the Sun is just as hot and luminous in the summer months, and the day is still 24 h long. This constancy is a consequence of our perception of time. The Sun may seem constant to our senses and memory, but it is undergoing continuous change and struggle. Each second, its composition alters ever so slightly (as we shall see in Chap. 4), and it fights against the ever-present threat of gravitational collapse.

All may seem calm and constant to the human eye, but the inside of the Sun, and that of the stars in general, is a seething battleground of checks and balances, atomic clashes, photon interactions and nuclear explosions. Arthur Eddington described the inside of a star as being akin to the chaos of a "jolly crockery smashing stall". He was entirely right. The inside of star is no place for the faint of heart or for those in search of a quiet life. All is in turmoil and motion. And yet, out of all this thrashing of atomic interactions and heaving of turbulent gas, stability—of a sort—is found. The stability is dynamic, and the Sun continually teeters on the brink of dramatic change. Just one slip, and the result would be devastation. If, for example, gravity were to triumph over the outward pressure forces acting within the Sun, not only would it collapse and plunge us into darkness and death, but it would do so in the same amount of time that we take for our daily lunch-hour. Rapid and continuous change is the name of the game within stellar interiors, but the net result of all this give and take is stability. How can this be?

Eddington, our historical guide, presented a very good minds-eye picture for a star. We can think of a star as two superimposed domains. One domain is concerned with physical stability, while the other is concerned with energy generation and the

© Springer Nature Switzerland AG 2019   69
M. Beech, *Introducing the Stars*, Undergraduate Lecture Notes in Physics,
https://doi.org/10.1007/978-3-030-11704-7_3

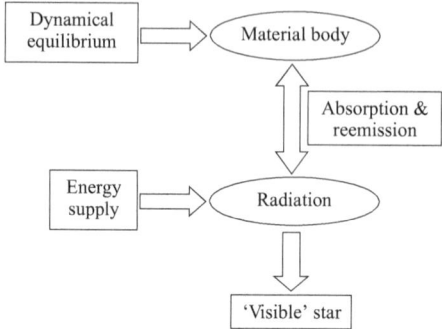

Fig. 3.1 Schematic diagram of the two superimposed domains of a star. The material body is made up of atoms, ions and electrons, while the radiative body is composed of electromagnetic radiation. The two domains remain stable through a continual cascade of absorption and reemission interactions between the electromagnetic radiation and the star's constituent atomic nuclei and electrons. (Credit: Author)

outward flow of radiation (Fig. 3.1). The two domains must interact, and neither can exist without the other, but we can treat them as mathematically separate. This chapter is concerned with the domain of stability, and the forces that act upon a star's constitutional atoms.

As argued in Chap. 1, the observed properties of a star of total mass M are described according to its radius R, surface temperature T, luminosity L and surface composition X_i, $i = 1, 2, 3, \ldots$. The theory of stellar structure takes us inside of the star, and the problem of stellar structure, as Chandrasekhar argued, is concerned with finding a mathematical solution to the equations $F(L, M, R, X_i) = 0$. Stated in words, the problem of stellar structure is:

Given a star of mass M and chemical composition X_i, $i = 1, 2, 3, \ldots$, what must the conditions within its interior be so that it has a specific radius R and luminosity L?

Note that we need not specify a specific surface temperature T in this condition, since it is given by the Stefan-Boltzmann law (recall Chap. 1): $L = 4\pi R^2 \sigma T^4$ To express the conditions within the interior of a star we introduce the following quantities, all expressed as a function of the distance r from the star's center:

Radius	$0 \leq r \leq R$
Mass	$M(r)$
Temperature	$T(r)$
Density	$\rho(r)$
Composition	$X_i(r), i = 1,2,3, \ldots$
Pressure	$P(r) = P(r, \rho, T, X_i)$

Later on in Chap. 4 we will introduce additional equations for describing the energy generation and the opacity. The pressure term $P(r)$ is additionally described by an equation of state, and this in turn will be a function of the density, temperature and composition that apply at a particular position r within the star. For a perfect gas

$$P_{gas}(r) = n(r) k T(r) \tag{3.1}$$

Where $n(r) = \rho(r)/\mu m_H$ is the number density of particles at position r, μ is the mean molecular weight (to be explained shortly), m_H is the mass of the hydrogen atom and k is the Boltzmann constant. At this stage we introduce a bit of nomenclature and express the composition of a star according to its hydrogen mass fraction X, its helium mass fraction Y and the mass fraction of all the other elements Z. By definition, these three terms must sum to unity: $X + Y + Z = 1$. In this manner we have

X = hydrogen mass fraction = X_1
Y = helium mass fraction = X_2
Z = 1 – X – Y = mass fraction of all other elements X_i, $i = 3, \ldots, 94$

Astronomers typically refer to those elements other than hydrogen and helium as being metals, and the Z term is generally called the metallicity. The mean molecular weight μ can now be determined upon the basis that all of the atoms within a star are fully ionized—they are fully stripped of all their electrons. In the interior of a star, full ionization is typically a good approximation, but it does not hold true in a star's upper envelope (as described in Sect. 1.6). Each hydrogen atom will contribute two particles, an electron and a proton, to the internal mix. Each helium atom of atomic mass 4 m_H will contribute three particles, two electrons and its nucleus, to the internal mix. The ion contribution from Z elements is typically going to be small, firstly because Z itself is small and secondly because the atomic mass A of such atoms is generally large—lithium, the third element in the periodic table, has an atomic mass of $A = 7$, and accordingly the ion contribution from Z elements is usually ignored. The number of electron contributed by Z elements, however, is typically going to be half the atomic mass $A / 2$. The number density of particles in a fully ionized gas can now be written as

$$n = \frac{\rho}{m_H}\left(2X + \frac{3}{4}Y + \frac{Z}{2}\right) = \frac{\rho}{\mu \, m_H}$$

and this provides an expression for the mean molecular weight as

$$\frac{1}{\mu} = 2X + \frac{3}{4}Y + \frac{Z}{2} \tag{3.2}$$

A star composed entirely of hydrogen $X = X_1 = 1$ will accordingly have a mean molecular weight of $\mu = 0.5$. For the Sun, it is found that the various mass fraction terms are $X = 0.71$, $Y = 0.27$ and $Z = 0.02$, indicating a mean molecular weight of $\mu = 0.613$.

(☺) **Exercise 3.1** Convince yourself that ½ is the appropriate multiplier of the Z element term in Eq. (3.2). Consider the following as an outline argument: For some metal X_i, with $i > 2$ (that is an element heavier than helium), the atomic mass will be A_i the atomic number will be z_i and the number of electrons will be z_i. Since for most metals the number of protons in the nucleus is roughly the same as the number of neutrons, so $A_i \approx 2\,z_i$. When such a metal is fully ionized the contribution to the number density of particles will be $(1 + z_i)/A_i \approx (1 + z_i)/2z_i = 1/2z_i + 1/2 \approx 1/2$ since the first term is always going to be smaller than 1/6, this being the $1/2z_i$ term for lithium ($i = 3$) for which $z_i = 3$. Carbon, corresponding in our notation to $i = 6$, has $A_i = 12.011$ and $z_i = 6$, and accordingly $(1 + z_i) / A_i = 7/12.011 = 0.58$. Iron has $A_i = 55.845$ and $z_i = 26$, and accordingly $(1 + z_i) / A_i = 27/55.845 = 0.48$.

When radiation pressure is dominant within the interior of a star, the pressure term will vary as:

$$P_{rad}(r) = \frac{1}{3} a\,T^4(r) \tag{3.3}$$

where a is the radiation density constant. Progress towards our first stellar models can now be made by developing two differential equations: one to describe the mass distribution within the star $dM(r) / dr$ and the second to describe the pressure gradient $dP(r) / dr$.

3.1 Hydrostatic Equilibrium

The stability of a star is expressed in terms of hydrostatic equilibrium. This condition requires that at every location r within the interior of a star, the sum of forces acting upon a given mass element δm must vanish. If the inward and outward forces acting upon the mass element did not balance, then the mass element would move either upwards or downwards according to which force dominated. The inward directed force acting to pull the mass element towards the star's center is gravity, while the outward directed force acting upon the same mass element is derived from the pressure gradient. Using Fig. 3.2 as a guide, imagine that we are examining a small cylindrical element of mass δm situated between r and $r + \delta r$. The inward gravitation force acting on the mass element will be entirely due to the mass $M(r)$ contained within radius r, and accordingly $F_{grav}(r) = G\,M(r)\delta m / r^2$, where G is the gravitational constant. In hydrostatic equilibrium the force due to the pressure exactly equals that due to gravity. The outward acting force at the base of our mass elemental at r is $F_{out}(r) = A\,P(r)$, where A is the base area of our mass element. The inward acting force at the top of our mass element at $r + \delta r$ is $F_{in}(r + \delta r) = A\,P(r + \delta r)$. The requirement that the inward and outward forces that

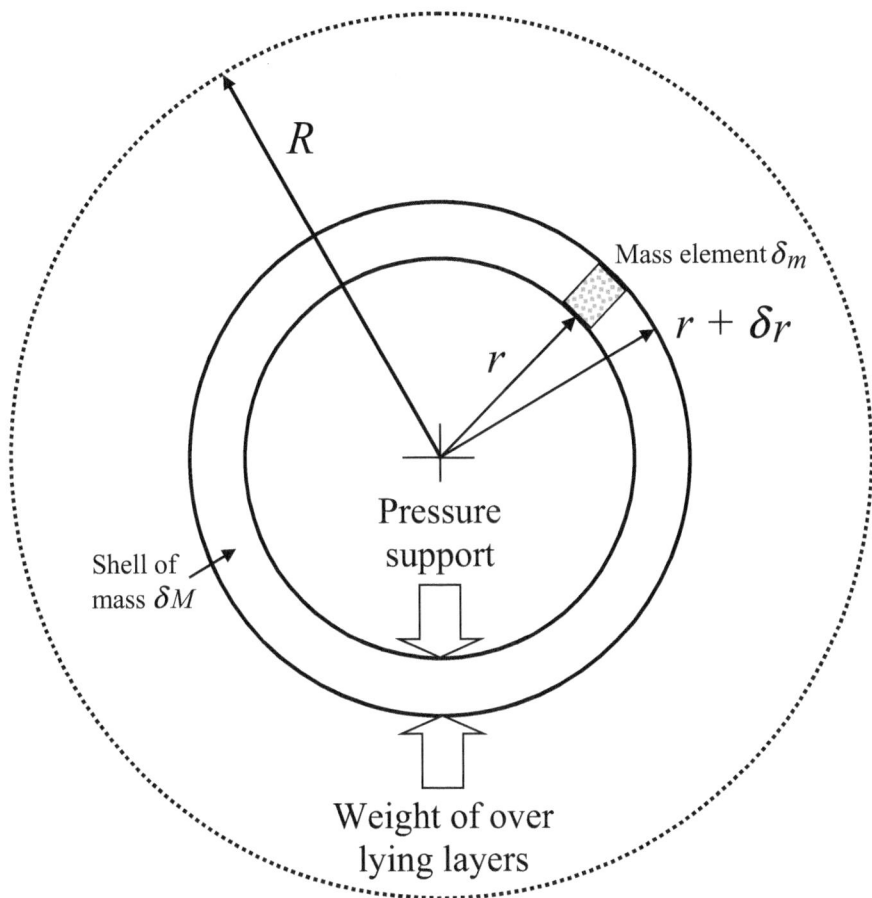

Fig. 3.2 At each layer of radius r within a star, the outward pressure balances the weight of overlying layers. (Credit: Author)

act upon the mass element can now be cast as: $F_{in}(r + \delta r)$ - $F_{out}(r) + F_{grav}(r) = 0$. Substituting for terms we have

$$AP(r + \delta r) - AP(r) + GM(r)\delta m / r^2 = 0 \tag{3.4}$$

Since we are considering very small elements, we can express the pressure difference across our mass element as $P(r + \delta r)$ - $P(r) = (dP / dr) \, \delta r$, where dP / dr is the pressure gradient. Additionally, the mass of our small element can be written as $\delta m = A \rho(r) \, \delta r$, and this allows us to rewrite Eq. (3.4) as

$$\frac{dP}{dr} = -G\frac{M(r)}{r^2}\rho \tag{3.5}$$

(☺) **Exercise 3.2** Complete the algebra that takes Eq. (3.4) to Eq. (3.5).

Equation (3.5) is the first of our equations for describing the interior structure of a star, and it is appropriately called the equation of hydrostatic equilibrium. The differential equation expressed in (3.5) has the independent variable r and three dependent variables P, M and ρ. Accordingly, if we are to solve for the whole system, we will need to find at least two more equations. The next equation to be derived is that corresponding to mass continuity, and again Fig. 3.2 acts as our visual guide. In this case, the mass increment δM that is associated with a complete shell of material situated between r and $r + \delta r$ is related to the density of material ρ at radius r and the volume of the shell V_{shell}. The volume of the shell can be expressed in terms of its surface area $4\pi r^2$, and its thickness δr. Now, we have $\delta M = 4\pi r^2 \rho \, \delta r$, which when we take to the limit gives our second equation of internal structure:

$$\frac{dM}{dr} = 4\pi r^2 \rho \qquad (3.6)$$

We immediately see from Eq. (3.6) that if the density is taken as being $\rho = $ constant, then a simple integration of Eq. (3.6) between $0 \leq r \leq R$ brings the familiar result that the density is the total mass divided by the volume, where in this case the volume is that of a sphere of radius R. Indeed, this latter example reminds us that we have assumed that the star under consideration is spherically symmetric (it is a sphere). This of course need not be the case, and it is a topic that will require a little more discussion later on.

Equations (3.5) and (3.6) can be solved for provided we can find one more equation linking together the pressure and the density. In general, this can be done through the equation of state, of which two simple forms are given in Eqs. (3.1) and (3.3). Before considering how the differential equations for hydrostatic equilibrium and mass continuity can be solved for analytically (see Sect. 3.2 below), we should first consider what will happen to a star if it no longer finds itself in a state of hydrostatic equilibrium. If the net force acting upon our small δm mass element is not zero, then it will begin to move with some acceleration a. The acceleration of the small mass element can be described by Newton's second law of motion, with the result of introducing a non-zero term $a \, \delta m$ on the right-hand side of Eq. (3.4). With this addition, the equation of motion becomes

$$\rho a = \rho \frac{d^2 r}{dt^2} = G \frac{M\rho}{r^2} + \frac{dP}{dr} \qquad (3.7)$$

Technically, we should use partial derivatives for the pressure derivative on the right-hand side of Eq. (3.7), since it is now considered to be a function of both position and time, but we shall let this mathematical detail slide for the moment. To further ruffle the feathers of the mathematicians, we shall in addition approximate the derivatives in Eq. (3.7) by dimensionally equivalent forms. Accordingly, we have $(d^2 r / dt^2) \Rightarrow R / t^2$, where R is the radius of the star and t is a characteristic time, and

$(dP / dr) \Rightarrow <P> / R$, where $<P>$ is a characteristic pressure term averaged across the star. With these approximations in place, we can now consider two limiting cases of (3.7): one case corresponds to the sudden disappearance of the outward pressure gradient, and the other corresponds to the sudden 'turning off' of gravity. Neither of these two limiting cases can actually be realized, but they do provide a characteristic maximum response timescale to deviations from hydrostatic equilibrium. In the first approximation the pressure gradient is imagined to vanish, with $(dP / dr) = 0$, and Eq. (3.7) accordingly reveals that

$$\rho \frac{R}{t_{coll}^2} = G \frac{M}{R^2} \rho \tag{3.8}$$

where t_{coll} is the characteristic timescale of collapse. A little more algebra reveals a characteristic collapse time of

$$t_{coll} = \frac{\eta}{\sqrt{G\rho_{bulk}}} \tag{3.9}$$

where $\eta = (3/4\pi)^{1/2} \approx 0.5$ and ρ_{bulk} is the bulk density of the star. The collapse time is determined by the inverse square root of a star's bulk density. The higher the density of the star, the more rapidly it will collapse under gravity if the outward pressure support becomes insignificant. The collapse time is often referred to as the dynamical timescale t_{dyn}, and it is a measure of the characteristic period with which the interior of a star will vibrate in response to some small mechanical disturbance, for example in response to the impact of a large asteroid or cometary nucleus.

(☺) **Exercise 3.3** Complete the algebra that takes Eq. (3.8) to Eq. (3.9).

(☺) **Exercise 3.4** Determine the bulk density for the Sun and calculate its dynamical collapse time.

The second limiting case to consider is that in which we imagine gravity to suddenly 'switch off'[1] In this case, Eq. (3.7) gives us

$$\rho \frac{R}{t_{exp}^2} = \frac{<P>}{R} \tag{3.10}$$

where $<P>$ is now a characteristic pressure term averaged across the star. With gravity no longer effectively holding the internal matter in place, the star will expand and disperse into its surroundings on a characteristic timescale given by

---

[1]There is, of course, no prospect of this ever happening, but remember this is a limiting case that allows us to find a timescale. In actuality, the outward pressure forces can under certain conditions overpower gravity and a star can expand, for example in the giant stage of evolution.

$$t_{exp} = \frac{R \, \gamma^{1/2}}{V_{sound}} \qquad (3.11)$$

where $V_{sound} = (\gamma < P/\rho>)^{1/2}$ is the characteristic sound speed within a star's interior, $\gamma \approx 1$, is the adiabatic gas constant, and we have adopted a star averaged value of the ratio P / ρ. This expansion time makes physical sense in that it corresponds to the sound crossing time across the star, and sound waves are basically pressure variation waves. For the Sun, the sound speed steadily decreases from a maximum of about 350 km/s at its center outward. Taking the maximum as a characteristic speed, however, yields $t_{exp} \approx 2000$ s, or some 33 min.

(☺) **Exercise 3.5** Complete the algebra that takes Eq. (3.10) to Eq. (3.11).

Having shown that the gravitational collapse and expansion times for the Sun in the limiting cases of hydrostatic equilibrium failure are shorter than 1 h, it is clear that at no time during the 4.5-billion-year age of the Solar System has the gravitational force dominated over that of the outward pressure or the outward pressure force dominated over gravity. In other words, the Sun must have maintained hydrostatic equilibrium to a very high order of accuracy. Some measure of how well the Sun has managed to maintain its state of hydrostatic equilibrium can be gauged if we imagine that the inward gravitational force is only very slightly greater than that of the outward pressure. If we assume that the net imbalance between the inward and outward acting forces is given by some factor λ of the acceleration due to gravity at the surface of a star, $g = G \, M/R^2$, then after a time t the inward displacement S will be of order: $S = \frac{1}{2}\lambda g \, t^2$. The time for the Sun, therefore, to decrease in size by say 25% (that is $S = R / 4$) will be $t_{25} = (\frac{1}{2}R/\lambda g)^{1/2}$, which upon taking $R = 1 \, R_\odot$ and $g = 273.4$ m/s^2 gives $t_{25} \approx 19 / \lambda^{1/2}$ min. Given that the geological record[2] effectively rules out any such change in the size of the Sun on a timescale of at least a billion years, this indicates that λ can be no larger than about 10^{-27}. Under these conditions, it seems safe to conclude that the equation of hydrostatic equilibrium—Eq. (3.5)—holds to a very high order of accuracy within the Sun and by inference within stars in general.

---

[2] One could also make recourse here to the detailed numerical models of stellar evolution, which tell us that the Sun's radius is not expected to decrease as it ages. The argument with respect to the geological record is built upon the idea that the Sun's luminosity has been effectively constant for billions of years; if its output had varied by more than say ±20% then the oceans would have either evaporated or frozen solid, which is not consistent with the fossil record or geological stratigraphy.

3.2   First Stellar Models

The differential equations describing hydrostatic equilibrium (Eq. (3.5)) and mass continuity (Eq. (3.6)) can be solved for analytically if we adopt some mathematically straightforward rule for the variation of density within a star, and assume that the pressure and density are related through the ideal gas equation (Eq. (3.1)). Given these conditions, the variation of the density, temperature and pressure within a model star of total mass M as a function of the radial distance from the center can be determined. By far the simplest approximation for the density variation is that in which it is held constant, $\rho(r) = \rho$. After this, we can assume that the density decreases linearly from the center to the surface, with $\rho(r) = \rho_c (1 - r / R)$, where ρ_c is the central density and R is the radius of the star. And, as a final approximation, we adopt an exponential law for the pressure gradient.

3.2.1   A Zeroth Order, Constant Density Stellar Model

In this highly simplified model, it is assumed that the density is constant $\rho(r) = \rho$ within a star's interior (see Fig. 3.3). Under this assumption, as noted before, the mass continuity Eq. (3.6) immediately provides the result that the mass interior to each location r within the interior is

$$M(r) = (4\pi\rho/3)r^3 \tag{3.12}$$

and this indicates that the total mass M of a star will vary according to the cube of its radius R, or $R \sim M^{1/3}$. This is something that we can test against the observations. Recall that in Chap. 1 (Fig. 1.11b specifically),we have seen a plot of measured stellar radii against stellar mass, and that the results indicate a relationship of the form $R \sim M^{\alpha}$, where $\alpha \approx 0.6$ for $M > 1\ M_{\odot}$ and $\alpha \approx 0.8$ for $M < 1\ M_{\odot}$. At first glance, the zeroth order model thus is not doing so badly with respect to predicting

Fig. 3.3 Density profile for the constant density stellar model. (Credit: Author)

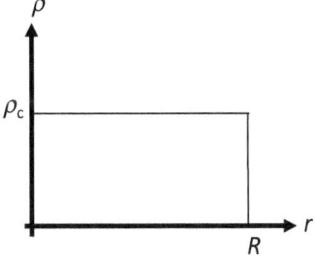

the radius-mass relationship for at least those stars more massive than the Sun[3]- So, let us persevere a little more.

The pressure variation within our constant density star is determined from the equation of hydrostatic equilibrium (Eq. (3.5)), and upon substitution from Eq. (3.12), this reveals:

$$\frac{dP}{dr} = -G\frac{M(r)\rho}{r^2} = -G\frac{(4\pi\rho_c/3)\ r^3\rho}{r^2} \tag{3.13}$$

Upon integration, from the center to the surface, Eq. (3.13) indicates that the variation of the pressure $P(r)$ inside of the star varies as

$$P(r) = P_c\left(1 - x^2\right) \tag{3.14}$$

where $x = r/R$, and P_c is the central pressure, found by adopting the boundary condition that the pressure is zero at the surface of the star, $P(r = R) = 0$. Accordingly,

$$P_c = \frac{2\pi}{3}G\rho^2 R^2 \tag{3.15}$$

(☺) **Exercise 3.6** If the density ρ is taken to be the bulk density of the star, show that

$$P_c = \frac{3}{8\pi}G\left(\frac{M^2}{R^4}\right) \tag{3.16}$$

The temperature variation with the constant density model can be determined if we assume that the interior corresponds to an ideal gas. In this manner, the equation of state is the ideal gas Eq. (3.1). The correspondence relation between pressure and temperature is accordingly

$$T(r) = (\mu\, m_p/k\, \rho)\, P(r) \tag{3.17}$$

Since the pressure goes to zero at the star's surface, $P(R) = 0$, so too does the temperature, $T(R) = 0$. This latter condition seems somewhat strange, since stars most definitely have hot surfaces, but here the idea is one of order of magnitude: across the Sun, the temperature varies from 16 million degrees at the center to 5800 degrees at the surface. Accordingly, $T(R) / T_c = 0.0004$, which gives us our approximation that the surface temperature is certainly very small compared to the central density, being effectively zero.

---

[3]The reason for the changeover in α for low-mass stars ($M < 1\ M_\odot$) is that these stars have extensive convective envelopes.

Fig. 3.4 Density (solid line) and pressure (dotted curve) variation within a constant density stellar model. (Credit: Author)

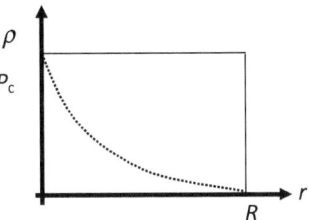

When ρ is taken as the bulk density, the central temperature T_c is

$$T_c = \frac{1}{2}\left(\frac{G\mu m_H}{k}\right)\left(\frac{M}{R}\right) \tag{3.18}$$

A schematic plot of the variation in the density (solid line) and pressure (dotted curve) for our constant density model is shown in Fig. 3.4.

The temperature variation will follow that of the pressure variation through Eq. (3.17). Taking solar values for the mass and radius of our model star, the central pressure and temperature turn out to be

$$P_c = 1.3 \times 10^{14}\text{Pascal}$$

$$T_c = 5.8 \times 10^{6}\text{Kelvin}$$

where we have adopted a mean molecular weight of $\mu = \frac{1}{2}$ (which corresponds to the Sun being entirely composed of hydrogen). Detailed numerical calculations indicate that the pressure and temperature at the center of the Sun are $P_c = 2.3 \times 10^{16}$Pascal, and $T_c = 1.6 \times 10^{7}$Kelvin respectively. From these values, the constant bulk density model underestimates the central pressure by a factor of 177 and the central temperature by a factor of 3. We do have the freedom to adjust the value of the density in this model, so rather than taking $= 1408$ kg/m^3, we can simply choose $\rho(r) = 6.65 \times 10^4$ kg/m^3 (an increase over the bulk density by a factor of 46). In this manner, the central pressure P_c can be brought into agreement with the detailed numerical calculations. By doing this, however, the central temperature for the constant density model becomes $T_c = 2.1 \times 10^7$ K, which is a factor of about 1.3 too high. The Sun's central density derived from the detailed numerical models is of order 1.6×10^5 kg/m^3, which is a factor of 110 times larger that the bulk density. We do not want to use this actual central density in our constant density model, since the central pressure and central temperature estimates would become larger overestimates. Additionally, we note that by changing the assumed density of our model, the derived model radius R for a given mass will also change, as indicated by Eq. (3.12). If we take $\rho = 6.65 \times 10^4$ kg/m^3, then the deduced radius for our constant density solar mass model with $M = 1$ M$_\odot$ is $R \approx 0.3$ R$_\odot$. Hence the model, while getting the central pressure right, overestimates the central temperature by a factor of some 30% and underestimates the radius by a factor of 70%.

(☺) **Exercise 3.7** Determine P_c, T_c and R when the density is taken as $\rho = 1.6 \times 10^5$ kg/m^3, and compare the results against the Sun's actual values.

At this point, we have encountered one of the inherent problems associated with simplified stellar models: they can only be pushed so far. Simplified models are highly useful, pedagogically informative and worth persevering with, but only up to some point. If one wants to understand how stars work, then simplified models are an important developmental tool, but if one wants to use stellar models to understand the actual observations relating to, say, a star cluster, then the only option is to employ detailed numerical calculations.

There is a useful story, possibly apocryphal, concerning the late and great astrophysicist Fred Hoyle that illustrates the point about stellar models. Upon being asked about the properties of stars during a conference discussion section, Hoyle remarked, 'stars are basically simple objects'. A voice in the audience then chirped in, 'you'd look pretty simple too Fred, from a distance of ten parsecs'. The comment by Hoyle and that of the unknown audience member are interesting in that they are both right and wrong. Hoyle is correct in that stars are basically simple, in the sense articulated by J. R. R. Tolkien—that they are just stable spheres of hot gas. The audience member is also correct in that the simplicity is in the general description only, and that the complexity is hidden from direct view, residing as it does in the fiendish subtleties of hydrodynamics and nuclear physics. The Devil is in the details, while the simplicity is in the visual appearance. With respect to a human looking simple at a distance from 10 parsecs, the audience member is also right and wrong. Indeed, from such a distance an individual may very well look entirely miniscule and insignificant, but the work and genius of many such insignificant individuals combined can make for a blazing radio signal that can be intercepted much farther than 10 parsecs away. This of course is the very foundation for the present-day search for extraterrestrial intelligence (SETI).

Returning to Fred Hoyle, now in a publication written with R. A. Lyttleton in 1949, we find the following sage comments: 'So far no one has succeeded in obtaining stellar models wholly free from both mathematical and physical approximations. Some form of simplification has always to be introduced, and the decision that invariably has to be made is whether to maintain a closely correct physical picture and make mathematical approximations to deal with the resulting equations, or to make considerable simplifications of the physical picture and thereby maintain as strict a mathematical treatment as possible'. To this statement, Hoyle and Lyttleton added, '. . . There is an ever-present difficulty in adopting the first method, and this is that only considerable familiarity with the problem shows what mathematical approximations can safely be made. Thus it has taken about thirty years for general confidence to be felt about what approximations are valid for the theory of stars of uniform composition'.

In developing the constant density model, we have certainly simplified the physics of a star, without applying any great mathematical rigor, but the important point is that the model is predicated upon a quantity that changes only moderately from center to surface when compared to the pressure and temperature. A detailed model of the Sun indicates a change by a factor of roughly 3000 in the central-to-surface temperature, and a change of 16 orders of magnitude between the central and surface pressures. In the case of density, the ratio $\rho_c/\bar{\rho}(R) = 115$, and it is this latter smaller variation across the interior of a star that allows for the zero-order model to work so unreasonably well.

Such simplified models have applicability, since stars genuinely do have some quantities that don't vary greatly with radial distance. Density is one term, composition is another; so too is the quantity $\beta = P(r) / P_{\text{gas}}(r)$ (to be described more fully below) nearly constant across a star. The other key factor that enables the simplified models to provide for good approximations is that the star's characteristics are largely determined by what is taking place at either its center or its surface, and that the variation within most variables (e.g. the temperature and pressure) is monotonically decreasing outward. The argument that the density cannot increase outwards is encapsulated in the stability condition of a star, since if a higher density region develops over a low density one, a turnover instability must arise. Given that the density cannot increase outward and that the pressure decreases outward, the temperature of a star must decrease from the center to the surface. Returning to Fred Hoyle's comment on the inherent simplicity of stars, we find he is assuredly right in that given that a model (zeroth-order or otherwise; see later) provides for a good order of magnitude description of central and surface values, the rest of the more complex story of stardom is effectively hidden within the more subtle details of the physics that has been approximated.

There is a theorem, first established in 1935, by British mathematician and pioneering astrophysicist E. A. Milne (modified by Chandrasekhar shortly thereafter) that encapsulates the various characteristics of the constant density model, and likewise that for the other model stars that we shall encounter. The theorem states:

In any equilibrium configuration in which the mean density $\bar{\rho}(r)$ interior to radius r decreases outwards,

$$\frac{1}{2}\left(\frac{4\pi}{3}\right)^{1/3} G\bar{\rho}^{4/3}m^{2/3} \leq P_c - P(r) \leq \frac{1}{2}\left(\frac{4\pi}{3}\right)^{1/3} G\rho_c^{4/3}m^{2/3} \tag{3.19}$$

What this theorem is getting at is that the central pressure of any star for which the density does not increase with radius is bounded below by the central pressure derived for a homogeneous star with density $\rho(r) = \bar{\rho}(R)$ and bounded above by the central pressure derived for a homogeneous star with density $\rho(r) = \rho_c$ (see Fig. 3.5).

The proof of Milne's theorem is straightforward and begins with Eq. (3.5), our erstwhile equation of hydrostatic equilibrium. Integrating this equation from the center $r = 0$ to some point r in the star's interior gives

$$P_c - P(r) = G\int_0^r \frac{M(r)}{r^2}\rho(r)\,dr \tag{3.20}$$

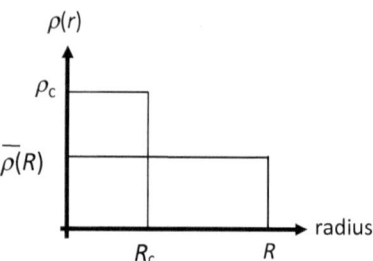

using the mass continuity equation, Eq. (3.6), we can change the variable of
integration from dr to $dM(r)$ and accordingly write

$$P_c - P(r) = \frac{G}{4\pi} \int_0^r \frac{M(r)}{r^4} dM(r) \qquad (3.21)$$

the r^4 term in Eq. (3.21) can be written in terms of the mass and mean density at
radius r as follows:

$$r^4 = \left[\frac{M(r)}{\frac{4}{3}\pi \overline{\rho}(r)} \right]^{4/3} \qquad (3.22)$$

Substituting for (3.22) in (3.21) now gives us

$$P_c - P(r) = \frac{1}{4\pi} \left(\frac{4\pi}{3} \right)^{4/3} \int_0^r \overline{\rho}^{4/3}(r) M^{-1/3}(r) dM(r) \qquad (3.23)$$

We are now nearly there, and the next step builds upon the statement that the
density decreases outwards, which enables the inequality statement that

$$P_c - P(r) \geq \frac{1}{4\pi} \left(\frac{4\pi}{3} \right)^{4/3} \overline{\rho}^{4/3}(r) \int_0^r M^{-1/3}(r) dM(r) = \frac{1}{2} \left(\frac{4\pi}{3} \right)^{1/3} \overline{\rho}^{4/3}(r) M^{2/3}(r)$$

In this final step, we have taken the mean density term outside of the integral and
set it to the mean density at radius r, and then integrated the $M^{-1/3}$ term. This proves
the left-hand side of the inequality in Eq. (3.19). The right-hand side of Eq. (3.19)
follows by again taking the mean density term outside of the integral in Eq. (3.23)
and using the conditional statement that the density decreases outward so that
$\overline{\rho}(r) \leq \rho_c$, where ρ_c is the central density, thereby completing the proof. In the
case where $r = R$, where R is the radius of the star, so $P(R) = 0$, and $\overline{\rho}(R)$ corresponds

to the bulk density of the star, and inequality (3.19) now provides upper and lower bounds upon the central pressure of a star. We note here that if the inequalities in Eq. (3.19) do not hold true, then this would imply that the density must be increasing outwards in some region of the star, and such a condition would suggest some form of instability. In this latter sense, inequality (3.19) is a necessary condition for a star to be stable.

Returning now to Eqs. (3.1) and (3.3), the ratio of the gas and radiation pressures, P_{gas} / P_{rad}, can be evaluated at the center of the constant density Sun model. Indeed, we have

$$\frac{P_{gas}}{P_{rad}} = \frac{nkT}{\frac{1}{3}aT^4} = \frac{\Re\rho T/\mu}{\frac{1}{3}aT^4} = \left(\frac{3\Re}{a\mu}\right)\left(\frac{\rho}{T^3}\right) \tag{3.24}$$

where $\Re = k / m_H = 8263.47$ is the gas constant. Taking $\rho = 6.65\times10^4$ kg/m^3 and $T_c = 2.1\times10^7$ K, we have that $P_{gas} / P_{rad} \approx 471.4$ and accordingly deduce that in solar mass stars, the radiation pressure is going to be unimportant. Generalizing this result, we now ask over what range in stellar mass is it appropriate to ignore radiation pressure, and additionally determine in which mass domain the radiation pressure is dominant.

In general, the pressure term can be written as the sum of the gas and the radiation contributions, with $P = P_{gas} + P_{rad}$. Writing now that $P_{gas} = \beta P$, and $P_{rad} = (1 - \beta) P$, we have

$$P = \frac{1}{(1-\beta)}\frac{aT^4}{3} = \frac{1}{\beta}\left(\frac{\Re}{\mu}\right)\rho T$$

which can be rearranged in terms of the temperature and density as

$$T = \left(\frac{1-\beta}{\beta}\frac{3\Re}{a\mu}\right)^{1/3}\rho^{1/3} \tag{3.25}$$

Substituting (3.25) into the definition $P_{rad} = (1 - \beta) P$ accordingly reveals

$$P = \left(\frac{1-\beta}{(\mu\beta)^4}\right)^{1/3}\left[\frac{3\Re^4}{a}\right]^{1/3}\rho^{4/3} \tag{3.26}$$

(☺) **Exercise 3.8** Complete the algebra steps taking Eq. (3.25) to (3.26).

We are now in a position to introduce what is known as the Eddington quartic equation. This is achieved by substituting for the pressure term in Eq. (3.26) with the right-hand side inequality of Eq. (3.19). In this case, $P = P_c$ and $\rho = \rho_c$, and

$$\left(\frac{1-\beta}{(\mu\beta)^4}\right)^{1/3}\left[\frac{3\mathfrak{R}^4}{a}\right]^{1/3} \leq \left(\frac{\pi}{6}\right)^{1/3} GM^{2/3}$$

Upon substitution for the constant terms and expressing the mass M in solar units, the last expression reduces to

$$\left(\frac{1-\beta}{\beta^4}\right) \leq 0.034 \left(\frac{M}{M_{sun}}\right)^2 \mu^4 \qquad (3.27)$$

Equation (3.27) is a fourth-order equation for β in terms of the mass of star and its composition, the latter being expressed through the mean molecular weight μ. Eq. (3.27) is, in all but the constant,[4] the Eddington quartic, and this latter equation plays a central role in description of what is called the Standard Model (to be described in Chap. 4).

(☺) **Exercise 3.9** Verify the derivation of the constant term in Eq. (3.27).

Equation (3.27) allows us to introduce an interesting philosophical point made by Eddington in his *Internal Constitution of the Stars*, a point that highlights the power of physical intuition and exemplifies how simple relationships can hint at other more complex processes. Eddington specifically noted that if someone—even a person who had never seen a star—determined β values for objects of varying mass, then for all masses smaller than about 10^{29} kg, its value would be essentially one. For all masses larger than about 10^{33} kg, however, the value of β is effectively zero (see Fig. 3.6). Clearly, in the mass range between 10^{29} and 10^{33} kg, something interesting happens, and, as Eddington put it, 'what happens is the stars'.

Figure 3.7 is an alternative way of expressing Fig. 3.6. In this case, however, the diagram shows temperature plotted against density. This latter diagram is more informative in the sense that the various degeneracy, ideal gas and radiation pressure-dominated regions are more clearly delineated. The domains in which the gas is highly degenerate will be covered later in Chap. 5 and will be an important part of our discussion concerning white dwarfs. The boundary between the ideal gas and radiation pressure-dominated domain is described according to Eq. (3.25). Since Fig. 3.7 uses a logarithmic scale, the boundary is a line of slope 1/3. The boundary between the ideal gas and the non-relativistic degenerate gas domain has a slope of 2/3 (as will become apparent later). The variation of density and temperature within a set of stars having masses of 100, 10, 1 and 0.1 $M_\odot$ is also shown in Fig. 3.7 by a series of red dashed lines. The 100 $M_\odot$ star skirts the boundary between the ideal gas and radiation pressure-dominated domain, while the 0.1 $M_\odot$ star skirts the boundary between the ideal gas and non-relativistic degenerate gas domain.

---

[4]As will be seen later, the constant in the Eddington quartic is actually some ten times smaller than that found in Eq. (3.27).

Fig. 3.6 Solution curves for Eq. (3.27). A mean molecular weight of $\mu = 0.6$ has been adopted in the calculations. Along the top axis are shown the approximate domains where the gas can be considered as being degenerate, ideal and radiation pressure-dominated. The crossover point of the curves takes place at a mass corresponding to 40 solar masses. (Credit: Author)

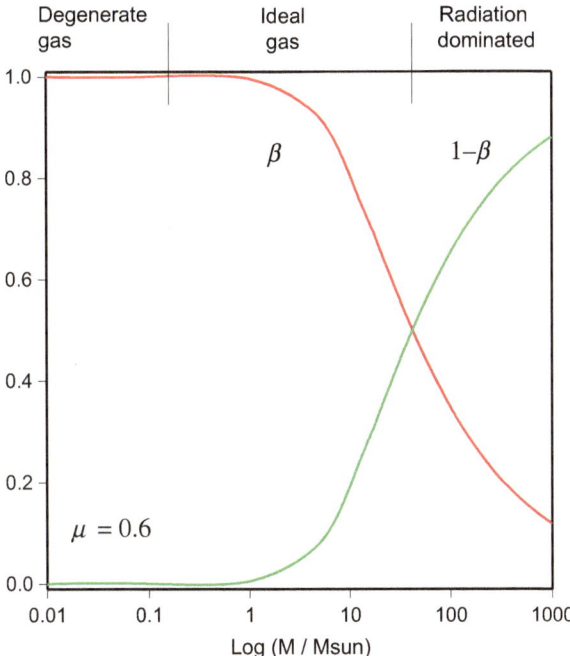

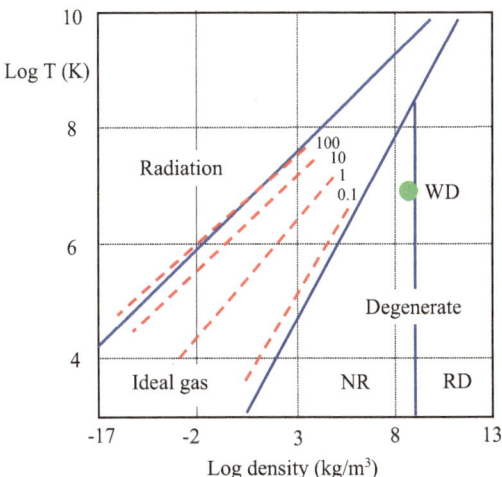

Fig. 3.7 Temperature versus density diagram. The degeneracy, ideal gas and radiation-pressure dominated regions are mapped out in this diagram (blue lines). The red dashed lines indicate the run of density and temperature for main sequence stars of varying mass (as labeled). Also shown is a data point corresponding to the central temperature and density of a 1 solar mass white dwarf. The region labeled NR corresponds to a non-relativistic degenerate gas; that labeled RD corresponds to a relativistic degenerate gas. (Credit: Author)

Fig. 3.8 Density profile for
the linear density stellar
model. (Credit: Author)

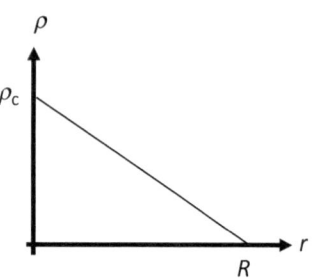

3.2.2   The Linear Density Stellar Model

Milne's theorem not only applies to constant density (zeroth-order) model stars, but
also to any stellar model in which the density decreases outward from the center. The
linearly decreasing density model is accordingly the next most straightforward
stellar model to examine. In this case, we are moving towards a better description
of the change in the density and accordingly write $\rho(r) = \rho_c (1 - r/R)$, where ρ_c is the
central density and R is the radius of the star (Fig. 3.8). Note that in general, we will
be using the parameter $x = r/R$ in what follows, with $0 \le x \le 1$ representing the
scaled distance between the center and the surface of the model. Building upon the
lecture notes of his colleague A. G. Cameron, Robert Stein first discussed the linear
density model in detail in 1966. Writing in the influential and exploratory text *Stellar
Evolution*, edited by Stein and Cameron[5] (published by Plenum Press in 1966), Stein
specifically developed the properties of the linear model and commented in the
Preface, 'An attempt is made to account for as much of the theory of stellar evolution
as is possible in simple analytic form'. We shall only look at a few of the many ideas
presented by Stein here. We begin with the equation for the conservation of mass
(Eq. 3.6) and obtain the result that

$$M(r) = \int_0^r 4\pi r^2 \rho_c (1 - r/R) \ dr = M \left(\frac{\rho_c}{\bar\rho} \right) x^3 \left(1 - \frac{3}{4}x \right) \qquad (3.28)$$

This equation determines a relationship between the mass, radius and central density
terms such that $\bar\rho/\rho_c = 1/4$. The variation of pressure through the interior follows by
integrating the equation of hydrostatic equilibrium (Eq. 3.5). Accordingly,

---

[5]The text is actually the Proceedings for a conference that was held at the Goddard Space Flight
Center in November 1963. The conference took place at the dawn of the electronic computation era.
The analytic models play a vital role in the understanding and the interpretation of the great mess of
numbers and phenomenon created by 'black box' computational methods. There is a possibly
apocryphal saying attributed to Martin Schwarzschild from this era, in which he noted that once the
money for computer time ran out, he had to start thinking again.

$$P(r) = P_c - \int_0^r G\frac{M(r)\rho(r)}{r^2}dr = P_c - GM\left(\frac{\rho_c}{\bar{\rho}}\right)\rho_c R \int_0^{r/R} x^3\left(1 - \frac{3}{4}x\right)(1 - x)dx$$

where P_c is the central pressure, and in the second integral we have changed from r to $r = Rx$ as the variable. This latter expression integrates to reveal:

$$P(r) = P_c - \frac{GM}{R}\left(\frac{\rho_c}{\bar{\rho}}\right)\rho_c \int_0^{r/R} x^3\left(1 - \frac{3}{4}x\right)(1 - x)dx$$

$$P(r) = P_c - \frac{GM}{R}\left(\frac{\rho_c}{\bar{\rho}}\right)\rho_c \left.\frac{x^2}{2}\left(1 - \frac{7}{6}x + \frac{3}{8}x^2\right)\right|_0^{r/R}$$

At the surface $r = R$ (that is, $x = 1$), it is required that the pressure goes to zero, with $P(R) = 0$. Accordingly, the central pressure is determined as

$$P_c = \frac{5}{48}\frac{GM}{R}\left(\frac{\rho_c}{\bar{\rho}}\right)\rho_c \tag{3.29}$$

and the run of the pressure through the interior of the stellar model is:

$$P(r) = P_c\left(1 - \frac{24}{5}x^2 + \frac{28}{5}x^3 - \frac{9}{5}x^4\right) \tag{3.30}$$

The run of the temperature through the model interior is determined by the ideal gas equation with $T(r) = (\mu / R) P(r) / \rho(r)$. The mathematically useful trick at this stage is to notice that the polynomial term in Eq. (3.30) can be rewritten as:

$$\left(1 - \frac{24}{5}x^2 + \frac{28}{5}x^3 - \frac{9}{5}x^4\right) = (1 - x)\left(1 + x - \frac{19}{5}x^2 + \frac{9}{5}x^3\right)$$

This rearrangement provides for the run of the temperature term as

$$T(r) = \left(\frac{\mu}{\Re}\right)\left(\frac{P_c}{\rho_c}\right)\left(1 + x - \frac{19}{5}x^2 + \frac{9}{5}x^3\right) \tag{3.31}$$

The equation for the internal run of temperature in the linear model satisfies the boundary conditions that at $x = 0$, the center, $T = T_c = (\mu / R) (P_c / \rho_c)$, and that at $x = 1$, the surface, $T = 0$. Interestingly, the equation for the temperature variation T (r) does not decrease monotonically, and it in fact yields a temperature maximum at $x \approx 0.15$ rather than at $x = 0$. The temperature at $x \approx 0.15$ is a factor of 1.07 times larger (that is 7% larger) than that at the center where $x = 0$. Accordingly, we have found an unexpected weakness in the linear density model. By improving the approximation for the variation in the density across the interior of a star, we have inadvertently lessened (albeit slightly) the accuracy with which the temperature variation is described. This result highlights one of the problems, as discussed by Hoyle and Lyttelton, that must sooner or later affect any analytic model—they are only approximations of the real physical situation. Just as in life, by making improvements in one area we tend to disrupt the harmony that was otherwise present in another. So be it.

(☺) **Exercise 3.10** Take the expression for the temperature variation in Eq. (3.31), differentiate it with respect to x and then set the differential to zero: $dT(x) / dx = 0$. Solve the resultant equation to show that a maximum occurs at $x = \left(19 - \sqrt{226}\right)/27 = 0.1469$.

A generalization of the linear density law is that in which the density varies as some power law, with $\rho(r) = \rho_c \left[1 - (r / R)^n\right]$, where the constant n can be taken as any positive real number. The linear density model is accordingly the $n = 1$ approximation, and for larger values of $0 < n < 1$, the density profile becomes more and more centrally peaked and the radius more and more extended. For larger and larger $n > 1$ values, the density profile begins to approach that of a highly compact, constant density model. The power law density expression in principle provides for a broad spectrum of possible stellar models and configurations, but nature, as we shall see below, is much more conservative in its choice for the density profiles that exist in real stars.

(☺) **Exercise 3.11** Given a density law of the form $\rho(r) = \rho_c[1 - (r/R)^n]$, with n being any real number greater than zero, and taking the bulk density to be $\bar{\rho} = 3M/\left(4\pi R^3\right)$, where M and R are the mass and radius of the star, determine and evaluate the ratio $\bar{\rho}/\rho_c$ for $n = 1, 2, 3$ and 4. What happens to this ratio as $n \Rightarrow \infty$? **Hints**: Start with the equation of mass continuity (Eq. 3.6) and show that $\bar{\rho}/\rho_c = 1 - 3/(n + 3)$; this is a fairly straightforward algebra and integration step. As n becomes larger and larger, so $\bar{\rho}/\rho_c \Rightarrow 1$ and the density profile becomes more and more broadly peaked. These latter high-n models describe a density profile that is approaching that of the extreme constant density model where $\rho(r) = \rho_c$. But the catch is that for such a configuration $\bar{\rho} \approx \rho_c$, and this, for a non-zero mass, can only come about if the radius is $R = 0$—not a very star-like object. Likewise, if we let $n \Rightarrow 0$, so $\bar{\rho}/\rho_c \Rightarrow 0$, which is equivalent to a constant density model in which $\bar{\rho} \approx 0$. In this case, the density profile becomes strongly peaked at the center, and the radius of the object must accordingly be very large, with $R \Rightarrow \infty$ —again, not a very realistic stellar model, although in this particular case it does start to resemble something that looks a little bit like a red giant star.

(☺) **Exercise 3.12** Use a computer graphing program to plot the variation of $\rho(r)/\rho_c = [1 - x^n]$ against x for $0 \leq x \leq 1$ and for various values of the power n, $n = \frac{1}{4}, \frac{1}{2}$, 1, 2, & 4. **Hint**: The plot should look something like that shown in Fig. 3.9. Note that the x-axis is scaled to the radius R which, as indicated earlier, will have a different value for each n.

power law in the density

$$\rho \,/\, \rho_c = 1 - x^n$$

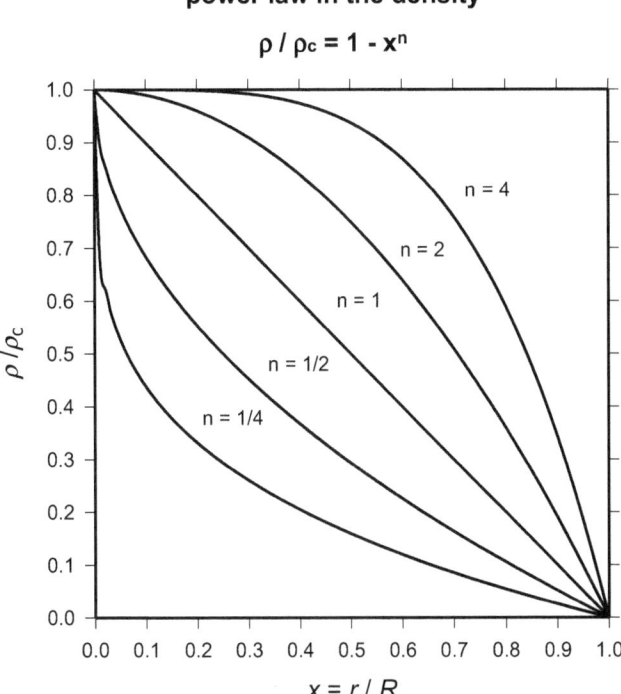

Fig. 3.9 Profiles for density ratio $\rho(r)/\rho_c = [1 - x^n]$ against x for $0 \leq x \leq 1$ for various values of n. As n increases towards infinity, the profile approaches that of the constant density model. As n decreases towards zero, the profile becomes more centrally peaked. (Credit: Author)

3.2.3 Clayton's Exponential Model

It is often found that many quantities in nature vary in some exponential fashion, and in the stellar model sense we might choose a density variation of the form $\rho(r) = \rho_c \exp\left[-(r/\lambda)^n\right]$ where ρ_c is the central density, and where $n > 0$ and $R >> \lambda > 0$ are constants to be chosen. With this kind of approximation, we do have to be careful with respect to surface boundary condition, since technically the density is now not going to zero at the surface where $r = R$. This perhaps is not so much of a problem since provided λ is chosen to be very small with respect to the radius R, then $\exp(-R/\lambda)$ will automatically be small as well, making $\rho(R) \approx 0$, as is required at the stellar surface. By far the biggest problem with such a choice for the density is in making any results reasonably tractable and analytic in final form, since integrals containing both powers and exponential terms are going to be encountered, and these may or may not have an analytic solution. In an intriguingly titled paper, *Stellar structure and the art of building boats*, Ludwik Celnikier (*American Journal of Physics*, 57 (2), 159–166, 1989) developed an analytic stellar model by assuming a set of "trial functions" for the run of mass, density and temperature. These functions were chosen since they have the correct limiting forms as $r \Rightarrow 0$ and $r \Rightarrow R$. To make progress, Celnikier chose to set $n = 2$ and then introduced the trial functions:

$$\rho(r) = \rho_c \exp\left[-\left(r/\lambda_\rho\right)^2\right]$$

$$\mathrm{T}(r) = T_c \exp\left[-\left(r/\lambda_\mathrm{T}\right)^2\right]$$

$$\mathrm{M}(r) = M \left[1 - \exp\left[-(r/\lambda_\mathrm{M})^2\right]\right]$$

The mathematical trick now is to place these expressions into the equation for hydrostatic equilibrium and the mass conservation equation, and determine appropriate values for the various λ-constants. The model can be made to work, but Celnikier has to insert 'by hand' values for the total mass, central density and central temperature. In spite of Celnikier placing an enthusiastic spin upon the results obtained in his article, the model is highly unwieldy and ultimately highly parameterized, leaving one with the uncomfortable feeling that physical details have been smothered and/or glossed over for the sake of pure mathematical convenience. With the Gaussian-like trial function model, we are beginning to move more deeply into the realm Hoyle and Lyttelton warned us against. Indeed, it is now the mathematical *tricks* and *niceties* that are driving the investigation, rather than the model providing us with any improved physical insight.

For all of the concerns just raised, let us persevere a little more with the topic. As with any approximate model, we need to make sure that there is a finite amount of mass M contained within some finite radius R over which the density is specified. Walking where angels fear to tread, one can certainly insert a density variation of the form, say, $\rho(r) = \rho_c \exp\left[-(r/\lambda_\rho)^2\right]$ into the equations for hydrostatic equilibrium and mass conservation, and after some considerable mathematical tinkering produce a set of rather unsightly analytic expressions for the pressure and mass distributions, as a function of central displacement r.

Rather than proceed directly to such derivations, however, we will backtrack slightly and ask if there is a better choice of variable term that might be employed than that of the density and the mass (as per Celnikier's approach). Indeed, this is exactly what Donald Clayton (Rice University) did in an article entitled *Solar structure without computers* [*American Journal of Physics*, 54 (4), 354–362, 1986]. To begin, let us consider the central boundary conditions that must apply within a star. As has already been argued, under the conditions of hydrostatic equilibrium, the pressure must attain its greatest value at the center, and in terms of its derivative we must accordingly have $dP / dr \Rightarrow 0$ as $r \Rightarrow 0$. Likewise, we also require that the temperature reaches its maximum value at the center, with $dT / dr \leq 0$ as $r \Rightarrow 0$. Now, under the conditions that the stellar interior is described by an ideal gas, so $T(r) = \text{constant } P(r) / \rho(r)$, and upon differentiating this equation we have:

$$\frac{dT}{dr} = \frac{1}{\rho(r)}\frac{dp}{dr} - \frac{P(r)}{\rho^2(r)}\frac{d\rho}{dr} \qquad (3.32)$$

Accordingly, if $dT / dr \leq 0$ as $r \Rightarrow 0$, we must have, from the right-hand side of Eq. (3.32), that $d\rho /dr \geq 0$ as $r \Rightarrow 0$. As pointed out by Clayton, 'To avoid an increasing temperature with radial distance from the center it is necessary that $\rho(r) \Rightarrow$ constant $= \rho_c$ near the center, i.e. that it not have a negative derivative'. This requirement placed upon the density derivative immediately informs us as to why the linear density approximation is a less physically correct model (for all of its appeal) than that of the zeroth-order constant density model, since for the linear density model $d\rho /dr = - \rho_c / R$, and as we have seen above the temperature does not attain its maximum value at the center, but rather at a point located away from the center. Likewise, we see that the Gaussian trial function provides no better an approximation to conditions at the center of a star than the linear model, since $d\rho/dr = - \rho_c(2r/\lambda_\rho) \exp [-(r/\lambda_\rho)^2] \approx - \rho_c(2r/\lambda_\rho) [1 - (r/\lambda_\rho)^2] < 0$. Clayton described the linear density model (with the same argument applying to the generalized power law form and the Gaussian trial function) as being 'diseased', and mathematically speaking he is correct. Such laws fail at the very first hurdle, at the very center of the model. However, Clayton's use of the term "diseased" is somewhat misplaced in terms of Eddington's motto, where in the making of any analytically useful model of a star, we are essentially making a sacrifice to the lesser shrine of plausibility. Nonetheless, and in spite of Clayton's concerns, the situation is not entirely lost. Rather than assuming some expression for the variation of the density within a stellar model, Clayton suggested that a useful, physically more realistic way forward is to adopt some trial function that describes the characteristic run of the pressure within the interior. Moreover, the choice of function is specifically adopted on the basis that it provides correct boundary values for $dP / dr \Rightarrow 0$ as $r \Rightarrow 0$ at the center, and at the surface where $r \Rightarrow R$. Accordingly, Clayton suggested that a physically reasonable expression for the variation of the pressure gradient that satisfies the limiting central and surface boundary conditions is:

$$\frac{dP}{dr} = -K\,r\exp\left[-\left(\frac{r}{a}\right)^2\right] \qquad (3.33)$$

where K and a are constants to be chosen appropriately. Indeed, to satisfy the surface boundary condition we are forced (see below) to choose $a << R$, where R is the adopted stellar radius. Indeed, Eq. (3.33) can be integrated directly to yield:

$$P(r) = \frac{Ka^2}{2}\left\{\exp\left[-\left(\frac{r}{a}\right)^2\right] - \exp\left[-\left(\frac{R}{a}\right)^2\right]\right\} \qquad (3.34)$$

where, as indicated earlier, a is eventually chosen so that the second exponential term in the right-hand side bracket is effectively zero (or at least very small). The form of Eq. (3.34) indicates that at the center the pressure is a maximum, with P $(r = 0) = K a^2 / 2$, and at the surface $P(r = R) = 0$. So far, we have not been required to do much work. However, to find the mass $m(r)$ enclosed within radius r, we are going to need some patience. The equation of hydrostatic equilibrium is our starting point, and we have:

$$G \int_0^r M(r) \; dM = -4\pi \int_0^r r^4 \frac{dp}{dr} dr$$

If we now substitute for the pressure gradient (Eq. (3.33)),

$$M^2(r) = \frac{4\pi K}{G} \int_0^r r^5 \exp\left[-\left(\frac{r}{a}\right)^2\right] dr \tag{3.35}$$

In order to make progress, we must now evaluate the integral on the right-hand side of Eq. (3.35). Following Clayton, we write

$$M^2(r) = \frac{4\pi K}{G} \Phi^2(r) \tag{3.36}$$

where

$$\Phi^2(r) = \int_0^r r^5 \exp\left[-\left(\frac{r}{a}\right)^2\right] dr \tag{3.37}$$

The trick now is to obtain an analytic expression for $\Phi(r)$. Here we can make use of the result that if $dV = s \exp(-s^2) \; ds$, so $V = -\frac{1}{2} \exp(-s^2)$, and accordingly the integral in Eq. (3.37) can be evaluated through integration by parts three times. In this manner, we find

$$\Phi^2(r) = a^6 \left[1 - \frac{1}{2}\left(s^4 + 2s^2 + 2\right)\exp\left(-s^2\right)\right] \tag{3.38}$$

where $s = r/a$ and where we have used the central boundary condition $\Phi(r = 0) = 0$. We now have the result, combining (3.36) with (3.38), that the mass enclosed within radius r is:

$$M(r) = a^3 \left(\frac{4\pi K}{G}\right)^{1/2} \left[1 - \frac{1}{2}\left(s^4 + 2s^2 + 2\right)\exp\left(-s^2\right)\right] \tag{3.39}$$

(☺) **Exercise 3.12** Complete the steps taking (3.37) to (3.38). This is not a technically difficult exercise, but it is one in which it is very easy to make an algebraic mistake. So, the only hint is to take a large sheet of paper and keep your wits about you.

To determine the density variation within the Clayton model, we can proceed from the equation for the conservation of mass. Accordingly,

$$\rho(r) = \frac{1}{4\pi r^2} \frac{dM}{dr}$$

Substituting for $s = r/a$ and using Eq. (3.36), we obtain the expression

$$\rho(r) = \frac{1}{a^3} \left(\frac{K}{4\pi G} \right)^{1/2} \frac{1}{s^2} \frac{d\Phi}{ds} \tag{3.40}$$

Now we can use the mathematical trick that

$$\frac{d\Phi}{ds} = \frac{1}{2\Phi} \frac{d}{ds} \left(\Phi^2 \right)$$

and accordingly derive an expression for the density variation as

$$\rho(r) = \left(\frac{K}{4\pi G} \right)^{1/2} \frac{r^3}{2\Phi(r)} \exp\left(-(r/a)^2 \right) \tag{3.41}$$

(☺) **Exercise 3.13** Once again, take a large piece of paper and complete the algebra steps taking (3.40) to (3.41). *Hint:* apart from straightforward substitutions and cancellations, the key step in the algebra is to note that from Eq. (3.37):

$$\frac{d}{ds} \left(\Phi^2 \right) = \frac{d}{ds} \left(\int r^5 \exp\left(-(r/a)^2 \right) dr \right) = a^6 \frac{d}{ds} \left(\int s^5 \exp\left(-s^2 \right) ds \right) = a^6 s^5 \exp\left(-s^2 \right)$$

To see that the density expression has the correct limiting form at the center, the next mathematical task is to show that $\Phi(r) \Rightarrow r^3 (1 - \tfrac{3}{4} r^2)^{\frac{1}{2}}$ as $r \Rightarrow 0$. Accordingly, with this limiting form, we find that $\rho(r) \Rightarrow 0$ as $r \Rightarrow 0$.

(☺) **Exercise 3.14** Show that $\Phi(r) \Rightarrow r^3 (1 - \tfrac{3}{4} r^2)^{\frac{1}{2}}$ as $r \Rightarrow 0$. Once again, we are back to our large piece of paper and the careful slog of substitutions and algebra. In this case, the key step is to approximate the exponential term in Eq. (3.38) as a polynomial, keeping only the first two terms. That is, $\exp(-s^2) \approx 1 - s^2$ when s is small.

So far, we have simply argued that the constant a must be very much smaller than the radius R, but if we consider the conditions that must apply near to the surface, then an analytic expression for a can be found. At the surface where $r \Rightarrow R$, we have

$$\left.\frac{dP}{dr}\right]_{r=R} \approx -\frac{GM}{R^2}\rho(R) = -KR\exp\left[-\left(\frac{R}{a}\right)^2\right] \tag{3.42}$$

where M is the total mass of the star $m(R)$. If we now substitute for $\rho(R)$ from Eq. (3.41) and work through the algebra, we find that

$$a^3 = M\sqrt{G/16\pi K} \tag{3.43}$$

Recall that the constant K is related to the central pressure through Eq. (3.34), and by choosing this term appropriately, the value of a is fully determined through Eq. (3.42). Clayton suggests using $K = (4\pi/3)G\rho_c^2$, which is just two times larger than the central pressure term P_c derived for the constant density model (recall Eq. (3.15)).

(☺) **Exercise 3.15** Derive the result (3.43). The essential mathematical trick in this exercise is to show from Eq. (3.38) that $\Phi(R) \Rightarrow a^3$ as $r \Rightarrow R$. Additionally, recall that we have $a << R$ and accordingly as $r \Rightarrow R$, so $\exp(-s^2) = \exp[-(R/a)^2] \Rightarrow 0$.

Clayton has made us work, albeit without too much sweat, in order to explore his model. However, in spite of its improved boundary condition formulation, Clayton's approach (like that adopted by Celnikier) is not overly satisfactory. It pushes the limits of what one might reasonably call an analytic model, and it does not provide for any order of magnitude estimates to key physical quantities—i.e. the central pressure and temperature. Rather, expressions for these terms have to be put into the model. Indeed, for all of the mathematical work required of the Clayton model, it is hardly an improvement over the zeroth-order constant density model, and it is certainly more unwieldy in its form and applicability. With the models of Clayton and Celnikier we have reached (even overshot) the limits of what reasonable constitutes analytic effort. Progress can only be made now by allowing for detailed numerical results to be included in our analysis. This we shall do in the following sections where the so-called polytropic model will be discussed.

Before finishing off this section, it is worthwhile to compare the various density approximations that have been described so far against those numbers derived for a detailed solar model. Such a comparison is shown in Fig. 3.10. Indeed, it is seen that in the innermost core region $x < 0.1$ the linear model provides a reasonably good fit to the density profile; the Clayton model, however, provides a much better fit in the entire range $0 \leq x \leq 1$.

Fig. 3.10 Comparison diagram showing various density approximation laws against that derived from the full numerical integration of the equations of stellar structure (red line). In each case, the density is scaled according to its central value. The dotted circles correspond to a Clayton model with $a = 11\,R_\odot/2$. (Credit: Author)

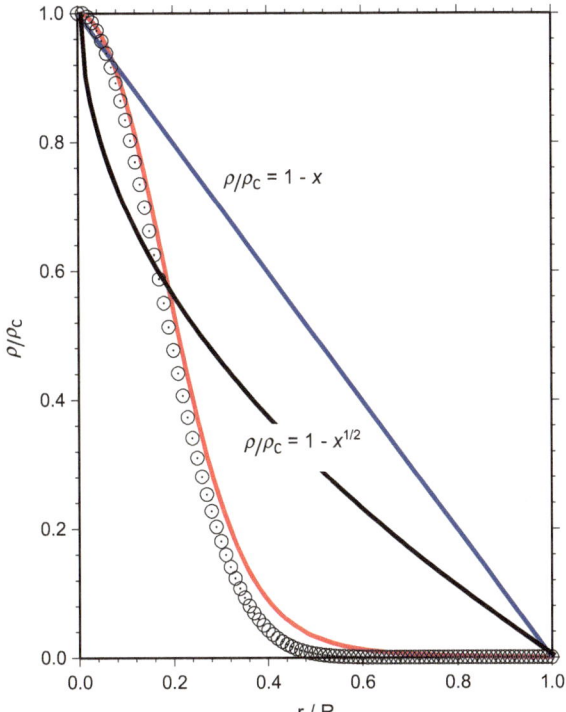

3.3 Polytropic Models

The intent of the last three sections has been to develop and explore simple analytic models that describe the run of pressure, mass and density with respect to radial displacement from the center of a star. Such models showing varying degrees of success, but all having some weakness or other in their physical formulation. By far the most successful semi-analytic approach, at least in terms of an ability to mimic detailed numerical calculations, is that in which a polytropic equation of state is assumed. In this case, the pressure is taken to be dependent upon the density only, with

$$P = K\rho^{(n+1)/n} \tag{3.44}$$

where K is a constant and n is the polytropic index. A vast amount of literature has been published upon polytropic stellar models, but our aim in the following section will be to develop the so-called Lane-Emden equation and then look at a few special solutions and applications.

To derive the Lane-Emden equation, we begin with the equation for hydrostatic equilibrium and the equation of mass continuity. Starting with the equation for hydrostatic equilibrium:

$$\frac{1}{\rho}\frac{dP}{dr} = -G\frac{m}{r^2}$$

Differentiating this equation with respect to r, we obtain

$$\frac{d}{dr}\left(\frac{1}{\rho}\frac{dP}{dr}\right) = -2G\frac{m}{r^3} - \frac{G}{r^2}\frac{dm}{dr}$$

Substituting for the mass continuity equation and reintroducing the equation for hydrostatic equilibrium, we find

$$\frac{d}{dr}\left(\frac{1}{\rho}\frac{dP}{dr}\right) = -\frac{2}{\rho r}\frac{dP}{dr} - 4\pi\ G\rho$$

Multiplying both sides of the above equation by r^2, we further obtain

$$r^2\frac{d}{dr}\left(\frac{1}{\rho}\frac{dP}{dr}\right) + \frac{2r}{\rho}\frac{dp}{dr} = \frac{d}{dr}\left(\frac{r^2}{\rho}\frac{dP}{dr}\right) = -4\pi\ Gr^2\rho$$

This equation is generally written as

$$\frac{1}{r^2}\frac{d}{dr}\left(\frac{r^2}{\rho}\frac{dP}{dr}\right) = -4\pi\ G\rho \qquad (3.45)$$

Equation (3.45) is effectively the Lane-Emden equation, although it is generally written in a dimensionless form, as the following exercise will reveal:

(☺) **Exercise 3.16**   Take Eq. (3.45), set $\rho = \rho_c\theta^n$ and $P = K\rho^{(n+1)/n}$, where ρ_c is the central density, and show that

$$\frac{1}{\xi^2}\frac{d}{d\xi}\left(\xi^2\frac{d\theta}{d\xi}\right) + \theta^n = 0 \qquad (3.46)$$

where the following substitutions have been made: $r = \alpha\,\xi$, and

$$\alpha^2 = (n+1)\frac{K}{4\pi G}\rho_c^{(1-n)/n} \qquad (3.47)$$

The only hint for exercise 3.16 is to take a large sheet of paper and slog away at the algebra.

Equation (3.46) is the Lane-Emden equation, named after Jonathan Homer Lane and Robert Emden, who first explored its characteristics in the late 1800 to early 1900s, respectively. Using a polytropic relation of the form $P \sim \rho^{\gamma}$, where $\gamma = C_p / C_v$ is the ratio of specific heats, Lane published in 1870 the first model discussion of the Sun considered to be a hot gas sphere (rather than a hot fluid sphere, as was the common assumption at that time). German engineer August Ritter further developed Lane's ideas in the early 1800s, but it was Emden's classic text *Gaskulgeln* (*Gas balls*), published in 1907, that brought the concept of the polytopic model to the full attention of mainstream astronomers.

In general, the polytopic index n can be taken as any positive or negative number. In astronomical applications, the most commonly used values are either $n = 1.5$ or $n = 3.0$. These two cases are applied according to the dominant mode of energy transfer within the interior of a star, with the $n = 1.5$ situation corresponding to convective energy transport, and the $n = 3$ case corresponding to radiative energy transport. We shall have more to say about these modes of energy transport in the next chapter. The $n = 1.5$ polytrope also applies to the non-relativistic white dwarf configuration (see Sect. 5.9). Neutron stars have additionally been approximated by polytropic models with $n = 0.5$.

Unfortunately, the Lane-Emden equation has no closed analytic form for the main polytopic indices of astrophysical interest (although see the Appendix). There are, however, special analytic solutions to Eq. (3.46) when $n = 0$, 1 or 5. The $n = 0$ solution corresponds to the constant density condition with $\rho = \rho_c$ and $P = P_c \theta$. The $n = 1$ solution has some limited application to the study of gaseous planets (such as Neptune; see below), while the $n = 5$ solution has some (possibly fundamental) application to the evolution of red giant stars (see sect. 5.6). The variation of θ with ξ for the $n = 0$, 1 and 5 polytropes are shown in Fig. 3.11.

(☺) **Exercise 3.17** Show that when $n = 0$, the solution to Lane-Emden equation is $\theta = 1 - \xi^2/6$. *Hint:* Proceed by substitution. *Note:* This result indicates a radius of $R = \alpha \, \xi_1$ for the $n = 0$ polytrope, with $\xi_1 = 6^{1/2} \approx 2.45$ being determined by the condition that $\theta(\zeta_1) = 0$.

(☺) **Exercise 3.18** Shown that when $n = 1$, the solution to Lane-Emden equation is: $\vartheta = \sin(\xi)/\xi$. In this situation, the resultant polytropic model has a constant radius determined purely by the constant K, with $R = \alpha \, \pi$, and $\alpha = (K / 2\pi \, G)^{1/2}$, which is independent of the total mass. *Hint:* proceed by substitution.

In order to obtain general solutions, the Lane-Emden equation has to integrated numerically outward from the center to the surface (see exercise 3.22 below and the Appendix). The surface is defined by the boundary condition $\rho = 0$ which corresponds to $\xi = \xi_1$ where $\theta(\zeta_1) = 0$. The central boundary conditions are $r = 0$, corresponding to $\xi = 0$; $\rho = \rho_c$, corresponding to $\theta = 1$; and $d\rho / dr = 0$, corresponding to $d\theta / d\xi = 0$. Accordingly, the surface radius of a polytropic model is given by $R = \alpha \, \xi_1$, and the total mass M is given as:

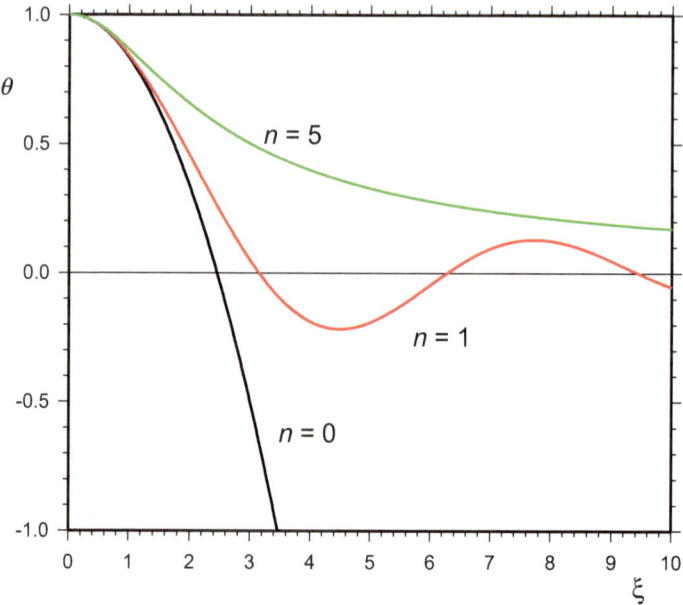

Fig. 3.11 Variation of the variables θ and ξ for $n = 0$, 1 and 5 polytropic configurations. The $n = 0$ case corresponds to the constant density model as discussed in Sect. 3.2.1. The solution for the $n = 5$ polytrope is discussed in Sect. 5.6. (Credit: Author)

$$M = \int_0^R 4\pi r^2 \rho \, dr = 4\pi\alpha^3\rho_c \int_0^{\xi_1} \xi^2\theta^n d\xi = -4\pi\alpha^3\rho_c\xi_1^2 \left(\frac{d\theta}{d\xi}\right)_{\xi_1} \qquad (3.48)$$

where the integral in (3.48) has been evaluated through substitution with Eq. (3.46). In general, the value of ζ_1 and $(d\theta / d\zeta)$ at $\zeta = \zeta_1$ can only be found through the numerical integration of Eq. (3.46), and historically such integration results have been published in tabulated form—see Table 3.1. For the analytic, constant density, $n = 0$, however, we have the simple result that

$$M = -4\pi\alpha^3\rho_c 6\left(-\frac{\sqrt{6}}{3}\right) = 8\pi\alpha^3\rho_c\sqrt{6} = \frac{4\pi}{3}\rho_c(\alpha\xi_1)^3 \qquad (3.49)$$

where $\alpha^2 = P_c/4\pi G\rho_c^2$. Eliminating the α term gives $P_c = (3/8\pi)(G M^2/R^4)$, and this is equivalent to Eq. (3.16) seen earlier. We additionally determine that the run of the pressure through the model interior is $P(\xi) = P_c(1 - \xi^2/6)$.

(☺) **Exercise 3.19** Complete the algebra for the central pressure term in the $n = 0$, constant density, polytrope when $\rho_c = \bar{\rho}$.

Starting with Eq. (3.48), we can eliminate the central density term ρ_c by substitution from Eq. (3.51), and this reveals a radius-mass relationship for polytropic models, with:

$$M^{n-1}R^{3-n} = \frac{1}{4\pi}\left(\frac{(n+1)K}{G}\right)^n \left(-\xi^2\frac{d\theta}{d\xi}\right)^{n-1}_{\xi_1} \xi_1^{3-n} \qquad (3.50)$$

Also, defining the term $D_n = \bar{\rho}/\rho_c$, we have

$$D_n = -\left[\frac{3}{\xi_1}\left(\frac{d\theta}{d\xi}\right)_{\xi_1}\right] \qquad (3.51)$$

(☺) **Exercise 3.20** No prize for guessing what to do here: Take a large piece of paper and derive the expressions for Eq. (3.50) and Eq. (3.51). **Comments:** Deriving (3.50) is just a hard slog, but Eq. (3.51) follows fairly easily from Eq. (3.48) with the substitution that $R = \alpha\,\xi_1$, and remember $\bar{\rho} = M/(4\pi R^3/3)$.

The radius-mass relationship revealed in Eq. (3.50) indicates several special situations where the mass and radius decouple from each other. Indeed, for $n = 3$ the mass is determined independently of the radius—that is, for a given value of K there is just one mass that satisfies the condition of hydrostatic equilibrium. For $n = 1$, the radius is determined independently of the mass. For $1 < n < 3$, we have $R \sim M^{-(n-1)/(3-n)}$, which indicates that the overall radius decreases with increasing total mass—that is, a more massive a star will be smaller and more dense than a lower mass star described by the same given polytropic index. The values for ξ_1, $-\xi^2(d\theta / d\zeta)$ and D_n, for a range of $0 \le n \le 5$ values, are given in Table 3.1. In this table we see that as n approaches 5, so the value of ξ_1 and hence the configuration radius tends to infinity. In this same limit, however, the mass remains constant but the ratio of the average to central density D_n tends to zero. As will be discussed in more detail in the next chapter, Eddington's standard model corresponds to the $n = 3$ polytrope. For this choice of n, not only is the total mass of the model independently determined with respect to the overall radius, but the average density is some 54 times smaller

Table 3.1 Solutions to the Lane-Emden equation for selected values of n. Data from Hordedt (2004)

n	ξ_1	$-\xi^2(d\theta / d\zeta)$	D_n
0	2.44948974	4.898979	1.000000
0.5	2.75269805	3.788651	0.544917
1	3.14159265	3.141593	0.303964
1.5	3.65375374	2.714054	0.166925
2	4.35287360	2.411046	0.087700
3	6.89684862	2.018236	0.018456
4	14.9715463	1.797230	0.001607
5	∞	1.732051	0.000000

than the central density (see Table 3.1: $D_3 = 0.018456 \approx 1/54$), indicating a distinctly peaked central density distribution.

(☺) **Exercise 3.21** For a given constant mass and central density, which polytope will have the larger radius, that for $n = 3$ or that for $n = 1.5$? *Hint:* Use the definitions for D_n and average density to show that $R(n = 3) / R(n = 1.5) = (D_{1.5} / D_3)^{1/3} = 2.08$. That is, the $n = 3$ polytrope will be nearly twice as large as the $n = 1.5$ polytrope for the same mass and central density.

(☺) **Exercise 3.22: Short Term Paper/Project** Write a computer program to numerically integrate the Lane-Emden equation (Eq. (3.46)). There are many internet sites that will take you through the process of writing such a program. The most straightforward numerical integration method to look at is that of the fourth order Runge-Kutta scheme (see Appendix). Test your results against the analytic solutions illustrated in Fig. 3.11, and against the terms given in Table 3.1.

The next two sections take us well beyond the exploration of straightforward models, and the results to be presented are largely based upon numerical rather than analytic developments. I would encourage the reader to work through these sections (although they may be safely skipped during a first read through of the text), and I would especially encourage the reader to tackle the exercises where a computer code needs to be developed, as per exercise 3.22 above. Perhaps these last two sections might be best read as outline notes for an extended term paper and/or term computing project.

3.4 The Bonnor-Ebert Sphere

Table 3.1 and Fig. 3.11 reveal that as the polytropic index n approaches 5, the radius of the configuration (through the ξ_1 term) approaches infinity. Interestingly, however, the mass of such a configuration (via column 3 in Table 3.1) remains finite. Having reached $n = 5$, there thus might seem to be little astrophysical application for polytropes with a higher index. This, however, is not the case, and indeed the $n = \infty$ polytrope affords a good approximation to an isothermal (constant temperature) sphere and mimics the situation within a molecular cloud core hovering on the edge of collapse before undergoing gravitational collapse to form a star (recall Sect. 2.2). In the isothermal sphere case, the pressure P is related to the density ρ and the speed of sound c_S, with $P = c_S^2 \rho$, where $c_S = (R\,T/\mu)^{1/2}$, where T is the temperature, μ is the mean molecular weight and R is the gas constant. Comparing the formulation for the pressure in an isothermal gas to that of the polytropic approximation (Eq. (3.44)), we have compatibility provided that $(n + 1) / n = 1$, which is the requirement that n be very large. Following the same algebraic steps required to derive Eq. (3.46), we have for an isothermal cloud in hydrostatic equilibrium:

$$\frac{1}{r^2}\frac{d}{dr}\left(\frac{r^2}{\rho}c_s^2\frac{d\rho}{dr}\right) = -4\pi G\rho \tag{3.52}$$

Introducing the variable $u = -\ln(\rho\,/\,\rho_0)$, where ρ_0 is the central density, Eq. (3.52) can be rewritten as:

$$\frac{1}{\xi^2}\frac{d}{d\xi}\left(\xi^2\frac{du}{d\xi}\right) = e^{-u} \tag{3.53}$$

where the substitution $r = \alpha_I\,\xi$, has been made and

$$\alpha_I = \frac{c_S}{\sqrt{4\pi\,G\rho_0}} \tag{3.54}$$

The boundary conditions are $u = 0$ and $du/d\xi = 0$ at $\xi = 0$. Formally, there is no specific surface to the isothermal model, and the outer radius ξ_1 is chosen so that the internal pressure matches that of the external pressure P_{ext} (note that at $\xi = \xi_1$ the density will not in general be zero). The mass interior to $\xi = \xi_1$ will be given in similar fashion to Eq. (3.48), with:

$$M(\xi_1) = 4\pi \int_0^{\xi_1} \rho_0\alpha_I^3 e^{-u}\xi^2 d\xi \tag{3.55}$$

Equation (3.53) has one special-case analytic solution corresponding to the so-called singular isotherm solution (SIS). This solution follows upon the assumption that Eq. (3.52) admits a power law solution of the form $\rho = A\,r^m$, where A and m are constants to be determined. Proceeding therefore by substitution, Eq. (3.52) becomes:

$$\frac{1}{r^2\rho}\frac{d}{dr}\left(\frac{r^2}{\rho}\frac{d\rho}{dr}\right) = -\frac{4\pi}{c_S^2}\frac{G}{A\,r^{(m+2)}} = \frac{m}{A\,r^{(m+2)}} \tag{3.56}$$

Accordingly, equating the last two terms in Eq. (3.56), the solution required corresponds to $m = -2$ and $A = c_S^2\,/\,(2\pi\,G)$, giving

$$\rho = \frac{c_S^2}{2\pi\,G}\frac{1}{r^2} \tag{3.57}$$

the mass interior to radius r is additionally

$$M(r) = \int_0^r 4\pi\,r^2\rho\,dr = 4\pi\frac{c_S}{2\pi\,G}\int_0^r dr = 2\frac{c_S^2}{G}r \tag{3.58}$$

That the power law assumption yields the SIS is evident from Eq. (3.57), which is singular (infinite density) at the origin $r = 0$. Sidestepping this problem, Eq. (3.58) further indicates that the mass of an isothermal sphere tends to infinity (since the mass increases linearly with the radius).

(☺) **Exercise 3.23** (1) Imagine that orbiting at some radius r from the center of an extended isothermal cloud is a small blob of gas. Given that the blob has a circular orbit, how does the velocity V_{orb} vary with distance r from the center of the cloud? **Hints:** Recall from Newton's shell theorem that the gravitational influence of the isothermal cloud will be entirely dependent upon the mass interior to radius r. On the basis that the gas blob has a much smaller mass than the cloud, then the orbital velocity will be $V_{orb} = (G\,M(r)\,/\,r)^{1/2}$ (2) Can you think of another circumstance where a similar such velocity relationship with distance holds true? **Hint:** investigate the properties of galaxy rotation curves, and see Sect. 6.3.

The SIS solution for the density variation in an isothermal cloud (Eq. (3.57)) provides a better and better approximation as the radius tends towards larger and larger values, but it is far from a satisfactory solution towards the cloud center, where the density must assuredly be finite. For solutions to Eqs. (3.52) or (3.53) with a finite central density ρ_0, some numerical integration scheme will need to be employed (see Fig. 3.12 and exercise 3.24). Such numerical schemes enable the exploration of more physically realistic situations, where, for example, an isothermal

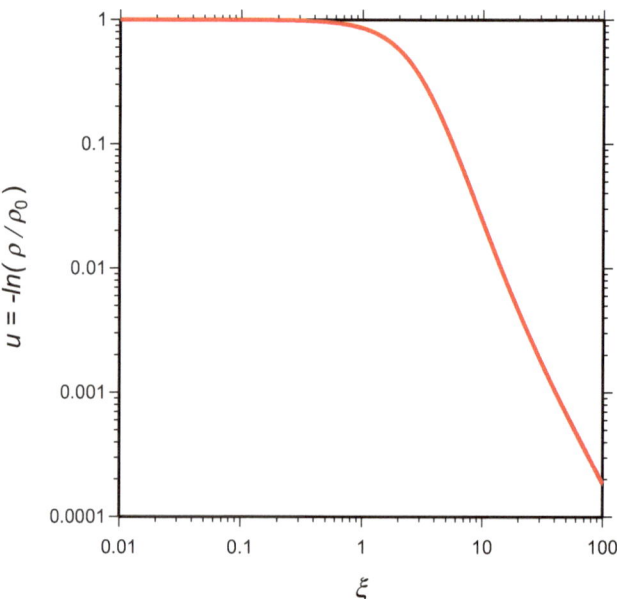

Fig. 3.12 The solution curve for Eq. (3.53) showing u verses ξ. The boundary conditions adopted are $u = 0$ and $du/d\xi = 0$ at $\xi = 0$, and the integration was arbitrarily stopped at $\xi = 100$. (Credit: Author)

cloud has a finite radius R_c and is constrained by some non-zero external pressure P $(R_c) = P_{ext} > 0$. Such isothermal clouds are known as Bonner-Ebert spheres after William Bonnor and Rolf Ebert, who first studied such model structures in 1956.

(☺) **Exercise 3.24: Short Project** Write a computer program (by adapting your code for evaluating the Lane-Emden equation developed in exercise 3.22) to solve for the isothermal model (Eq. (3.53)). Test your results against Fig. 3.12.

A Bonnor-Ebert sphere corresponds to a stable, finite-sized, finite-mass isothermal cloud with a non-zero pressure at its outermost boundary. It turns out that a Bonnor-Ebert sphere in effect determines, for a given mass of material, the limiting conditions beyond which an isothermal cloud must undergo gravitational collapse and begin to produce stars (as described in Chap. 2). To see how this limiting condition comes about, we need to consider an isothermal cloud of some given mass M_c. Accordingly, for a given ξ_1 the mass Eq. (3.55) determines the central density ρ_0. Furthermore, with ρ_0 determined, Eq. (3.54) reveals the radius of the cloud as $R_c = \alpha_1 \xi_1$. We therefore have

$$\rho_0 = \left(\frac{c_S^6}{4\pi G M_c^2}\right) \left[\xi_1 \left(\frac{du}{d\xi}\right)_{\xi_1}\right]^2 \tag{3.59}$$

and

$$R_c = \left(\frac{G \, M_c}{c_S^2}\right) \left(\frac{1}{\xi_1 (du/d\xi)_{\xi_1}}\right) \tag{3.60}$$

The external pressure at the cloud boundary R_c is given by $P_{ext} = c_S^2 \, \rho(\xi_1)$, and using the relation $u = -\ln(\rho \, / \, \rho_0)$, it is found that

$$P_{ext} = \left(\frac{c_S^8}{4\pi G^3 M_c^2}\right) \xi_1^4 \left(\frac{du}{d\xi}\right)_{\xi_1}^2 \exp(-u) \tag{3.61}$$

Figure 3.13 shows a plot of the scaled pressure, P_{ext} divided by the first bracketed term in Eq. (3.61) versus ξ_1. Here we see something interesting happening, in that there is a well-defined maximum at $\xi_{crit} = 6.45$. This critical value for ξ_1 sets a stability limit and determines the conditions for cloud collapse. Inserting numerical terms at $\xi_1 = \xi_{crit}$, the corresponding critical radius and critical external pressure for an isothermal cloud of mass M_c are

$$R_{crit} = 0.41 \frac{GM_c}{c_S^2} \quad \text{and} \quad P_{crit} = 1.40 \frac{c_S^8}{G^3 M_c^2} \tag{3.62}$$

Should the radius of a cloud or the external pressure exceed one (or both) of these critical values, then cloud collapse will occur. Alternatively, the critical mass M_{crit} of a Bonnor-Ebert sphere, for a given sound speed c_S and external pressure P_{surf}, is

$$M_{crit} = 1.18 \frac{c_S^4}{G^{3/2}} \frac{1}{\sqrt{P_{surf}}} \qquad (3.63)$$

Given a 1 $M_{\odot}$ molecular cloud core with a uniform temperature of T = 10 K, the characteristic sound speed will be $c_S = (R\,T\,/\,\mu)^{1/2} \sim 200$ m/s, and accordingly $R_{crit} \approx 1.5 \times 10^{15}$ meters $\approx 10^4$ AU (or 0.05 pc), and $P_{crit} \approx 3 \times 10^{-12}$ Pa. This latter critical pressure is about 10 times larger than the typical pressures exhibited within interstellar clouds (recall Sect. 2.2).

(☺) **Exercise 3.25: Short Project** Using the computer code developed in exercise 3.22 to verify the results shown in Fig. 3.13. Convince yourself that the maximum in the curve falls at $\xi_{crit} = 6.45$ and that at that point, $(du\,/\,d\xi) = 0.377$.

The cloud collapse conditions (3.62) and (3.63) come about due to a relative interplay between the importance of gravity and external pressure. For $\xi_1 < \xi_{crit}$, it is the external pressure rather than gravity that is important in determining the

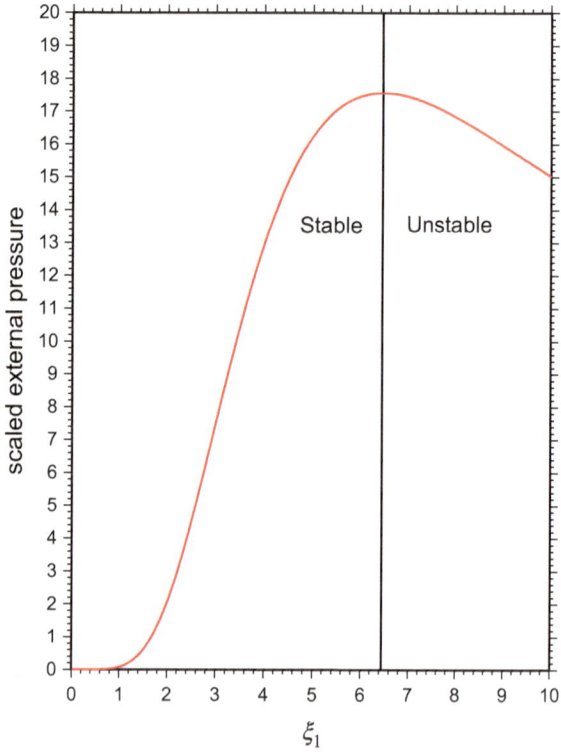

Fig. 3.13 The scaled external pressure $[P_{ext}\,/\,(c_S^8\,/\,4\,\pi\,G^3\,M_c^2)]$ versus ξ_1 for a Bonnor-Ebert sphere of mass M_c. (Credit: Author)

equilibrium condition. Indeed, an increase in the external pressure (for a given M_c and c_S) results in cloud compression and a concomitant increase in the internal pressure to match that at the cloud's surface, and the sphere oscillates about its equilibrium radius. For $\xi_1 > \xi_{crit}$, the external pressure is no longer important in determining the cloud's equilibrium condition, and a small decrease in the cloud radius now results in the gravitational force becoming stronger than the restoring internal pressure force—a situation that is inevitably unstable, and one in which cloud collapse will quickly follow. This being said, a critical Bonnor-Ebert sphere need not collapse if other internal pressure support mechanisms such as turbulence, rotation and magnetic fields are in play, as was explored earlier in Chap. 2.

(☺) **Exercise 3.25** Show that the ratio of the central density to the average density in a Bonnor-Ebert sphere is $\rho_0/\bar{\rho} = \frac{1}{3}\xi_1/(du/d\xi)_{\xi_1}$ and that in the critical case $\rho_0/\bar{\rho} = 5.7$. The latter result indicates that a critical Bonnor-Ebert sphere has just a modest enhancement in its central density, and that it is accordingly not too far removed from the constant density, isothermal sphere model that resulted in the critical Jeans mass derivation (recall Sect. 2.2).

3.5 Multiple Polytropes

There is no specific reason that a stellar or planetary model has to be based upon a single choice of polytropic index. Indeed, multilayered polytropic models can be constructed provided that appropriate care is taken to match boundary conditions. While August Ritter considered such multiple polytrope models in the 1880s, the first utilitarian double polytrope model was developed by E. A. Milne in the early 1930s, when he conjoined an $n = 1.5$ polytrope envelope to a fully degenerate, relativistic white dwarf core of polytropic index of $n = 3$. More recently, Archibald Henry (Indiana University) developed in the early 1990s a multilayered polytropic model for the Sun's interior, and Saul Rappaport (MIT) and coworkers in the early 1980s developed a set of multiple polytrope models of very low-mass stars to be used in a detailed study of cataclysmic variable and compact X-ray binaries. Other examples of double polytrope models include an $n = 3$ core, $n = 3$ envelope description of massive main sequence stars in which radiation pressure is dominant; an $n = \infty$ (isothermal) core, $n = 1$ envelope model for exploring the so-called Chandrasekhar-Schoenberg limit (to be discussed latter); Peter Eggleton (University of Cambridge) and coworkers further developed an $n = 5$ core, $n = 1$ double polytrope model to describe in part the internal structure and physical size of red giants (see Sect. 5.8). Planetary structure can also be described according to multiple polytrope models; in the early 1980s, George Horedt (DFVLR, Germany) and William Hubbard (University of Arizona) discussed a number of increasingly complex such models to describe the properties of Jupiter, Saturn and Uranus.

Key to making progress in the construction of a multilayered polytropic model is to match the boundary conditions across the boundary at which the polytropic index changes. At this boundary there must be continuity in the mass, radius and pressure descriptions. Indeed, Chandrasekhar in his classic *An Introduction to Stellar Structure* provides us with the appropriate matching conditions. Before seeing these boundary matching conditions, let us first introduce some notation. In the following, the core is to be described by a polytrope of index $n = n_1$, and the envelope is to be described by a polytrope of index $n = n_2$. The core is described by the Lane-Emden equation (Eq. (3.46)) with variables θ and ξ, while the envelope is described by the Lane-Emden equation but with variables ϕ and η. Also (for brevity), a dash notation is to be used to indicate differentiation, with $\theta' = d\theta / d\xi$ and $\phi' = d\phi / d\eta$. Chandrasekhar shows (see exercise 3.27 below) that the required matching conditions at the core-envelope interface located at $\xi = \xi_c$ are

$$\frac{\xi \theta^{n_1}}{\theta'} = \frac{\eta \phi^{n_2}}{\phi'} \tag{3.64a}$$

and

$$(n_1 + 1)\frac{\xi \theta'}{\theta} = (n_2 + 1)\frac{\eta \phi'}{\phi} \tag{3.64b}$$

At the core interface, taking $\phi(\eta_c) = \theta(\xi_c)$, these matching conditions indicate that

$$\phi'(\eta_c) = \frac{(n_1 + 1)}{(n_2 + 1)} \left(\frac{\xi}{\eta}\right)_c \theta'(\xi_c) \text{ and } \left(\frac{\xi}{\eta}\right)_c = \theta(\xi_c)^{(n_2 - n_1)} \left(\frac{\theta'}{\phi'}\right)_c \tag{3.64c}$$

and matching the pressure across the interface we also have

$$\frac{P_e}{P_c} = \theta(\xi_c)^{n_1 - n_2} \tag{3.64d}$$

where P_c is the central pressure and P_e is the pressure at the base of the envelope. By way of an example of a double polytrope, we might consider an $n = 0$, constant density core, matched to an $n = 1$ envelope. This model was studied in some considerable detail by J. O. Murphy (Monash University) in the early to mid-1980s. Both of the chosen indices in this model provide for analytic solutions, and while somewhat unwieldy and physically oversimplified, the model does approximate that of a Jovian planet with a solid, incompressible core overlain by a gaseous envelope. The core of our planet will have a radius R_c given by $R_c = \alpha_c \xi_c$, where $\alpha_c^2 = P_c / (4\pi G \rho_c^2)$, where P_c and ρ_c are the central pressure and the (constant) core density respectively. The mass of the core will be $M_c = (4\pi/3)R_c^3\rho_c$. From exercise 3.17, we have $\theta = 1 - \xi^2/6$, and accordingly, at the core-envelope interface where $\xi = \xi_c$, we have $\theta(\xi_c) = 1 - \xi_c^2/6$ and $\theta'(\xi_c) = -\xi_c/3$. The interface matching conditions now dictate that

$$\eta_c = \xi_c / \sqrt{2\,\theta(\xi_c)} \text{ and } \varphi'(\eta_c) = \left(\frac{\theta_c}{2}\right)^{1/2} \theta'_c \qquad (3.65)$$

For an $n = 1$ envelope, the solution to the Lane-Emden equation is such that

$$\varphi(\eta) = A\frac{\sin(\eta - B)}{\eta} \qquad (3.66)$$

where A and B are determined by the boundary conditions (Eq. (3.65)). In the single polytrope case, $A = 1$ and $B = 0$. In the $n = 0$ core, $n = 1$ envelope double polytrope situation, we have:

$$A = \theta(\xi_c)\eta_c / \sin(\eta_c - B) \qquad (3.67)$$

$$B = \eta_c - \cot^{-1}\left(\frac{\varphi'(\eta_c)}{\varphi(\eta_c)} - \frac{1}{\eta_c}\right) \qquad (3.68)$$

With the last two equations the model is fully resolved, and for a set choice of P_c and ξ_c, a complete model mass and radius can be determined. Indeed, the radius of the complete model will be $R = \alpha_e\,\eta_1$, where $\eta_1 = \pi + B$ is the first zero of (3.66), and $\alpha_e^2 = P_e / (2\pi G\,\rho_e^2)$, where $P_e = P_c / \theta(\xi_c)$ and $\rho_e = \rho_c$ are the pressure and density at the base of the envelope. The total mass of the model follows as per Eq. (3.48) with the integral running from η_c to η_1. Table 3.2 provides an evaluation of the intercept constants for various values of $0 \le \xi_c \le \sqrt{6}$.

By way of an example calculation for our double, $n = 0$ core, $n = 1$ envelope polytrope planet model, let us look at the planet Neptune. Observations reveal that Neptune has a radius and mass of 2.5×10^7 meters and 10^{26} kg respectively. For the sake of argument in this example, let us assume that the core-envelope boundary is

Table 3.2   Intercept constants for the $n = 0$ core, $n = 1$ envelope double polytrope

ξ_c	η_c	A	B	η_1	$-\eta^2(d\phi/d\eta)$
0	0	1	0	π	π
$\sqrt{6}/10$	0.2449	-1.0147	0.3447	3.4863	3.5377
$2\sqrt{6}/10$	0.4898	-1.0560	0.6808	3.8223	4.0365
$3\sqrt{6}/10$	0.7348	-1.1161	1.0049	4.1465	4.6279
$4\sqrt{6}/10$	0.9798	-1.1846	1.3217	4.4633	5.2870
$5\sqrt{6}/10$	1.2247	-1.2500	1.6435	4.7851	5.9814
$6\sqrt{6}/10$	1.4697	-1.3005	1.9926	5.1342	6.6769
$7\sqrt{6}/10$	1.7146	-1.3228	2.4114	5.5530	7.3453
$8\sqrt{6}/10$	1.9596	-1.3005	3.0030	6.1446	7.9908
$9\sqrt{6}/10$	2.2045	-1.2090	4.1731	7.3147	8.8435
$\sqrt{6}$	–	–	–	$\xi_1 = \sqrt{6}$	$2\sqrt{6}$

The first row corresponds to a pure $n = 1$ (all envelope) polytrope, while the last row corresponds to a pure $n = 0$ (all core) polytrope located at $\xi_c = 5\sqrt{6} / 10$, and accordingly the radius is expressed as $R_{Nep} = 2.5 \times 10^7 = \alpha_e \eta_1$, and from Table 3.2 we have $\eta_1 = 4.7851$. This relation gives $\alpha_e = 5.2 \times 10^6$. The total mass (via Eq. (3.48)) is additionally given by $M_{Nep} = 10^{26} = 4 \pi \alpha_e^3 \rho_e \times 5.9814$, where the last numerical term is taken from Table 3.2. This relationship indicates that $\rho_e = 9330$ kg/m^3, and this sets the density of the $n = 0$ (constant density) core. Using the expression $\alpha_e^2 = P_e / (2\pi G \rho_e^2)$, we now determine that $P_e = 9.94 \times 10^{11}$ Pa. Furthermore, the pressure at the core-envelope boundary is $P(\xi_c = \sqrt{6} / 2) = P_e = P_c(1-0.25) = 0.75\ P_c$, giving P_c, the central pressure as $P_c = 1.33 \times 10^{12}$ Pa. From the expression $\alpha_c^2 = P_c / (4\pi G \rho_c^2)$, we now find that $\alpha_c = 4.3 \times 10^6$, and the core mass now follows from Eq. (3.49) as $M_{core} = M(\xi_c) = 5.6 \times 10^{24}$ kg $\approx 1\ M_{Earth}$. The core-mass fraction is $q = M_{core} / M_{Nep} = 0.06$ and the core radius fraction is $Q = r(\xi_c) / R_{Nep} = 0.21$.

The model is now fully determined, and a quick check against the more detailed numerical models for the internal structure of Neptune reveals that the double polytrope model does a reasonably good job, and indeed current detailed numerical models predict that Neptune has an inner rocky core of about one Earth mass occupying the inner 25% by radius. The range in central density and central pressure vary from one numerical model to the next depending upon the various input assumptions, but the double polytrope model values ($\sim$ 9000 kg/m^3 and $\sim$ 1000 GPa respectively) are both within the predicted limits. Clearly, the double polytrope model could be tweaked to yield a range of central pressure, density, core mass, q and Q values by systematically exploring other values for ξ_c. We shall leave this, however, as an exercise for the student (see exercise 3.27). Indeed, we have worked rather hard to develop the equations for even the simplest (analytic) double polytrope model, and while the algebra can be pushed further (see Sect. 5.8) it is probably simpler to proceed numerically. It is in fact the computational simplicity (once the equations and code have been developed, as per exercise 3.22) that underscores the real utility of the multiple-polytrope approach. This situation is especially so when a large parameter space is to be explored.

(⊗) **Exercise 3.27: Longer Term Paper/Extended Term Project.** Investigate the history of multilayered polytropes and their application to stellar / planetary structures. Research and discuss the advantages and reasoning behind the construction of such models, derive the interface matching eqs. (3.64a) to (3.64d) and develop one specific model example in detail—say an $n = 3$ core with an $n = 1.5$ envelope. Extend the computer code developed in exercise 3.22 to enable the investigation of a double polytrope model with arbitrary core and envelope indices n_1 and n_2. In terms of the historical literature, the internet will be your most useful resource, but see specifically the SAO/NASA Astrophysics Data System webpage (absabs.harvard. edu/) for actual research papers. For additional details and reference material, see:

M. Beech and R. Mitalas, 1990. Effect of mass loss and overshooting on the width of the main sequence of massive stars. *Astrophysical Journal*, 352, 291–299.

M. Beech, 1988.The Schoenberg-Chandrasekhar limits: a polytropic approxima-tion. *Astrophysics and Space Science*, 147, 219–227.

3.6   Answers to Exercises

Following are the answers to all exercises in this chapter. The exercises for which no answer is provided are those in which the student is asked to check a result or complete the algebra steps between two equations given in the text.

Exercise 3.4 The bulk density is $\rho_{\text{bulk}} \approx 1408$ kg/m^3, and $t_{\text{coll}} \approx 1632$ s (or about 27 min).

Exercise 3.7 $P_c = 3.2\,P_{c\odot}$, $T_c = 1.7\,T_{c\odot}$, and $R = 0.2\,\text{R}_\odot$.

Exercise 3.23 The oddity concerning the mass distribution in an isothermal cloud is that the circular velocity is the same irrespective of the distance from the center. In the case of galactic rotation curves, this same constant velocity with distance relationship is also observed but explained in accordance to the presence of a dark matter halo (see Chap. 6).

Chapter 4
From the Inside Out

Take another look at Fig. 3.1. Chap. 3 was largely concerned with the material body shown in the diagram, while this chapter is largely concerned with the flow of radiation, the second critical component in Eddington's two-part star model. Recall that the material body is described according to the interactions between the protons, electrons and ions that make up the material mass of a star. In contrast, the radiative body is described according to the passage of radiation and energy through the material body. Neither component can exist for long without the other, and as we have seen, the material body of a star would collapse on a timescale of hours if its interior was not hot and supported by a pressure gradient. The radiative component would dissipate even more rapidly if it did not interact with a star's material body. Traveling at the speed of light, a photon would take just 2 s to cross the Sun's radius, and yet for all its haste, the passage of radiation through a star is a tortuous affair, and the escape time is stretched out to many hundreds of thousands of years. As Arthur Eddington descriptively put it in his 1920 address to the British Association, 'The star is like a sieve, which can retain them [photons] only temporarily; they are turned aside, scattered, absorbed for a moment, and flung out again in a new direction. An element of energy may thread the maze for hundreds of years [this is an underestimate; see below] before it attains the freedom of outer space'.

The light crossing time of star is a mere few seconds, and if photons were free to fly outward from the center unimpeded, a star's internal energy would rapidly drain away into space. Not only this, but given that the temperature at the center of a star is typically many millions of degrees, the radiation would have a characteristic wavelength corresponding to that of hard X-rays. This is in contrast to what we actually see, which is radiation having a characteristic wavelength corresponding to visible light.[1] What controls the photon transformation and dramatically slows the journey time from the center to the surface are myriad atomic interactions. The photon

---

[1]Wien's law [equation (1.12)] dictates that between the center and surface of the Sun, the wavelength λ_{max} will increase from 2×10^{-10} m to 5×10^{-7} m—an increase by a factor of 2500.

© Springer Nature Switzerland AG 2019
M. Beech, *Introducing the Stars*, Undergraduate Lecture Notes in Physics,
https://doi.org/10.1007/978-3-030-11704-7_4

Fig. 4.1 The random walk
path of a photon through the
interior of a star. (Credit:
Author)

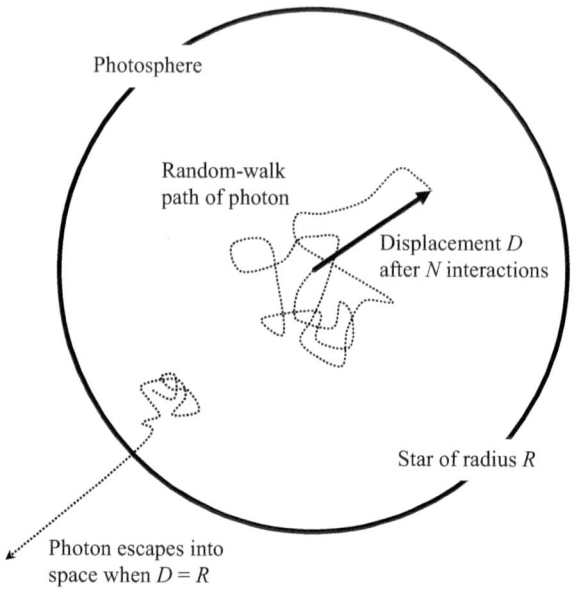

Photosphere

Random-walk
path of photon

Displacement D
after N interactions

Star of radius R

Photon escapes into
space when $D = R$

literally zigzags its way outward, taking a random walk from one interaction to another, traveling no more than about a tenth of a millimeter from one encounter and the next. The mean free path l between photon interactions is related to the density of material ρ and the opacity κ, with $l = 1 / \kappa\rho$. The opacity we have yet to define (see below), but it is effectively a measure of how much the stellar material hinders the passage of radiation through it. A photon's journey outward through a star is accordingly pictured as a series of short sprints, each of step length l, between interactions, with the direction of motion changing at random each time an interaction with a material component takes place (Fig. 4.1). This is a random walk model, and the radial displacement D form the center of a star after n interactions is $D^2 = n$ l^2. Accordingly, the typical number of random walk-steps experienced by a photon in crossing a star will be $n = (R / l)^2$, where R is the star's radius. Each step is accomplished at the speed of light c, with a travel time $T_S = l / c$, and accordingly the photon travel time from the center to the surface will be of order $T_{PR} = n (l / c) = R^2 /$ $(l \ c)$. In the case of the Sun, the number of photon interactions will be $n \approx (7 \times 10^8 /$ $10^{-4})^2 \sim 5 \times 10^{25}$, and the associated diffusion time will be of order 52,000 years. Compared to the direct exit travel time, the random walk journey of a photon through the Sun is a factor of some 7 million times slower. It is remarkable to think that the light that we see emanating from the Sun's surface today started its outward journey long before human civilization even existed; indeed, at a time when Mastodons and Sabertooth tigers walked the Earth.

It is the temperature gradient $\Delta T / \Delta R$ within a star that allows the energy generated at the center to flow outward. Characteristically, for the Sun $\Delta T / \Delta R \approx (T_C - T_S)/R_\odot = 0.02$ K/m, where T_C and T_S are the central and surface temperatures, respectively. Accordingly, the temperature within the Sun drops by just two 1/100ths of a degree for every meter moved outward towards the surface. This small temperature gradient is enough to allow energy to flow from the center to the surface, but importantly, since the typical distance between photon interactions is less than a millimeter, the photon essentially senses at any given moment a constant temperature environment. The temperature change experienced by a photon during one mean free path will be $\delta T = l \, \Delta T / \Delta R = 2 \times 10^{-6}$ K (taking $l = 10^{-4}$ m). At the center of the Sun, the fractional change in the temperature in different directions is thus $\delta T / T_C = 2 \times 10^{-13}$ (taking the central temperature to be 10^7 K). Even close to the surface of the Sun, $\delta T / T_S \sim 10^{-9}$. Under such conditions, the radiation field is clearly very close to being isotropic, and the very frequent collisions between particles maintain, at each radius, results in a local thermodynamic equilibrium. This is why the Sun radiates like a blackbody radiator corresponding to its surface temperature $T_S = 5780$ K rather than one of $T_C = 15$ million Kelvin.

Given that a star is in a steady state—neither heating up nor cooling down with time—the energy radiated into space at its surface must be exactly compensated for by the energy generated deep in its interior. This balanced state, described as thermodynamic equilibrium, does not always apply (as we shall see), but when it does, it enables an estimate to the surface temperature of a star to be made. Indeed, we can equate the central and surface energy fluxes. At the surface, the photon leaves the star at the speed of light c and the temperature is T_S and the energy flux is $(a \, T_S^4) \, c$, where a is the radiation density constant. This surface energy flux is matched by the energy flux at the star's center, which has moved through the star at an effective speed V_E determined by the photon diffusion process. The steady state condition now requires that $(a \, T_S^4) \, c = (a \, T_C^4) \, V_E$, which gives the surface temperature as

$$T_S = T_C \left(\frac{V_E}{c} \right)^{1/4} \tag{4.1}$$

Since V_E is going to be significantly smaller than the speed of light, so we are informed by Eq. (4.1) that the surface temperature must be significantly smaller than the central temperature. The effective diffusion speed of the photon will be of order $V_E = c / n$, where n can be no smaller than the direct number of mean-free path lengths between the star's center and surface i.e., $n > R / l \sim 10^{13}$ in the case of the Sun. Accordingly, given a central temperature of some ten million Kelvin, the surface temperature of the Sun is expected, when in a steady state, to be of order $T_S = 10^7 \, (1 / 10^{13})^{1/4} \approx 5600$ K, which is essentially as observed.

With the above initial study in place, we shall next develop a series of equations to more fully account for the temperature gradient within a star, and look more closely at the opacity term κ that enters into the expression for the mean-free photon path length.

4.1   The Temperature Gradient

It was argued in the last chapter that the temperature of a star must increase inwards from its surface. This temperature gradient is required in order to maintain the pressure gradient that keeps a star in dynamical equilibrium – that is, the outward pressure at each radial shell within a star capable of supporting the weight of its overlying layers. The temperature gradient is in turn determined by the energy flux per square meter per second F at each point within a star. This quantity is itself related to the way in which energy is being transported outward, at each radial location within a star, and there are just three ways in which this can be done. The modes of energy transport are the classical conduction, radiation and convection. Effectively, the first two modes entail collisions (interactions) in which energy is transferred from high energy particles to lower energy particles. The third mode, convection, entails the bulk motion of the material body. In the case of radiative energy transport, photon interactions with matter are important, whereas with conduction, typically collisions between electrons are being considered. Since collisions (interactions) occur repeatedly throughout the random walk followed by an electron (or photon), the outward energy flux is described by a diffusion equation involving the temperature gradient and a coefficient of thermal conductivity. Accordingly, we can write:

$$F = F_{rad} + F_{cond} = -(K_{rad} + K_{cond})\frac{dT}{dr} \qquad (4.2)$$

where K_{rad} and K_{cond} are coefficients of thermal conductivity (with units of energy per second per meter per Kelvin). If the energy is carried entirely by radiation and conduction, then the luminosity (energy per second) at each location within a star is given by the relationship $L = 4\pi r^2 (F_{rad} + F_{cond})$, and accordingly,

$$L = 4\pi r^2 (F_{rad} + F_{cond}) = -4\pi r^2 (K_{rad} + K_{cond})\frac{dT}{dr} \qquad (4.3)$$

Rather than work in terms of thermal conductivity coefficients, astronomers since Arthur Eddington (who first derived the equation for the temperature gradient in the early 1920s) work in terms of what is called the opacity. This term, as described earlier, is a measure of the resistance of a material body (gas) to the flow of energy through it, and it is defined as the cross section of interaction per unit mass (giving it units of square meters per kilogram). Accordingly,

$$K = \frac{4}{3}\frac{acT^3}{\rho \kappa} \qquad (4.4)$$

where K is the appropriate thermal conductivity term. A derivation of Eq. (4.4) takes us well beyond the range and topic of this book; suffice to say that it is derived from detailed kinematics calculations that result in the conductivity coefficient being defined as $K = \frac{1}{3}\bar{v}lC_V$, where the $\bar{v}$ is the average particle velocity, l is the mean free path between collisions and C_V is the specific heat per unit volume. For a photon gas $\bar{v} = c$, the speed of light, the energy density is $u = aT^4$, giving a specific heat of $C_V = du/dT = 4aT^3$, and the mean free path (as seen earlier) is $l = 1/\kappa\rho$. Eq. (4.4) emerges from the substitution of the various terms.

Using Eq. (4.4) in Eq. (4.3) now enables us to write an expression for the temperature gradient, with

$$\frac{dT}{dr} = -\frac{3\kappa\rho L}{16\pi a c r^2 T^3} \qquad (4.5)$$

where

$$\frac{1}{\kappa} = \frac{1}{\kappa_{rad}} + \frac{1}{\kappa_{cond}}$$

The dominant opacity term is the smaller one of κ_{rad} or κ_{cond}, and typically this is the κ_{rad} term. In other words, energy transfer by conduction is generally negligible within a star because κ_{cond} is large. The opposite situation can be true, however, when the gas is degenerate, as is the situation with white dwarfs (see Sect. 5.9).

(☺) **Exercise 4.1** (1) Complete the algebra steps taking Eq. (4.3) to (4.4). (2) Show that Eq. (4.5) can be rewritten as:

$$\frac{dP_{rad}}{dr} = -\frac{1}{4}\frac{\kappa}{4\pi r^2}\frac{\rho}{c}\frac{L}{c} \qquad (4.6)$$

Hint: Start with Eq. (4.5) and recall Eq. (3.3) for the radiation pressure. The relevance of Eq. (4.6) will become clear below when we consider Eddington's standard model and when we derive the so-called Eddington luminosity.

Working in terms of the opacity rather than the thermal conductivity does not actually save the astrophysicist from having to work with and solve a highly complicated set of equations. There is no simple way to determine the opacity appropriate to the gas inside of a star, and when such calculations are made in the modern era, they are usually presented as tables parameterized in terms of composition, density and temperature. More will be said about the opacity term in the next section.

Equation (4.5) is our third fundamental equation of stellar structure, and it takes its place alongside the mass continuity Eq. (3.6) and the equation of hydrostatic equilibrium (3.5).

(☺) **Exercise 4.2** It turns out that for very massive stars in which the surface temperature is very high, the opacity is approximately constant. (1) Using the condition that $\kappa = $ constant, show that in the outer layers of such a massive star that the pressure and temperature vary as $P = \Psi(T^4 - T_S^4)$, where $\Psi = 4\pi a c G M / 3\kappa L$ is a constant with L and M being the surface luminosity and total mass of the star, and where T_S is the surface temperature. **Hint:** Combine Eqs. (3.5) and (4.5), note which terms will be constant in the outer regions of the star, and then integrate and adopt the boundary condition that $P = 0$ at the star's surface. Note also that as we are considering what is happening in the outer layers of the star, the mass and luminosity terms will be essentially constant and equal to their total and surface values respectively. (2) Now show that if the outer layers of the star are composed of an ideal gas, the temperature variation near the surface is given by

$$T = \Theta \left(\frac{1}{r} - \frac{1}{R} \right), \text{ where } \Theta = \frac{\mu G M}{4 \Re}$$

and where R is the radius of the star. **Hint:** In this case, substitute the pressure term $P = \Psi(T^4 - T_S^4)$ into the equation of hydrostatic equilibrium Eq. (3.5) and use the ideal gas formula Eq. (3.1) to eliminate the density term and then integrate. Also, in this case, take the surface temperature boundary condition to be $T_S = 0$. (3) **Optional exercise.** Part (2) becomes much more complicated if T_S is not set to zero—you might like to investigate some of the complications that arise when T_S is some finite value. While an analytic solution (of sorts) is still available, this one small change in the boundary condition transforms our emoji label from ☺ into ☹.

(☺) **Exercise 4.3** Show that in the outermost layers of a star the equation of hydrostatic equilibrium can be rewritten in the form:

$$\frac{dP}{d\tau} = \frac{g}{\kappa} \tag{4.7}$$

where g is the gravitational acceleration at the star's surface, κ is the opacity and τ is a dimensionless quantity, called the optical depth, defined as $d\tau = -\kappa \rho \, dr$. The optical depth is a measure of the transparency of a medium, with an opaque medium (with a high opacity) having a large optical depth, and a transparent medium (with a low opacity) having a low optical depth. The optical depth can be used to determine the actual radius R of a star where $T = T_S$. Indeed, the integral $\int \kappa \rho dr$, from R to *infinity*, must be of the order of one (a more detailed set of physical arguments and detailed calculations reveals that the radius of a star is actually located at $\tau = 2/3$).

It was noted earlier that there is a third way in which energy can be transported within the interior of a star, and this is by convection, a process that involves the bulk motion of the material within a star. There is no complete hydrodynamic theory of turbulence and convection under stellar conditions —which is not to say that there are not highly complex theories and simulations available. The topic is fraught with mathematical difficulties, and some form of approximate theory has invariably to be employed. Even in the modern era (with state-of-the-art supercomputers), there are no first-principle stellar models that treat convection in a self-consistent manner, and this is perhaps the greatest area of uncertainty in all of astrophysical research. For all this, however, there is much that can be said and successfully modeled, albeit with parameterized theories.

Under convective conditions, Eq. (4.5) for the temperature gradient needs to be modified. Before seeing what this modification is, we will first consider the conditions under which convection might arise within the interior of a star. Since convection involves fluid motion, let us imagine a blob of matter at some radius r within a star. This blob of matter, as previously described, will have the same temperature and density as its surroundings. Suppose now that the blob is displaced upwards (away from the center to cooler surroundings) by some distance δr. Also suppose that the displacement is sufficiently fast that the blob does not exchange energy with its surroundings—this is known as an adiabatic change, and accordingly the pressure and density of the rising blob must vary according to the relationship $P / \rho^{\gamma} = $ constant, where $\gamma = C_P / C_V$ is the ratio of specific heats. In an ideal gas environment $\gamma = 5/3$, but when radiation pressure begins to dominates, $\gamma = 4/3$. Invoking pressure balance between the blob and its new surroundings at $r + \delta r$, we can write:

$$P(r + \delta r) = P(r) + \frac{dP}{dr} \delta r \qquad (4.8)$$

where dP/dr is the pressure gradient in the surrounding material. In moving upwards, however, the density of the blob has changed from $\rho(r)$ to $\rho(r) + \delta\rho$, where from the adiabatic relation we derive

$$\frac{d}{dr}[\ln P - \delta \ln \rho] = \frac{1}{P}\frac{dP}{dr} - \gamma\frac{1}{\rho}\frac{d\rho}{dr} = 0$$

giving

$$\delta\rho = \frac{1}{\gamma}\frac{\rho}{P}\frac{dP}{dr}\delta r \qquad (4.9)$$

In turn, the density of the new surroundings in which the blob finds its self has changed to

$$\rho(r + \delta r) = \rho(r) + \frac{d\rho}{dr}\delta r \qquad (4.10)$$

Since the blob will have (in general) a different density to that of its new surroundings, it will now do one of two things. If the density of the blob is smaller than that of its surroundings, it will keep rising upwards. If the blob has a greater density than its surroundings, it will sink downwards (back to its initial position). If the density of the blob is smaller than its surroundings, then it will continue to move upwards under a buoyancy force $(\rho_{surroundings} - \rho_{blob})\, g(r)$, where $g(r)$ is the gravitational acceleration at radius r. The material under these latter conditions is unstable and convective. The condition for convective instability can now be set as:

$$[\rho(r) + \delta\rho_{blob} < [\rho(r + \delta r)]_{surroundings} \tag{4.11}$$

Bringing together expressions (4.8) and (4.9) and inserting these in (4.11), we can rewrite the instability condition as

$$\frac{d\rho}{dr} > \frac{1}{\gamma}\frac{\rho}{P}\frac{dP}{dr} \tag{4.12}$$

in physical terms, the right-hand side of (4.12) is the density gradient if the density and the pressure of the surroundings varied in an adiabatic manner. The left-hand side of (4.12) is the actual density gradient in the surrounding gas.

(☺) **Exercise 4.4** Verify the derivation of (4.12).

(☺) **Exercise 4.5** Eq. (4.12) is usually cast in terms of the temperature gradient. This can be done via substitution from the ideal gas equation: $P = (\mathfrak{R}/\mu)\rho T$. Accordingly, show:

$$\frac{P}{T}\frac{dT}{dP} = \frac{\gamma - 1}{\gamma} \tag{4.13}$$

Equation (4.13) effectively says that a stellar gas will be unstable to convection if the actual temperature gradient is steeper than the adiabatic gradient.

In very general terms, convection is all about the motion and energy transport by hot blobs of gas moving upwards and cold blobs of gas falling downward. Again, in very general terms, the amount of energy carried by convection can be written as

$$F_{conv} = 2\rho\,V_{conv}\,l\,C_P\,\Delta T \tag{4.14}$$

where the factor of 2 accounts for rising and falling blobs of gas, V_{conv} is the characteristic velocity (typically some 5 to 50 m/s) of the convective blobs, C_P is the specific heat at constant pressure, ΔT is the temperature excess and l is known as the mixing length. The energy carried per second across radius r by convection will accordingly be $L_{conv} = 4\pi\,r^2\,F_{conv}$. The mixing length distance is the characteristic distance traveled by a convective blob before it gives up its energy to the surroundings (and then effectively dissolves into the surrounding gas). There is no actual theory to predict what the mixing length should be, and it is taken as a parameter to be found by comparing model calculations against the observations. It is customary to write the mixing length in terms of the pressure scale height $H_P = P(dr/dP)$, with $l = \alpha\,H_P$, with α, the mixing length parameter, characteristically being of order one.

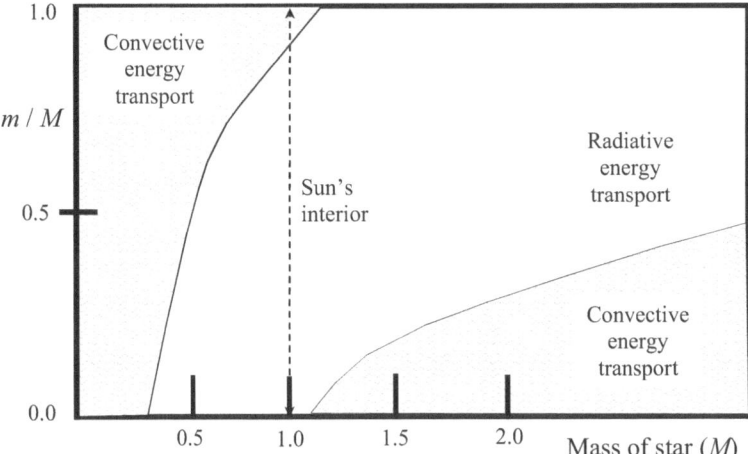

Fig. 4.2 Variation in the mass-fraction extent of convective zones within main sequence stars of varying mass. The shaded regions indicate those zones that are convective. (Credit: Author)

(☺) **Exercise 4.6** Proxima Centauri has a radius of 0.145 $R_\odot$. Given that it is also convective throughout its interior (see Fig. 4.2), estimate the convective turnover time $t_{conv} = R / V_{conv}$. How does this convective turnover compare to the star's rotation period of $P_{rot} = 82.6$ days? *Comments:* the convective turnover time is a measure of how well-mixed the interior of a star will be, and indeed, the implications are that Proxima is well mixed (chemically homogeneous) within its interior. The dimensionless ratio P_{rot} / T_{conv} is called the Rossby number (named after the Swedish meteorologist Carl-Gustav Rossby), and it is known to correlate with chromospheric activity, which in turn varies according to surface magnetic field activity. And, indeed, Proxima shows continuous flare activity and supports a reasonably well-ordered surface magnetic field.

The equation of energy generation (to be developed below) will dictate the total energy flux that needs to be carried across a given radius within a star, with $L = 4\pi r^2 F_{total}$. For a given temperature gradient, Eq. (4.5) derived earlier informs us how much of that specified total energy flux can be carried by radiation F_{rad}, and accordingly, the difference between the total and the radiative energy fluxes must be the energy flux carried by convection: $F_{conv} = F_{total} - F_{rad}$. In general, detailed numerical models reveal that the temperature excess entering into Eq. (4.14) is very small and only very slightly larger than that dictated for by the adiabatic gradient given in Eq. (4.13). In practice, therefore, the temperature gradient within a stellar model, when the convective instability condition (Eq. 4.12) is satisfied, is taken to be the adiabatic gradient. In this manner, the energy flow within a convective region is determined by Eqs. (4.13), (4.5) and (4.15). Specifically, Eq. (4.13) determines the temperature gradient, Eq. (4.5) then provides an expression for L_{rad}, the energy that is carried by radiation in that region, and finally, the energy carried by convection will be $L_{conv} = L - L_{rad}$.

In general, the convective instability condition (Eq. (4.13)) can be satisfied when either the ratio of the specific heats γ is close to unity (as in an extended region of partial ionization), or if the temperature gradient is very steep. The former conditions often holds true within the outer layers of low-mass stars, while the latter condition holds true in very hot regions where the energy generation rate is highly sensitive to the temperature. Detailed numerical models indicate that stars less massive than about 0.3 $M_\odot$ are fully convective throughout their interiors. Stars up to about the mass of the Sun have radiative cores, but convective envelopes. In contrast, stars move massive than about 1.3 $M_\odot$ have radiative envelopes and convective cores. Figure 4.2 shows the extent of the convective regions found through detailed stellar models over a range of masses.

4.2 Opacity

The opacity is a measure of the resistance to the flow of radiation. It is based upon detailed calculations concerning the probability of photon absorption by the stellar material through which it must pass while migrating towards a star's surface (the photosphere). The opacity is a complex function of the temperature, density and composition, $\kappa = \kappa(\rho, T, X_i)$, and is usually presented in a tabulated form rather than being directly calculated within a stellar model. There are many processes that contribute to the calculation of opacity tables, but the four main ones of interest are:

1. **bound-bound absorption**: in this situation, an electron moves from one bound orbit to another of higher energy through the absorption of a photon.
2. **bound-free absorption**: this is an ionization event where an electron in a bound orbit is moved through the absorption of energy into a free hyperbolic orbit.
3. **free-free absorption**: in this case, a free electron that chances to be close to a free ion can absorb energy by interacting with a photon.
4. **electron scattering**: in this case, a photon is scattered (its direction altered) by a close encounter with an electron. This process is also known as Thomson scattering (named after physicist J. J. Thomson, who discovered and identified the electron in 1897).

Figure 4.3 shows the variation of the opacity with temperature for various fixed values of the density. The opacity varies in a systematic manner according to temperature domain. At very high temperatures, the opacity is approximately constant and is predominantly due to scattering by electrons. In the electron scattering domain, the opacity takes on a simple analytic form, with

$$\kappa = \kappa_{scat} = 0.02(1 + X) - \left(units: m^2/kg\right) \tag{4.15}$$

where $X \equiv X_1$ is the mass-fraction of hydrogen. At lower temperatures, bound-free and free-free absorption are the dominant opacity-generating processes, and over a

Fig. 4.3 Opacity versus temperature for a selection of R values - see Exercise 4.8 below. The opacities are from the OPAL code. The red dots indicate the opacity and temperature for a standard solar model at radial fractions $r/R_\odot = 0.0 - 0.9$ in steps of 0.1. The electron scattering, constant opacity domain is indicated along with an arrow showing the characteristic power law variation in the Kramers' opacity domain. (Credit: Author)

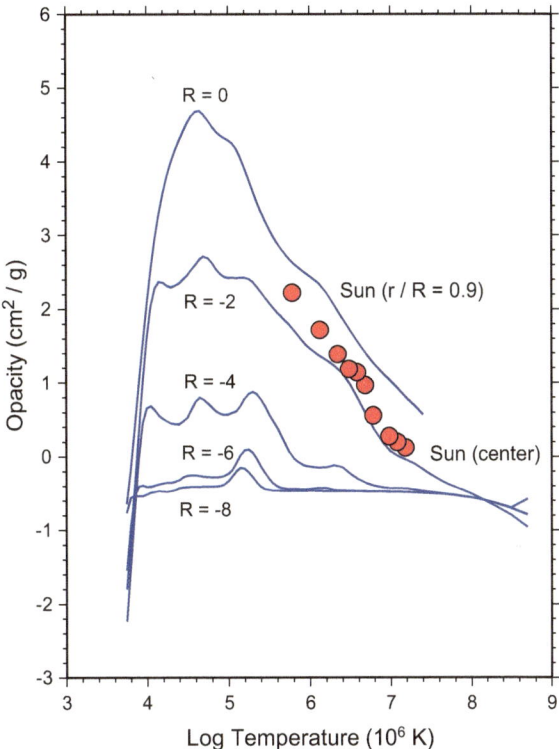

wide range of conditions the so-called Kramers' opacity law (named after Dutch physicist Henrik Kramers) holds true, with

$$\kappa = \kappa_{Kramers} = \kappa_K \rho T^{-3.5} \tag{4.16}$$

where κ_K is a constant related to the composition. At low temperatures, the opacity decreases with decreasing temperature and can be approximated by a power law of the form

$$\kappa = \kappa_1 \rho^{0.5} T^4 \tag{4.17}$$

where κ_1 is a constant related to the composition. In general, it will be convenient to express the opacity law as a power law of the form:

$$\kappa = \kappa_0 \rho^{\lambda-1} T^{\nu-3} \tag{4.18}$$

where κ_0 is a constant, and λ and ν are chosen according to the specific approximation formula being used for the opacity.

(☺) **Exercise 4.7** The density at the center of the Sun is 1.6×10^5 kg/m^3, while the opacity is 0.1 m^2/kg. What is the mean free path of a photon at the Sun's center?

(☺) **Exercise 4.8: Computer Project/Lab Exercise** Go to the OPAL opacity code page hosted by the Lawrence Livermore National Laboratory (https://opalopacity. llnl.gov/). Download the opacity table (#73) for a solar composition (X = 0.72, Y = 0.28, Z = 0.02). The table gives the total opacity (in units of cm^2/g) as a function of log T (rows) and log R (columns)—note that $R = \rho / (T_6)^3$, where ρ is the density (in units of g/cm^3) and T_6 is the temperature in units of 10^6 K. Construct a diagram showing the opacity as a function of temperature for a series of R values (say $R = -8, -6, -4, -2, 0$). Undertake a web search to find a numerical model that describes the run of temperature and density through the Sun and plot a few of these values in the opacity versus temperature diagram. *Hints:* To do this, you will need to interpolate between the various fixed R–value loci. The result should look something like Fig. 4.3. A 3-dimensional surface plot (Fig. 4.4) further illustrates the opacity *landscape* showing the precipitous drop-off in the opacity at low temperatures and the electron scattering *plane* at high temperatures. At intermediate temperatures, the descent from the peak of the opacity mountain is governed by the Kramers' opacity law relationship.

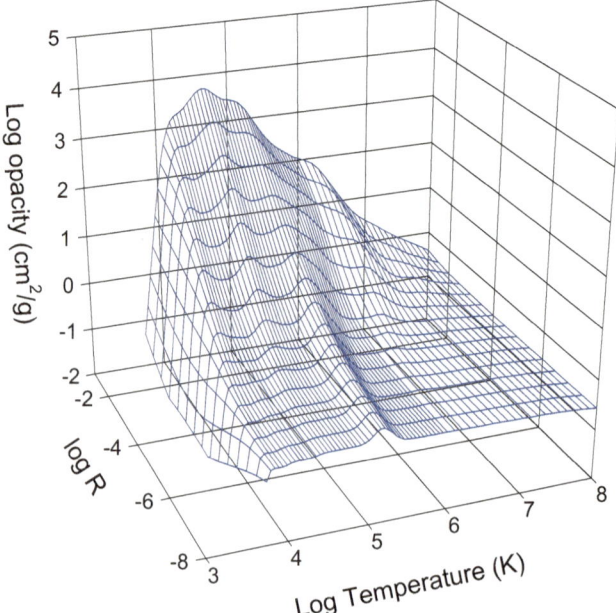

Fig. 4.4. 3-D surface plot of the opacity as a function of log R and log T. Data from table #73 of the OPAL opacity library. (Credit: Author)

4.3 Energy Generation

Stars generate energy within their interiors through nuclear fusion, which is a process that enables the conversion of one atomic element into another with an associated liberation of energy. It is the energy liberated during the conversion process that powers a star. Let the energy generated per kilogram per second at radius r within a star be $\varepsilon(r)$. Suppose energy flows across the sphere of radius r at a rate given by the luminosity $L(r)$. The difference between the energy entering the sphere and that crossing a sphere of radius $r + \delta r$ will correspond to the energy generated within the shell. Accordingly,

$$L(r + \delta r) - L(r) = 4\pi r^2 \delta r \rho \, \varepsilon$$

which reduces to the differential expression

$$\frac{dL}{dr} = 4\pi r^2 \rho \varepsilon \qquad (4.19)$$

In deriving (4.19), it has been assumed that none of the energy that is liberated within the shell of radius δr is used to heat up and/or change the volume of the shell—such a condition will apply when the star is in thermal equilibrium.

While the energy generation term in general is a relatively complex function of the temperature, density and composition, $\varepsilon = \varepsilon(T, \rho, X_i)$, it can be well approximated by a power law, with

$$\varepsilon = \varepsilon_0 \rho^\chi T^\eta \qquad (4.20)$$

where ε_0 is a constant, and where the powers χ and η will vary according to the temperature domain and the fusion reactions (see below) that are in play. For hydrogen fusion reactions in a star like the Sun, it is found that $\eta \approx 4$; for more massive stars, with higher central temperatures $\eta \approx 15$. For hydrogen fusion reactions, $\chi = 1$, while for the triple-alpha (helium fusion) reaction (see below), $\chi = 2$.

(☺) **Exercise 4.9** (1) Insert Eq. (4.20) into Eq. (4.19) and determine an expression for the luminosity when the interior is approximated as a polytrope of index n. *Hints:* Recall Sect. 3.3 on polytropes, and use the substitutions: $r = \alpha\xi$, $\rho = \rho_c\theta^n$, and $T = T_c\theta$. These steps give

$$L = 4\pi\varepsilon_0 T_c^\eta \rho_c^{\chi+1} \alpha^3 \int_0^{\xi_1} \xi^2 \theta^{(1+\chi)n+\eta} d\xi$$

In general, the integral will need to be evaluated numerically, but we can consider the special case when $n = 0$, corresponding to our zeroth-order, constant density

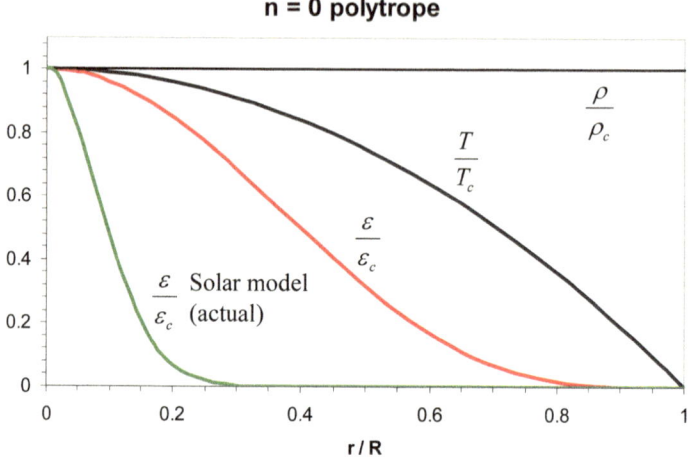

Fig. 4.5 Run of model parameters for an $n = 0$ polytropic Sun model. (Credit: Author)

model (Sect. 3.2.1). (2) When $n = 0$, we have seen that $\theta(\xi) = 1 - \xi^2/6$. Taking $\chi = 1$ and $\eta = 4$, evaluate the integral in part (1) between 0 and $\xi_1 = \sqrt{6}$. This is a rather tedious but straightforward integration, with

$$I = \int x^2 \left(1 - \frac{1}{6}x^2\right)^4 dx = \frac{x^{11}}{14256} - \frac{x^9}{486} + \frac{x^7}{42} - \frac{2x^5}{15} + \frac{x^3}{3}$$

which, when $x = \xi_1 = \sqrt{6}$, evaluates to $I = 0.54$. (3) Given $\alpha = 2.8 \times 10^8$, $\rho_c = 1400$ kg/m^3, $T_C = 10^7$ K and $\varepsilon_0 = 4.5 \times 10^{-34}$ W/kg, what is the resultant luminosity for the constant density Sun model, and how does this number compare to $L_\odot$? (4) Plot a graph showing the variation of $\varepsilon / \varepsilon_C$ (where ε_C is the energy generated per kilogram per second at the center) against the fractional radius $0 \leq r/R \leq 1$ for the $n = 0$ polytropic model. **Hint:** Check your results against Fig. 4.5. **Comment:** compared to the $n = 0$ polytropic model, the energy generated per kilogram per second is more centrally concentrated in a detailed solar model.

(☺) **Exercise 4.10** Given that the Sun has a central temperature of 15 million Kelvin, and that energy generation through hydrogen fusion reactions is only efficient provided the temperature is greater than ten million Kelvin, show that energy generation is only taking place within the inner ~1/3rd of the Sun's radius. **Hint:** Recall that the temperature gradient for the Sun is $\Delta T/\Delta R \approx (T_C - T_S)/R_\odot = 0.02$ K/m.

Up to this point, all that has been said is that a star can generate energy within its interior, and that the amount of energy generated per kilogram per second is given by some function $\varepsilon(r)$. Indeed, the energy generation rate per unit mass within the Sun is quite modest, with $L_\odot/M_\odot \approx 0.0002$ Watts per kilogram. This number can be contrasted against that for say a human being with $(L/M) = 12$ Watts / kg. Indeed, as we have seen in earlier discussions with respect to the Sun and with respect to stars in general, the problem is not so much the amount of energy that must be generated, but the amount of time over which the energy has to be generated—in the case of the Sun, at least 4.56 billion years. A human being, provided they are fortunate enough, gets to eat three square meals a day, and accordingly they can maintain a high luminosity-to-mass ratio. The stars must feed upon their own bodies (in some ghoulish fashion), and accordingly we must look to mechanism that not only enable the generation of energy within a star, but that also enable the long-term generation of energy.

(☺) **Exercise 4.11** Take a spherically symmetric human being of radius $r = 0.4$ m, having a mass of 75 kg and a temperature of 300 K, and show that $(L / M) \approx 12$ Watts / kg.

But what is the energy-generating mechanism, and how does it work? The idea that energy could be extracted from the conversion of four hydrogen atoms (protons in the ionized stellar interior) into a helium atom (the nucleus of which is composed of two protons and two neutrons) became clear in the early 1920s. Key to developing this idea was the discovery by Francis Aston, working at the Cavendish Laboratory in Cambridge, that the mass of the helium nucleus was not the same as the mass of four protons: $4m_P < 1m_{He}$. In atomic mass units (where 1 u $= 1.66 \times 10^{-27}$ kg), a hydrogen atom weighs in at 1.00794 u, while the helium atom weighs in at 4.002602 u. Accordingly, the mass of the helium atom differs from that of four hydrogen atoms by 0.029158 u, or by 4.842×10^{-29} kg. Indeed, the observational measurements appeared to indicate that H + H + H + H does not actually equal 4H, but rather 3.97H.

The question that immediately arises is where in the process of making a helium atom out of four hydrogen atoms does the missing mass go? Arthur Eddington realized as early as February 1920 that where the mass *goes* is into energy—indeed, the energy is given by Einstein's famous formula $E = 0.007(4m_p) c^2$. In this manner, some 4.4×10^{-12} Joules of energy is liberated per transformation of 4H $\Rightarrow$ He. While this is not very much energy per transformation, the Sun has a vast store of 10^{57} hydrogen atoms (protons) that it can potentially work with. In order to explain the Sun's energy output—its current luminosity—the amount of matter that must be converted into energy per second is of the order of $L_\odot = dE/dt = (dM/dt)c^2$ or $dM/dt = L_\odot/c^2 = 3.85 \times 10^{26}/9 \times 10^{16} \approx 4$ billion kilograms per second.

(☺) **Exercise 4.12** (1) How many conversions of 4H $\Rightarrow$ He take place per second within the Sun? (2) What fraction of the total number of protons within the Sun is involved in generating its energy each second? *Hints:* The Sun's luminosity is 3.85×10^{26} Joules per second, and each 4H $\Rightarrow$ He conversion generates 4.4×10^{-12} Joules worth of energy.

(☺) **Exercise 4.13** (1) Given that the Sun is 4.56 billion years old, and that its luminosity has remained constant over this time, how much matter has it converted into energy? (2) Compare the mass from part (1) to that of the Earth and to that of the Sun.

(☺) **Exercise 4.14** In his 1920 address to the British Association (see the journal *Nature*, 106, September 1920, p. 14), Eddington argued, 'If 5 per cent of a star's mass consists initially of hydrogen atoms, which are gradually combined to form more complex elements, the total amount of heat [energy] liberated will more than suffice for our demands, and we need look no further for the source of a star's energy'. (1) How much energy does this amount to in the case of the Sun, and (2) how long could the Sun maintain its current luminosity on this '5% of hydrogen' energy supply?

(☺) **Exercise 4.15** If the Sun were simply converting matter into energy by some form of annihilation process, how long could it survive for if radiating at its present luminosity?

Given that stars can be powered by fusion reactions, which enable the transformation of 4H $\Rightarrow$ He, what is the characteristic nuclear timescale T_{nuc}? Assuming that all of the hydrogen within a star is available for transformation, then the nuclear lifetime of a star of mass M and luminosity L will be of order

$$T_{nuc} = \frac{(M/4m_p)(4.4 \times 10^{-12})}{L} = (10^{11}) \left(\frac{M}{L}\right)_{Solar} \text{years} \qquad (4.21)$$

where, recall, 4.4×10^{-12} Joules is the amount of energy liberated per 4H $\Rightarrow$ He transformation. For the Sun, the nuclear timescale is smaller than the annihilation timescale (recall exercise 4.14) by a factor of about 100. Additionally, as will be described later on in Chap. 5, the Sun can only tap about 10% of its actually available hydrogen supply during its main sequence phase (recall Fig. 1.9). Taking a general form of the mass-luminosity relationship $(L/L_\odot) = (M/M_\odot)^\alpha$ with $\alpha > 1$ (see Eq. 1.24), and allowing for a 10% access of the hydrogen available, Eq. (4.21) reveals that the characteristic (now main sequence) lifetime of a star of mass $M/M_\odot$ will be

$$T_{ms}(years) = \frac{10^{10}}{M_{solar}^{\alpha-1}} \qquad (4.22)$$

Perhaps in a counterintuitive sense, Eq. (4.22) indicates that the more massive a star is, the shorter its main sequence lifetime—this is a key point that we shall come back to in Chap. 5. In the case of the Sun, Eq. (4.22) indicates that it is about middle-aged, and that it will continue to be a main sequence star for another five billion years. Eq. (4.22) also underlies the earlier discussion in Chap. 1 (and as enumerated in Table 1.3) concerning the age determination of star clusters. For a given value of $\alpha > 1$, the lower the mass of the stars at the turnoff point, the older the cluster.

(☺) **Exercise 4.16** What is the key physical reason for the decreasing main sequence lifetime with increasing mass? *Hint:* Think about the luminosity.

While much progress in understanding stars can be made by simply asserting that $4H \Rightarrow He$, we have thus far not actually explained how the transformation proceeds. This was a detailed physical problem that was only solved some 20 years after Eddington had first discussed the idea of powering stars through the conversion of hydrogen into helium. One of the initial problems that had to be dealt with concerned the physical ability of stars to initiate fusion reactions. Were they hot enough, even at 15 million degrees within their central regions, to allow two protons to come together in order to start the transformation process?

In order for two protons to interact, they must have enough energy to overcome the Coulomb barrier that exists between them. The Coulomb energy between two protons a distance r apart is $U = \zeta e^2/r$, where ζ is the Coulomb constant and e is the fundamental proton charge. The average temperature T of the gas required to overcome such a Coulomb barrier is $3kT/2 = U$, where k is the Boltzmann constant. Setting the separation distance between the protons to be that of the classical nucleus radius $r = 10^{-15}$ m, then the barrier penetrating condition requires $T > 9$ billion Kelvin—some 600 times hotter than the Sun's core temperature. Not only this, but even if two protons could be brought together to form a nucleus, it turns out the ^{2}He is unstable. The problem with building up a helium nucleus step by step thus appeared to be that it was impossible to get past—or even to—the first p + p step. For all this, Eddington wisely wrote in his *The Internal Constitution of the Stars* (published in 1926), 'The helium which we handle must have been put together at some time and some place. We do not argue with the critic who urges that the stars are not hot enough for this process; we tell him to go and find a hotter place'. And indeed, Eddington was right: the helium must have been put together somewhere and somehow, and it was done so at temperatures easily accessible to the stars.

The solution to the hydrogen fusion problem lay in the fundamental developments that emerged from quantum mechanics and elementary particle physics during the 1930s. In order to get the first p + p reaction to work, scientists first had to understand the concept of quantum tunneling and the process of β^+-decay. The quantum tunneling effect importantly allows a particle to instantaneously jump through an otherwise insurmountable energy barrier (this allows for close encounters between protons at stellar temperatures), and the β^+-decay enables the transformation of a proton into a neutron with the associated emission of a positron and an electron neutrino: $p \Rightarrow n + e^+ + \nu_e$. Provided one of the two protons being brought together can tunnel close enough to its companion and then undergo a β^+-decay, a deuterium nucleus $^2D_1$ will be formed (the deuterium nucleus contains one proton and one neutron). Accordingly, hydrogen fusion begins with the step: $p + p \Rightarrow D + e^+ + \nu_e$. Even at this stage, a star gets some energy, since the positron will very quickly interact with an electron to produce two gamma-ray photons: $e^- + e^+ \Rightarrow 2\gamma$, with the gamma rays carrying off 1.02 MeV (1.63×10^{-13} Joules) of energy. Energy is also lost from the star since the neutrino will exit the star at close to the speed of light, such particles only rarely interacting with matter. Once deuterium has formed, it can interact with another proton to build up ^{3}He. That is, $D + p \Rightarrow {}^3He + \gamma$, with the gamma-ray photon carrying off some 5.49 MeV in energy, this energy ultimately

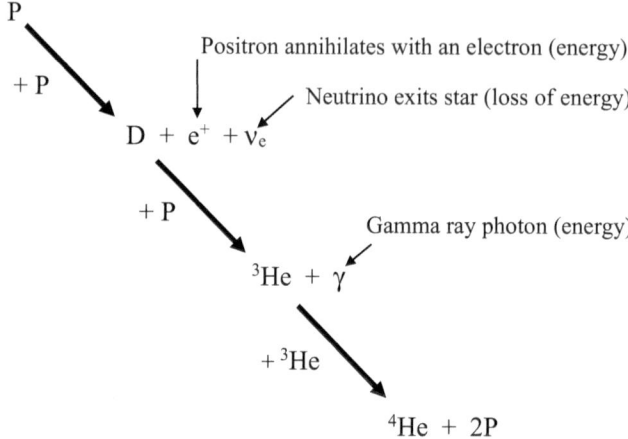

P

+ P

Positron annihilates with an electron (energy)

D + e^+ + v_e

Neutrino exits star (loss of energy)

+ P

^{3}He + γ

Gamma ray photon (energy)

+ ^{3}He

^{4}He + 2P

Fig. 4.6 A schematic flow chart for the PP I branch interactions of the proton-proton chain. (Credit: Author)

being incorporated into the star's internal energy reserves through interactions with its constituent protons and electrons.

The final step in building up ^{4}He is for two ^{3}He nuclei to be brought together, with ^{3}He + ^{3}He $\Rightarrow$ ^{4}He + 2p. The overall reaction is therefore 4p $\Rightarrow$ ^{4}He + $2e^+$ + $2v_e$, with a net energy release of 27.732 MeV (4.44×10^{-12} Joules) per complete transformation (2% of this energy is lost to the star because of the formation—and subsequent rapid exit—of the two electron neutrinos). What has just been described is called the proton-proton chain reaction— more technically, it is the PP I branch of the proton-proton reaction that has been described (Fig. 4.6). There are two other P-P branches that can be followed, but the end result in terms of energy released is pretty much the same. The main effect of these other P-P branches is to produce more energetic electron neutrinos. While such neutrinos are highly elusive and carry energy away from a star, they can nonetheless be detected at Earth. Solar neutrinos have been detected since the mid-1960s, and neutrinos were detected from the supernova outburst SN 1987A on February 23rd, 1987.

(☺) **Exercise 4.17 Review essay question.** Investigate the history of the solar neutrino problem and review the key role that neutrino detectors are playing in the development of modern astrophysical concepts.

In addition to the P-P chain, another way to forge a helium nucleus from four protons is via the CNO-cycle. This latter method is a catalytic cycle that operates upon a ^{12}C nucleus. The end result, however, is the same as in the P-P chain in that four protons are ultimately converted into a helium nucleus, and 27.732 MeV of energy is made available to the star for internal heating each time the transformation process is completed. Although the energy released per transformation 4p $\Rightarrow$ He is the same in both the P-P chain and the CNO cycle, the actual amount of energy generated per kilogram per second is strongly dependent upon a star's composition, central temperature and density. Detailed modeling indicates that the P-P chain is the dominant energy-generating mechanism in low-mass stars up to about 1.3 times the mass of the Sun. Above 1.3 $M_\odot$, the CNO cycle begins to dominate. In addition, for the CNO cycle, the exponent to the power law expression for the energy generation—Eq. (4.20)—is more appropriately $\eta = 17$ rather than the four appropriate to the P-P chain. The energy generation rate is therefore highly temperature sensitive and as a consequence of this, once energy generation by the CNO-cycle becomes dominant, the core becomes convectively unstable (recall Fig. 4.2).

With the exhaustion of hydrogen at the end of the main sequence phase, a star is left hunting for its next nuclear fuel: helium. In this case, three helium nuclei (alpha particles) are brought together in order to produce Carbon: Schematically, $3\,^4\text{He} \Rightarrow \,^{12}\text{C} + energy$, and once again it is the energy released through the fusion reactions that keeps the interior of a star hot and stable. The triple-alpha reaction yields a net energy release of 7.275 MeV (1.17×10^{-12} Joules) per reaction and characteristically runs at temperatures of order 100 million Kelvin—10 times higher than those associated with hydrogen fusion reactions. In terms of the energy generation power law (Eq. (4.20)), the appropriate parameters for the triple-alpha reaction are $\chi = 2$, $\eta = 40$, the later value indicating an extreme temperature sensitivity for this fusion reaction. The density power of $\chi = 2$ comes about since the triple-alpha reaction is essentially a three-particle reaction, wherein all three of the ^4He nuclei must come together at the same time. This is in contrast to the step-by-step, particle-plus-particle reactions that take place during hydrogen fusion (for which $\chi = 1$). To achieve the very high-density, very high-temperature conditions required of the triple-alpha reaction, a star has to radically alter its internal structure; the core shrinks dramatically to enable the helium fusion reactions, and in contrast the outer envelope expands, with the star moving into the red giant region of the HR diagram (recall Fig. 1.9 and see Sect. 5.7 later).

Once all the helium is used up within the central core, further fusion reactions involving carbon can take place, but now it transpires that the key factor in determining the onset of such advanced reactions is the mass of the star. Only the most massive stars, with initial masses greater than 8 $M_\odot$, can initiate fusion reactions beyond that of helium, although we shall leave this topic for a later discussion. Stars with an initial mass less than 8 $M_\odot$ enter into a final end state after the exhaustion of helium within their central regions—a planetary nebula phase with their cores eventually evolving a white dwarf structure.

4.4 What Is a Star (Again)?

In *The Voyage of the Dawn Treader*, written by C. S. Lewis (first published in 1952), Eustace is amazed to meet an actual star, and comments (somewhat rudely), 'In our world a star is a huge ball of flaming gas'. The star, called Ramandu, replies to Eustace with the understanding that, 'Even in our world, my son, that is not what a star is but only what it is made of'. In similar refrain to Ramandu, we are now in a position to say what a star is made of and how it is structured, but we are still left wanting for a good definition for what a star actually is (recall Chap. 2 and definition 2.1). For all this want of definitions, however, we do have a complete set of equations that describe how a star actually works. Indeed, all of the equations required to fully describe stellar structure have been presented, and for convenience they are brought together here:

Hydrostatic equilibrium: $\dfrac{dP}{dr} = -\, G\dfrac{M(r)}{r^2}\rho$ (3.5)

Mass conservation: $\dfrac{dM}{dr} = 4\pi r^2 \rho$ (3.6)

Energy generation: $\dfrac{dL}{dr} = 4\pi r^2 \rho \varepsilon$ (4.19)

Temperature gradient $\dfrac{dT}{dr} = -\dfrac{3\kappa \rho L}{16\pi a c r^2 T^3}$ radiative zone   (4.5)

$\dfrac{P dT}{T dP} = \dfrac{\gamma - 1}{\gamma}$ convective zone   (4.13)

To these we add the equation for the composition-dependent mean molecular weight and the power law expressions for the energy generation and the opacity:

$$\frac{1}{\mu} = 2X + \frac{3}{4}Y + \frac{Z}{2}$$ (3.2)

$$\varepsilon = \varepsilon_0 \rho^\chi T^\eta$$ (4.20)

$$\kappa = \kappa_0 \rho^{\lambda-1} T^{\upsilon-3}$$ (4.18)

The boundary conditions associated with this set of equations are such that, at the center $M = 0$ and $L = 0$ at $r = 0$, and $T = 0$, and $P = 0$ at $r = R_{\text{star}}$. The power law constants are such that:

	χ	η	λ	υ
PP I chain	1	4		
CNO-cycle	1	14		
Triple-alpha reaction	2	40		
Kramers opacity			2	- 0.5
Low temperature opacity			1.5	7
Electron scattering opacity			1	3

And finally, the equation of state is such that $P = P_{gas} + P_{rad} = F(\rho, T, X_i)$, where ideal gas:

$$P_{gas}(r) = n(r) \, k \; T(r) = \left(\frac{\Re}{\mu} \right) \rho(r) \, T(r)$$

radiation pressure:

$$P_{rad}(r) = \frac{1}{3} a T^4(r)$$

It is far from obvious that there is any solution to the collected equations of stellar structure, but the so-called Vogt-Russell theorem (really it is a statement about boundary conditions and has never actually been proved in any rigorous manner) dictates that:

VR Theorem   If the pressure P, the opacity κ and the rate of energy generation ε are functions of the local density ρ, temperature T and composition X_i, $i = 1, 2, 3 \ldots$ only, then the structure of a star is uniquely determined by the total mass M and the chemical composition X_i, $i = 1, 2, 3 \ldots$.

The *proof* of the VR theorem essentially resides in the fact that unique solutions are actually found when detailed numerical integrations of the equations of stellar structure are performed.[2]

(☺) **Exercise 4.18**   In mathematical terms, the VR theorem is a very bold and brash statement, and while there are mathematical techniques that can prove the existence of a solution to a given set of equations, no such proof has been developed for the VR theorem. First introduced as a *statement* in a paper by Heinrich Vogt in early 1926, and independently in a book coauthored by Henry Norris Russell in 1927, it was primarily Russell who pushed the idea that Vogt's *statement* was actual a *theorem*. (1) Complete a web search and write a one-side-of-paper essay on the mathematical difference between a *statement*, a *conjecture* and a *theorem*. (2) Take a look at the collected equations of stellar structure as given above. Count the number of differential equations, the number of independent variables, the number of dependent variables and the number of boundary conditions.

Putting the solution-existence and solution-uniqueness problems aside for the moment—they are, after all, a problem for the mathematician rather than an issue for the pragmatic astrophysicist (who can readily find numerical solutions to the equations of stellar structure)—Fig. 4.7 is a schematic representation of how the various equations link together to provide a specific stellar model. The figure also hints at the

---

[2] As always there are caveats. It transpires that multiple solutions to the equations of stellar structure can be *engineered* when special conditions are made to apply. Technically, this constitutes a numerical proof that the VR theorem is, in fact, not always true!

Fig. 4.7 A schematic
representation of how a star
works—see text for details.
(Credit: Author)

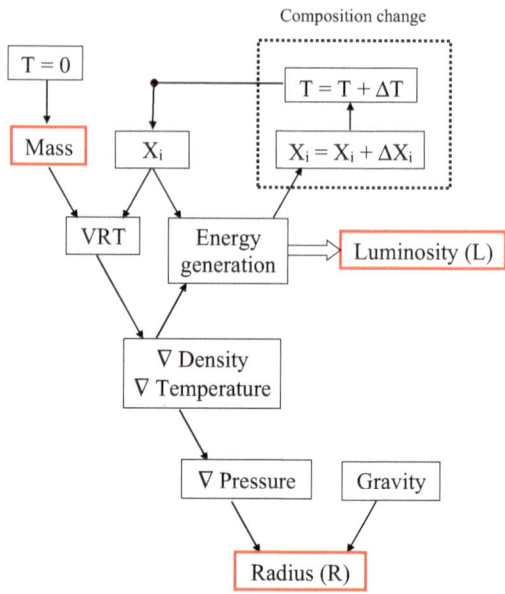

reasons why a star changes its physical characteristics (luminosity, temperature and radius) as it ages.

To interpret Fig. 4.7, we start at time T = 0. At this time, the accretion and formation stage has stopped and a *bona fide* star of initial mass M and composition $X_i, i = 1, 2, 3, \ldots$ has emerged. The VR theorem then dictates that for the given mass and composition, a star in hydrostatic equilibrium will come about with a specific luminosity L (predicated upon a specific energy generation mode (PP I or CNO-cycle) and an overall radius R. The density, temperature and pressure gradients (indicated by the symbol ∇) are additionally inter-adjusted so as to maintain a state of dynamic and thermal equilibrium; the former keeping the star stable and acting against gravitational collapse, and the latter ensuring an exact balance between the energy generated via fusion reactions in a star's core and the energy radiated into space at its surface.

A star is thus a giant negative feedback mechanism. If the radius decreases, the temperature goes up, and this changes the pressure support so as to return the star to its initial radius. Likewise, if the star expands, the temperature goes down and the pressure gradient changes so as to restore the star to its original size. Additionally, if a star begins to generate more energy within its core than it radiates into space at its surface, it heats up and the star expands, resulting in a temperature decrease that in turn lowers the energy generation rate, and the star returns to its equilibrium luminosity. The same negative feedback mechanism applies if the star suddenly begins to radiate more energy into space at its surface than it generates within its core. For a star, both dynamic and thermal stability can be maintained for as long as there is a central fuel that can sustain the fusion reactions at work in its interior. Once dynamic and thermal equilibrium are established, a star can remain in such a state for at least a nuclear timescale, the nuclear timescale being determined by the star's initial mass (Eq. (4.22)). The reason why stars must undergo changes in their observable (intrinsic) characteristics of luminosity, radius and temperature as they age is hinted at by the dashed box in Fig. 4.7. Specifically, as fusion reactions proceed within the core of a star, a change in its internal composition comes about (it is becoming increasingly helium $= X_2$ rich and hydrogen $= X_1$ poor), and this results in a new set of age-appropriate equilibrium conditions coming about. Likewise, a star may also lose mass as it ages, and this will further change its age-appropriate equilibrium conditions.

(☺) **Exercise 4.19: Essay Question** Imagine that some advanced alien civilization has placed a star within a massive box, and that the inner surface of the box acts as a perfect reflector—no energy escapes from the box and the box itself does not heat up over time (this is not a question about a Dyson Sphere). Ignoring all the engineering problems involved with the process, the question of concern is: What happens to the star? *Discussion points:* Let us narrow down the options for an answer. It seems that two essential outcomes are possible. (1) The star will continue to radiate more and more energy into the box and eventually the box will just explode—this is what would happen, for example, to a lightbulb wrapped up in tinfoil (don't try this experiment at home). Or (2), the star, being a feedback system, will *sense* that no energy is being lost into space at its surface, and accordingly evolve into a constant temperature state, with a temperature below that at which thermonuclear reactions will operate.

4.5 Back to Reality

Finding a solution to the equations of stellar structure for a given stellar mass and initial composition is all well and good, but the ultimate test of such solutions is their ability to describe what is actually observed. As discussed in Chap. 1, the key observations that require explanation are the HR diagram (where the temperature, radius and luminosity of star are displayed), the mass-luminosity relationship and the mass-radius relationship (recall Figs. 1.9, and 1.11). The ability of stellar models to fully describe the main sequence in the HR diagram will be addressed in the next chapter, and for the moment we concentrate on other features—specifically, a basic explanation of the mass-luminosity relationship will be developed. To achieve this goal, we return to the use of approximation methods, and with the appropriate warnings of Eddington, Hoyle and Lyttleton ringing in our ears (recall Sect. 3.2.1), the various structure equations will be brought together in order to yield a set of scaling-law relationships. We begin with Eq. (4.5), and boldly approximating the temperature gradient as T/R, the luminosity is expressed as

$$L = K_1 \frac{T^4 R}{\kappa \rho} \tag{4.23}$$

where K_1 is a constant. Substituting from Eq. (4.18) and letting $\rho \sim M / R^3$, Eq. (4.23) is rewritten as

$$L = K_2 \, T^{7-\nu} \, R^{1+3\lambda} \, M^{-\lambda} \qquad (4.24)$$

where K_2 is a constant. From Chap. 3, Eq. (3.18) indicates that $T \sim K_3 \, M / R$, where K_3 is a constant, and inserting this relationship into Eq. (4.24) we find that

$$L = K_4 \, M^{7-\nu-\lambda} \, R^{3\lambda+\nu-6} \qquad (4.25)$$

where K_4 is a constant. Now, for a Kramers' law opacity we have $\lambda = 2$ and $\nu = -0.5$, giving upon substitution in (4.25) the result that

$$L = K_5 \; M^{5.5} \, R^{-0.5} \qquad (4.26)$$

where K_5 is a constant. In the case of a dominant electron scattering opacity we have $\lambda = 1$ and $\nu = 3$, and upon substitution in (4.25) this gives

$$L = K_6 M^3 \qquad (4.27)$$

where K_6 is a constant. Eqs. (4.26) and (4.27) provide a good approximation to the mass-luminosity relationship described by Eq. (1.24) in the mass range $0.5 < M/M_\odot < 20$. Taking the mass-radius relationship to be $R \sim M$, we find for stars similar in mass to the Sun, for which the Kramers' opacity is dominant, that $L \sim M^5$, while for more massive stars in which the electron scattering opacity is dominant, the luminosity varies as $L \sim M^3$. Another change in the power of the mass-luminosity relationship for the most massive stars will be introduced in Sect. 4.6.3. In this latter case, it is found that $L \sim M$.

Although we have been mathematically fast and loose in our scaling law analysis, the end comparisons are in fact reasonably good, and this gives us some confidence in both the scaling argument method and the ability of the equations of stellar structure to actually explain what the stars are doing. In terms of a basic explanation of the main sequence, we first turn to the Stefan-Boltzmann law (Eq. (1.14)) that relates the surface temperature T_S to the luminosity and radius, with $T_S = K_7 (L / R^2)^{1/4}$, where K_7 is a constant. Substituting for L from Eq. (4.25) now indicates

$$T_S = K_8 \, M^{(7-\nu-\lambda)/4} \, R^{(3\lambda+\nu-8)/4} \qquad (4.28)$$

where (as per usual) K_8 is a constant. If we substitute for λ and ν values appropriate to Kramer's opacity law and to the electron scattering opacity, use the approximation that $R \sim M$ and adopt the mass-luminosity scaling results derived earlier, we find

$$L = K_9 \, T_S^{20/3} \text{ in the case of a Kramers' opacity} \qquad (4.29a)$$

$$L = K_{10} \, T_S^{12} \text{ in the case of electron scattering opacity} \qquad (4.29b)$$

where in (4.29a) and (4.29b), K_9 and K_{10} are constants. Effectively, Eqs. (4.29a and 4.29b) indicate that the slope of the main sequence essentially doubles, becoming much steeper when the electron scattering opacity becomes dominant. This is indeed what the observations reveal (recall Sect. 1.3, and Fig. 1.9).

(☺) **Exercise 4.20: Term Paper Topic** The scaling law arguments that have been used to explain the basic properties of the mass-luminosity relationship and the slope of the main sequence in the HR diagram can be developed much more rigorously (and with less apparent *fudging*) by the adoption of homology relationships. This is a topic that requires the application of some considerable algebraic manipulation, but is entirely practical given a good hour of work and a plentiful supply of scrap paper. For this term paper, begin with a look at Chapter 1.6 in Hansen and Kawaler's text, *Stellar Interiors—physical principles, structure and evolution* (Springer, 1994). Develop the full set of homology relationships and then check that the scaling argument approximations, Eqs. (4.26), (4.27), (4.29a) and (4.29b), are indeed correct in their power law form.

4.6 More Introductory Stellar Models

In this section, a number of reduced stellar models will be examined. By reduced, I mean that various physical simplifications, assumption and sometimes mathematical dodges will be employed. The models to be developed are classics, and their inventors were generally well known for their pioneering contributions to astrophysics and the study of stellar structure in particular. The first model to be looked at is the Standard Model developed by Arthur Eddington in the early to mid-1920s. The second model to be considered was developed by Thomas Cowling in the 1930s and is sometimes called the Point Source model. The final suite of models is concerned with the most massive stars, and our development will follow from the works first published in the late 1970s by Herbert Falk and Romas Mitalas (University of Western Ontario, Canada).

4.6.1 Eddington's Standard Model

With Eddington's standard model, our attention is drawn back to the polytropic approximation for the equation of state, discussed in Sect. 3.3, and specifically the $n = 3$ polytropic model that gives an equation of state having $P = K \rho^{4/3}$, where K is a constant. To see how this situation comes about, we begin with Eq. (4.6), which describes the radiation pressure gradient. Dividing this equation by that for hydrostatic equilibrium (Eq. (3.5)) yields at radius r

$$\frac{dP_{rad}}{dP} = \frac{\kappa}{4\pi cGM_r}\frac{L_r}{} \qquad (4.30)$$

From Eq. (4.19) divided by the mass continuity equation, we define the following quantities $<\varepsilon(r)>$ and $\eta(r)$ such that

$$< \varepsilon(r) > \equiv \frac{L(r)}{M(r)} = \frac{\int_0^r \varepsilon \, dM_r}{\int_0^r dM_r} \qquad (4.31)$$

and

$$\eta(r) \equiv \frac{< \varepsilon(r) >}{< \varepsilon(R) >} = \frac{L_r/M_r}{L/M} \qquad (4.32)$$

Note that at the surface, $\eta(R) = 1$. With these two quantities defined, Eq. (4.30) can be rewritten as

$$\frac{dP_{rad}}{dP} = \frac{1}{4\pi cGM}\frac{L}{} k(r)\eta(r) \qquad (4.33)$$

Having developed Eq. (4.33), and in order to make further analytical progress, Eddington next introduced a controversial idea. Specifically, he argued that $\kappa(r)\eta(r)$ is a constant throughout a star's interior. It is not obvious that this assumption has to be true, but later numerical modeling found that it is approximately correct. In this manner, $\kappa(r)\eta(r) = \text{constant} = \kappa_s$, where κ_s is the surface opacity. With this assumption in place, Eq. (4.33) can be directly integrated to yield

$$P_{rad} = \frac{\kappa_s}{4\pi cGM}\frac{L}{}P \qquad (4.44)$$

where the surface boundary condition $P = 0$ has been applied. Eddington's approximation now becomes clear, and it implies a constant ratio between the radiation pressure and the total pressure through the interior of a star. With (4.44) in place, and recalling the earlier definition that $P_{rad} / P = 1 - \beta$, we obtain a mass-luminosity relationship of the form

$$L = \frac{4\pi cG}{\kappa_s}M(1 - \beta) = L_{Edd}(1 - \beta) \qquad (4.45)$$

with the meaning of Eq. (4.45) being that as the radiation pressure becomes more and more dominant (that is $\beta \Rightarrow 0$), the luminosity approaches the so-called Eddington Luminosity L_{Edd}, where

$$L_{Edd} = \frac{4\pi c G}{\kappa_s} M \qquad (4.46)$$

In stars where the gas pressure is given by the ideal gas law, we additionally have

$$P = \frac{P_{rad}}{(1-\beta)} = \frac{P_{gas}}{\beta} = \frac{\mathfrak{R}}{\mu\beta}\rho T \qquad (4.47)$$

and writing $P_{rad} = aT^4/3$, this yields the important result that

$$T = \left[\frac{3}{a}\left(\frac{\mathfrak{R}}{\mu}\right)\frac{(1-\beta)}{\beta}\right]^{1/3}\rho^{1/3} \qquad (4.48)$$

and accordingly, the equation of state can be written

$$P = K\rho^{4/3}, \text{ where } K = \left[\frac{3}{a}\left(\frac{\mathfrak{R}}{\mu}\right)^4\frac{(1-\beta)}{\beta^4}\right]^{1/3} \qquad (4.49)$$

Equation (4.49) is the equation of state for a polytrope of index $n = 3$ (recall Eq. (3.40)).

(☺) **Exercise 4.21** Complete the algebra steps taking (4.48) to (4.49).

Furthermore, from Eq. (3.46), we recall that the constant K is related to the mass M alone in an $n = 3$ polytrope, and this enables a derivation of the Eddington quartic equation (recall Eq. (3.27)) to be made such that:

$$\frac{1-\beta}{\beta^4} = \left(\frac{\pi a G^3}{48\mathfrak{R}^4}\right)\frac{1}{M_3^2}\mu^4 M^2 = 0.003\,\mu^4(M/M_{sun})^2 \qquad (4.50)$$

where $M_3 = (-\xi^2\, d\theta/d\xi)$ evaluated at $\xi = \xi_1$ for an $n = 3$ polytrope (as tabulated in Table 3.1; also recall exercise 3.22).

(☺) **Exercise 4.22** (1) Complete the algebra leading to Eq. (4.50). (2) Verify the constant term is equal to 0.003.

With Eq. (4.50) in place, the Standard Model is now fully defined. For a given mass M, radius R and chemical composition (which, recall, will determine the value of the mean molecular weight μ), Eq. (4.50) can be solved to determine β. The polytropic constant K can then be determined via Eq. (4.49). With K determined, the run of the density ρ against radius $r = \alpha\xi$ can be evaluated from the $n = 3$ polytropic model (recall the configuration radius is given by $R = \alpha\,\xi_1$). With the density known, the run of the temperature and pressure can be evaluated via Eqs. (4.48) and (4.49). The luminosity at each point within the model interior can be evaluated in a similar manner to that outlined in exercise 4.8.

For a model of the present Sun, we set $M = 1\,M_\odot$, $R = 1\,R_\odot$ and $\mu = 0.6$. From Eq. (4.50) we derive a value of $\beta = 0.99892$, indicating that radiation pressure is not important. From the given mass and radius, we deduce that the Sun's average density is 1408 kg/m^3 (recall exercise 3.4) and the relationship between the central and bulk density for an $n = 3$ polytrope—Eq. (3.47) and Table 3.1—gives the central density as $\rho_c = 7.63 \times 10^4$ kg/m^3. Eq. (4.49) and the value of β indicate that $K = 5.366 \times 10^9$, giving a central pressure of $P_c = 1.736 \times 10^{16}$ Pascals. Eq. (4.48) further indicates that the central temperature is $T_c = 1.651 \times 10^7$ Kelvin. Combining the energy generation Eq. (4.20) with (4.19) allows for the luminosity to be determined through the integral:

$$L = 4\pi\varepsilon_0 T_c^4 \rho_c^2 \alpha^3 \int_0^{\xi_1} \xi^2 \theta^{10} d\xi \qquad (4.51)$$

where in the integral we have taken $n = 3$, $\chi = 1$ and $\eta = 4$. Adopting a typical value of $\varepsilon_0 = 4.5 \times 10^{-34}$ W/kg for the PP I fusion reactions, and evaluating $\alpha = 10^8$ from the given solar radius, the constant term in Eq. (4.51) is 2.4×10^{30} Watts. Using the computer program developed in exercise 3.22 to evaluate the integral I in Eq. (4.51), we find $I \approx 0.02$, and accordingly the luminosity is of order 125 $L_\odot$. In this latter estimate we have seemingly drifted well away from any reasonable value. The problem is that we have assumed that the power of the temperature term in Eq. (4.20), as well as the ε_0 term, are constants, and this is very much not the case. As the temperature decreases, it transpires that the value of ε_0 decreases and that the power of the temperature term becomes larger. To improve upon the luminosity estimate, we will have to reevaluate the functional form of Eq. (4.20)—this can in fact be done, and much more reasonable results for the luminosity term obtained— but we shall not follow such refinements here (see, however, Sect. 4.6.3 below).

Key to developing a Standard Model solution is the evaluation of Eq. (4.50). In principle, since it is a quartic equation, there is an analytic solution to Eq. (4.50), but it is typically easier (at least in the modern era) to write a computer program to solve for β given a specific value of the constant $0.003\mu^4(M/M_\odot)^2$. Fig. 4.8 shows just such a solution curve.

(☺) **Exercise 4.23: Lab/Computer Project** Write a computer program to solve Eq. (4.50) for β given a specific value of μ and M. Produce a graph of β versus $\mu^2(M/M_\odot)$ over the range $10^{-3} \le \mu^2(M/M_\odot) \le 100$. Base your program on the Newton–Raphson iteration method (there are plenty of webpages that will guide you through the development of this latter procedure). As a quicker alternative you could also use one of the web-based quartic equation solvers. Your results should look the same as those shown in Fig. 4.8.

Figure 4.8 illustrates a number of important points. First, for the Sun, the diagram indicates that radiation pressure is not important within its interior. Second, as the mass of a star increases, radiation pressure becomes increasingly important within its interior. Also, as indicated by Eq. (4.45), as the radiation pressure becomes

Fig. 4.8 Solution curve to the Eddington quartic— Eq. (4.50). Also highlighted are β values for the Sun and stars of mass 10, 50 and 100 $M_\odot$ (assuming a mean molecular weight of $\mu = 0.6$). (Credit: Author)

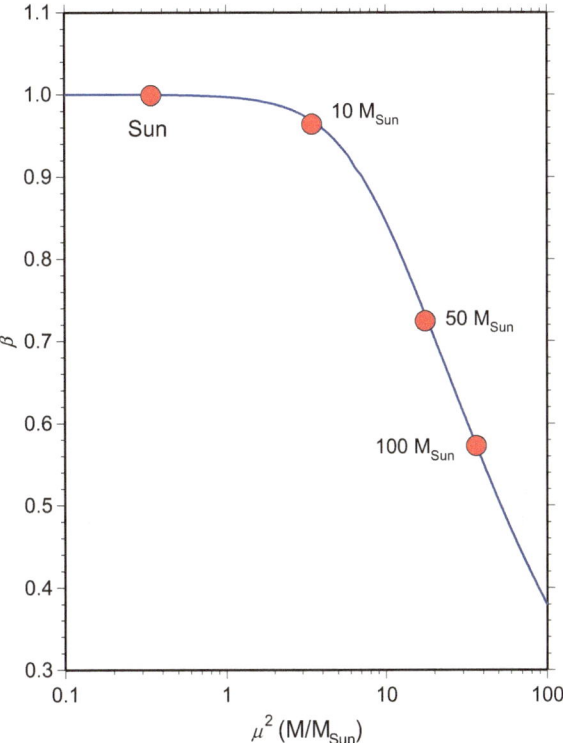

increasingly important, so the luminosity approaches the Eddington limit L_{EDD}. Recall that stars in which the radiation pressure becomes dominant will also become unstable to disruption (see Sect. 2.3).

4.6.2   Cowling's Model

In the Standard Model, Eddington adopted the stance that the ratio of the radiation pressure to total pressure remains constant throughout a star. In the early 1930s, Thomas Cowling adopted a power law expression for the opacity and assumed that all of the energy is generated within a very small region of the central core, and accordingly that the luminosity was constant throughout the interior with $L(r) = L$. This is a minimal model in the sense that it avoids the problem of energy generation (still an unknown process in the early 1930s) by sweeping the entire issue into a central boundary condition. Even with this simplification concerning the energy generation mechanism, as will be seen below, the Cowling Model only takes us a little ways forward in the understanding of stellar structure, and it does not actually yield any analytic results; indeed, it is far more difficult to solve for the equations of the Cowling Model than those for the Standard Model.

The Cowling Model begins by inserting the opacity law Eq. (4.18) into Eq. (4.6), giving

$$\frac{dP_{rad}}{dr} = -\frac{\kappa_0 L}{4\pi c} \left(\rho^{\chi+1} T^{\eta}\right) \frac{1}{r^2} \qquad (4.52)$$

Further to this, the equation for hydrostatic equilibrium can be written as

$$\frac{d}{dr}\left(P_{rad} + P_{gas}\right) = -\frac{GM\rho}{r^2} \qquad (4.53)$$

Dividing (4.53) by (4.52) then gives us.

$$1 + \frac{dP_{gas}}{dP_{rad}} = \left(\frac{L_{Edd}}{L}\right)\left(\frac{\kappa_s}{\kappa_0}\right) \rho^{-\chi} T^{-\eta} \qquad (4.54)$$

At this stage we essentially have to give up with respect to further general simplifications, and analytic progress (of a sorts) can only be made by assuming a constant opacity—setting $\chi = \eta = 0$. If we do this and then differentiate Eq. (4.54), we find

$$\frac{d^2 P_{gas}}{dP_{rad}^2}\left(\frac{dP_{rad}}{dr}\right) = \left(\frac{4\pi c G}{\kappa_0 L}\right)\left(\frac{dM}{dr}\right) \qquad (4.55)$$

substituting now for the mass continuity equation dM / dr and dP_{rad} / dr yields the second order differential equation

$$\frac{d^2 P_{gas}}{dP_{rad}^2} = -A r^4 \qquad (4.56)$$

where A is a constant. Going back to Eq. (4.6), we now want to develop an equation between P_{gas}, P_{rad} and r independent of the density ρ. First, by taking the equations for P_{gas} and P_{rad} [Eqs. (3.1) and (3.3)], we obtain the expression

$$\rho = \left[\frac{\mu}{\mathfrak{R}}\left(\frac{a}{3}\right)^{1/4}\right] P_{gas} P_{rad}^{-1/4} \qquad (4.57)$$

Substituting Eq. (4.57) into Eq. (4.6) now yields

$$\frac{d}{dP_{rad}}\left(\frac{1}{r}\right) = B \frac{P_{rad}^{1/4}}{P_{gas}} \qquad (4.58)$$

where B is a constant. Eqs. (4.56) and (4.58) now provide a set of differential equations that describe the variation of P_{gas} and P_{rad} as a function of the radius

r throughout a star. Here we stop, but the point historically was that the equations so developed could be integrated numerically (not without difficulty), and that once the run of the various pressure terms was known, the run of the density and temperature could be determined.

The Cowling Model was developed some 30 years before the first electronic computers became generally available for research, and of course, from the mid-1960s onwards there was no specific need to make the simplifications that characterized the Standard and/or the Cowling Models. Perhaps the best lesson to take away from the Cowling Model just described is that even under the most extreme regime of physical simplification, it is still very difficult to find solutions to the equations of stellar structure. Indeed, in his 1966 Presidential address to the Royal Astronomical Society, Thomas Cowling, commented on the early problems of integrating the equations of structure numerically, noting that, 'a slight computational error is enough to set the solution hurtling off on one side or the other, like a climber walking along a sharp ice-covered ridge between two precipices'. It was not until the late-1930s that Carl von Weizsäcker, Hans Bethe and coworkers developed the first clear descriptions of the energy generation process via the CNO and PP chain. Incorporating such knowledge into a pre-electronic computer model was far from straightforward. Returning to Cowling for a few final words, he specifically noted in his 1966 address that, 'with the advent of this new physical information, complete data for the construction of stellar models were for the first time available; it was like supplying the fourth leg of a chair which so far had had only one back leg'.

(☺) **Exercise 4.24: Lab/Computer Project**   Among the first researchers to begin developing computer programs to numerically solve the detailed equations of stellar structure (including opacity and energy generation equations) was Martin Schwarzschild, and he is credited (amongst others) with the probably apocryphal remark that, 'once the computer money ran out we had to start thinking'. In the modern era the idea that one has to *pay* to use a computer is strange enough, but for all this, his comments still hold relevance. Computer models are all well and good, and important, but they can also provide a somewhat blind set of results. For all the complexity of the problem, how do we have any idea that the numbers derived from a massive set of computations mean anything at all? This, of course, is exactly where the analytic, simplified models come in: they are the test cases for more general computer codes. In the modern era, there are a number of publicly available stellar evolution codes and/or websites that will provide a user with a detailed set of stellar models. By far the best webpage that I have encountered is that of *Mad Star* hosted by Richard Townsend (University of Wisconsin at Madison): www.astro.wisc.edu/~townsend/static.php?ref=ez-web. The website allows access to the EZ stellar evolution code that has been described by Bill Paxton (University of California, Santa Barbara) in http://arxiv.org/pdf/astro-ph/0405130.pdf. (1) Read through the information on the webpage and generate a solar model: set the metallicity to $Z = 0.02$, the maximum age to 5×10^9 years and the number of steps to 4—make sure you tick the box for the detailed structure files. (2) From the generated data files,

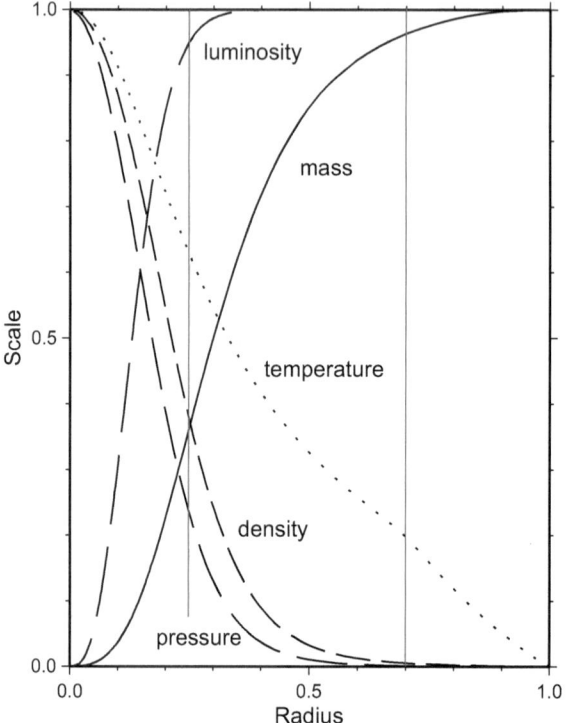

Fig. 4.9 The internal run of the scaled mass, temperature, density and luminosity for a solar mass model (age = 4.5 Gyr). Data from the EZweb star simulator. The maximum mass, radius and luminosity are 1 $M_\odot$, 1 $R_\odot$ and 1 $L_\odot$ respectively. The maximum pressure, density and temperature are 1.42×10^{16} Pa, 7.83×10^4 kg/m^3 and 1.34×10^7 K respectively. The vertical lines at $r/R = 0.25$ and 0.7 indicate, respectively, the extent of the energy generating core and the boundary at which the radiative interior transitions into a convective envelope (recall Fig. 4.2). (Credit: Author)

construct a diagram showing the variation of the scaled temperature T/T_c, the scaled density ρ/ρ_c, the scaled luminosity L/L_{star} and the scaled mass and scaled pressure, against the scaled radius r/R_{star}. See Fig. 4.9.

4.6.3   The Most Massive Stars

It was seen earlier that as the importance of radiation pressure increases within a stellar model, the value of β in the Eddington quartic approaches zero and the luminosity approaches the Eddington luminosity L_{Edd}. A star with a luminosity of L_{Edd} is supported entirely by radiation pressure $\beta = 0$, and if the luminosity exceeds this limit, then surface material will be ejected into space through the transfer of momentum from the outgoing photons. Certainly, significant mass loss is observed to take place from the most luminous (most massive) stars, with the rates varying from of order $10^{-7} - 10^{-4} M_\odot/yr$. Although such massive stars have relatively short main sequence lifetimes (recall Eq. (4.22)) amounting to a few millions of years, this is still time enough for their mass to be significantly reduced. As we have seen with respect to the Vogt-Russell theorem, any change in mass must have a ripple effect with respect to the determination of a star's luminosity, radius and surface temperature. As will be shown below, it turns out that the effects of mass loss can be modeled analytically, and especially so in the case of massive stars that tend to evolve in a near-homogeneous (fully mixed) manner. This latter result is illustrated in Fig. 4.2, where it is revealed that the mass of the convective core M_{cc} increases with increasing stellar mass.

The Eddington luminosity is given in Eq. (4.46), and if we take the opacity to be that due to electron scattering (which applies in the high temperature limit; see Fig. 4.4), then

$$\frac{L_{Edd}}{L_{Sun}} = \frac{4\pi c G}{0.02(1+X)} \left(\frac{M}{M_{Sun}}\right) \left(\frac{M_{Sun}}{L_{Sun}}\right) = 6.5 \times 10^4 \frac{(M/M_{sun})}{(1+X)} \qquad (4.59)$$

where, recall, $X \equiv X_1$ is the hydrogen mass fraction. If we now make use of the result that for massive stars the mass-luminosity relationship indicates that $(L/L_\odot) = (M/M_\odot)^3$, Eq. (4.59) tells us that the Eddington luminosity is achieved for stars having a mass of order $M_{Edd} = \sqrt{6.5 \times 10^4/(1+X)} M_{Sun}$, which for $X = 0.7$ indicates that $M_{Edd} \sim 195\ M_\odot$. Placing this mass back into the mass-luminosity relationship further indicates that such massive stars will have luminosities of order $7.5 \times 10^6 L_\odot$. Stars with such luminosities will occupy the very upper most part of the HR diagram and indeed, the Humphreys-Davidson exclusion region (recall Sect. 1.7) and the dramatic behaviors observed for the Luminous Blue Variable (LBV) stars (such as η Carina) are believed to be a direct consequence of the ravages of extreme mass loss in high-luminosity, high-mass stars.

To see that the mass M_{cc} of the convective core should increase as the stellar mass itself increases (as shown in Fig. 4.2) can be demonstrated by an analysis of the temperature gradient expression under convective condition, that is, via Eq. (4.13). Accordingly, we write:

$$\frac{dT}{dr} = \frac{\gamma - 1}{\gamma} \frac{T}{P} \frac{dP}{dr} = -\frac{\gamma - 1}{\gamma} \frac{T}{P} \frac{G M_{cc}}{r^2} \rho \qquad (4.60)$$

where in the second part of (4.60) we have substituted for the equation of hydrostatic equilibrium. At the boundary of the core and the envelope, the adiabatic temperature gradient (Eq. (4.60)) must equal the radiative temperature gradient. We invoke the continuity of the temperature gradient dT/dr across the convective core to radiative envelope boundary. Accordingly, equating the left-hand side of Eq. (4.60) with the radiative temperature gradient given in Eq. (4.5), we have:

$$\frac{\gamma - 1}{\gamma} \frac{T}{P} G M_{cc} = \frac{3}{16\pi a c} \frac{\kappa}{T^3} L \qquad (4.61)$$

In constructing (4.61) we have implicitly assumed that there is no energy generation outside of the convective core. A little more work in rearranging Eq. (4.61) yields the result

$$\frac{\gamma - 1}{\gamma} M_{cc} = \frac{1}{4(1 - \beta) 4\pi a c G} \kappa L$$

where we have substituted for the identity $P_{rad} / P = (1 - \beta) = (a T^4/3) / P$. This gives us the final result that (for constant opacity in the radiative envelope) the core mass fraction $q = M_{cc}/M$ is:

$$q = \left(\frac{M_{cc}}{M}\right) = \frac{\gamma}{4(1 - \gamma)(1 - \beta)} \left(\frac{L}{L_{Edd}}\right) \qquad (4.62)$$

If we look at the limit of Eq. (4.62) as $\beta \Rightarrow 0$, so $L \Rightarrow L_{Edd}$, and

$$\left(\frac{M_{cc}}{M}\right) \approx \frac{\gamma}{4(1 - \gamma)}$$

As discussed earlier, $4/3 < \gamma < 5/3$ for dynamical stability, and since $\gamma \Rightarrow 4/3$ in the radiation dominated limit, so $M_{cc} / M \Rightarrow 1$, meaning the convective cores of the most massive stars must takeup more and more of the interior volume, justifying the trend shown in Fig. 4.2.

So far we have deduced that for massive stars, the luminosity will vary approximately as $L = (1 - \beta) L_{Edd}$, where β is given by the Eddington quartic, and that the interiors of such stars will be well mixed with $M_{cc} \sim M$; in other words, they will tend to evolve in a homogeneous (fully mixed) manner. Working within the framework of the relatively good approximations that apply as $\beta \Rightarrow 0$, Fred Hoyle and William Fowler in a classic research paper published in 1963 derived a relationship linking the central temperature T_c of a massive star to its surface temperature T_s. For our future discussion, this will be a useful formula to have.

The surface temperature of a star can be expressed through the Stefan-Boltzmann law (Eq. (1.14)), and accordingly $T_s = (L / 4\pi R^2)^{1/4}$. Eventually we will set $L = L_{Edd}$ $(1 - \beta)$, but first we look for a relationship between the radius R and the central temperature T_c for a polytrope of index $n = 3$. This is rather lengthy and twisting derivation, so take a deep breath and find a good quantity of scrap paper to work with. The radius of a polytrope is given by the relation $R = \alpha \xi_1$, where for $n = 3$, $\xi_1 = 6.897$ (as given in Table 3.1). Equation (3.43) further indicates that for $n = 3$,

$$\alpha^2 = \frac{K}{\pi G} \rho_c^{-2/3}$$

Furthermore, for an $n = 3$ polytrope we have $P_c = K \rho_c^{4/3}$, and using $P_c = a\, T_c^4 / [3 (1-\beta)]$ the K term can be eliminated to yield

$$\alpha^2 = \frac{a}{3(1-\beta)\pi\, G}\frac{T_c^4}{\rho_c^2}$$

Once more, we can use the basic expression $P = P_{gas} + P_{rad}$ and the definition that $P_{gas} = \beta P$ to obtain a relationship between ρ_c and T_c such that

$$\rho_c = \frac{a\mu\beta}{3\mathfrak{R}(1-\beta)}T_c^3 \tag{4.63}$$

Substituting for ρ_c in the equation giving α^2, we find

$$\alpha^2 = \frac{3\pi\,\mathfrak{R}^2}{G a}\frac{(1-\beta)}{(\mu\beta)^2}T_c^{-2}$$

we can now substitute for the $(\mu\beta)^2$ term from the Eddington quartic equation (Eq. (4.50)) to yield a final expression for α^2 such that:

$$\alpha^2 = A(1-\beta)^{1/2}\frac{M}{T_c}$$

where A is constant. The final step is to insert appropriate terms into the Stefan-Boltzmann relation $T_s = (L/4\pi R^2)^{1/4}$. Writing $L = (1-\beta)\,4\pi c\, G\, M/[0.02(1+X)]$, where we have adopted an electron scattering opacity term, we derive:

$$T_{s4} = (1.010)\,(1-\beta)^{1/8}(1+X)^{-1/4}T_{c6}^{1/2} \tag{4.64}$$

where the temperatures are now expressed in units of 10^4 K in the case of the surface temperature T_{s4} and 10^6 K in the case of the central temperature T_{c6}. With Eq. (4.64) in place, the straightforward analytic work ends, since at present we have no specific method of linking a value of the central temperature to any specific given stellar mass. This gap in our knowledge can be closed if we are prepared to do some numerical integration and adopt a better approximation for the energy generation term $\varepsilon(r)$. In what follows, I will be drawing heavily on two research papers[3]:

---

[3]See also the research paper, Very Massive Stars: evolution with mass loss by Jaime Klapp (*Astrophysics and Space Science*, 93, 313–345, 1983). In this paper, stars in the mass range from 500 to 10,000 $M_\odot$ are studied. Such stellar models are of interest from a cosmological viewpoint, since the very first Population III stars are thought to form with such very high masses.

(P1): R. Mitalas and P. W. Manuel. 1987. *Relation between mass and central temperature in supermassive stars.* Astronomy and Astrophysics, 173, 244–246.
(P2): M. Beech and R. Mitalas. 1989. *The homogeneous evolution of massive stars.* Astronomy and Astrophysics, 213, 127–132.

These two papers can be found and downloaded from the SAO/NASA Astrophysics Data System webpage at: http://adsabs.harvard.edu/. In P1, the authors obtain a specific set of numerical results that link the central temperature T_c of an $n = 3$ polytrope to the mass M, luminosity L and radius R for a specific energy generation formula derived for the CNO cycle—recall, that the CNO cycle will dominate in terms of energy generation in stars more massive than about 1.5 the mass of the Sun. Specifically, the energy generation expression is taken from Donald Clayton's classic text, *Principles of Stellar Evolution and Nucleosynthesis* (McGraw Hill, New York, 1968). Clayton gives the following:

$$\varepsilon_{CNO} = \varepsilon_0 \rho T_6^{-2/3} \exp\left(-B/T_6^{1/3}\right) \tag{4.65}$$

where $\varepsilon_0 = 8 \times 10^{27} \, XX_{CN}$, where X is the hydrogen mass fraction and X_{CN} is the mass fraction of carbon and nitrogen respectively, and $B = 152.31$. This particular expression can be contrasted against the simple power law expression for ε used earlier in Eq. (4.20). Just as explored in exercise 4.8, Eq. (4.65) can be integrated numerically over a polytrope of index $n = 3$.

(⊛) **Exercise 4.25: Lab/Computer Project** Adapt the computer code developed in Exercise 3.22 to solve the integral $I(T_{c6}, 16/3)$ over a polytrope of index $n = 3$, where,

$$I\left(T_{c6}, \frac{16}{3}\right) = \int_0^{\xi_1} \xi^2 \theta^{16/3} \exp\left[-\frac{B}{(T_{c6}\theta)^{1/3}}\right] d\xi \tag{4.66}$$

Solve for this integral over a range of values for $10 \leq T_{c6} \leq 100$.

In paper P2 the results from a set of calculations for the integral (4.66) were tabulated, and for ease of use a fitting formula was found such that

$$I\left(T_{c6}, \frac{16}{3}\right) = \exp\left(a T_{c6}^{-1/3} + b\right) \tag{4.67}$$

where $a = -156.58656$, and $b = -1.33061$. With this result in place, we are in a position to develop a formula describing the luminosity L, and accordingly with

$$L = \int 4\pi r^2 \rho \, \varepsilon_{CNO} \, dr$$

we obtain the result that

$$L = \varepsilon_0 \left(\frac{\rho_c}{T_{c6}^{2/3}} \right) M \, \frac{I\left(T_{c6}, \frac{16}{3}\right)}{I(3)} \qquad (4.68)$$

where M is the mass of the star and $I(3) = \int \xi^2 \theta^3 d\xi = 2.018236$ (recall Eq. (3.44) and see Table 3.1). The ρ_c term in the brackets of Eq. (4.68) can be substituted for from Eq. (4.63) to yield an L-T_c-M relationship. With Eq. (4.68) in place, we can further substitute from Eqs. (4.45), (4.46) and (4.50) to find a relationship between the mass M and the central temperature T_{c6}—the full formula for this is presented in P1, Eq. (9). We now have a complete set of formula (none of them particularly straightforward) to link together the mass, surface temperature, radius and central density of a massive star in terms of its central temperature and core hydrogen mass fraction.

In P2, these same formulae are used to generate evolutionary tracks for homogeneous (fully mixed) massive stars—in these calculations, the decreasing (towards zero) hydrogen mass fraction X_1 is used as a proxy for model age. Figure 4.10 shows

Fig. 4.10 The evolution of a fully mixed 30 M$_\odot$ star in the HR diagram. The various evolutionary tracks are developed by decreasing the value of the hydrogen mass fraction from $X = 0.71$ to 0.01. The dashed lines show contours of constant hydrogen abundance for $X = 0.1$ and $X = 0.01$. The green dots correspond to equal mass locations corresponding to 15 and 5 M$_\odot$ (for the $N = 400$ and 500 tracks) respectively. See the text for details. (Credit: Author)

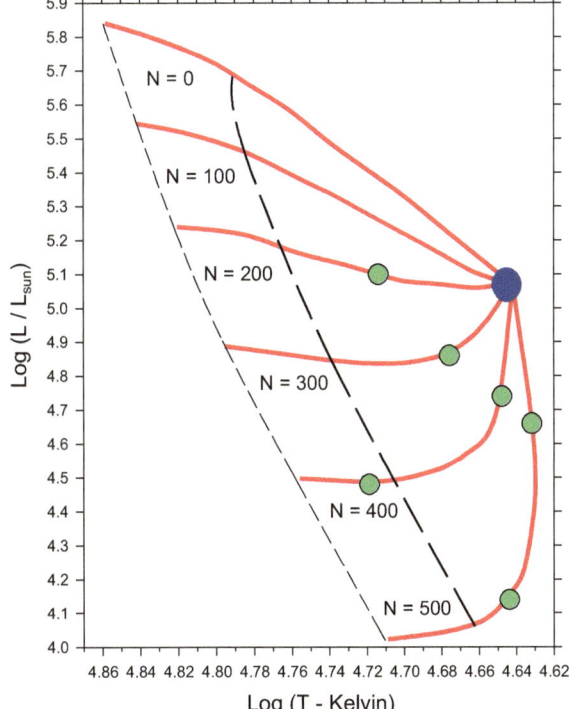

the evolutionary tracks for an initially 30 $M_\odot$ star undergoing mass loss parameter-ized according to $N = 0$, 100, 300 and 500 (see the next section and Sect. 5.6), where $dm/dt = N L / c^2$, where L is the luminosity and c is the speed of light. The various tracks in Fig. 4.10 are parameterized according to the mass loss parameter N. In all cases, even the $N = 0$ case, the evolution is towards higher surface temperature (an issue to be discussed further in Sect. 5.8). The luminosity evolution, however, is highly dependent upon the adopted mass loss rate, and we will discuss this behavior in the next chapter.

Taking a well-earned breath at this time, we can now reflect upon what the last two sections have told us. Just as we found with the Cowling Point Source Model, as soon as some effort is made to account for the energy generation term within a particular stellar model, life becomes mathematically difficult. There really is no easy way of solving this problem.

4.6.4   The Evolution of Massive, Fully Mixed Stars

Those readers who have been brave enough to look over paper P2 by Beech and Mitalas (as referenced in the last section) will have seen that the paper uses the mass-central temperature relationship to construct a series of evolutionary models for massive homogeneous stars evolving with mass loss (Fig. 4.10). The basics of this approach were outlined in another research paper[4]:

(P3). R. Mitalas and H. Falk. 1984. *On the main-sequence evolution of massive stars with mass loss.* Monthly Notices of the Royal Astronomical Society, 210, 641–653.

The key idea behind the analytic models developed by Mitalas and Falk was that the detailed numerical models do reveal that the value of β is very nearly constant throughout the interior of a massive star, and that such stars are well approximated by $n = 3$ polytropes. Accordingly, the luminosity for a given hydrogen mass fraction X within the central core is well described by Eqs. (4.45) and (4.46) when the opacity is taken to be that for electron scattering: $\kappa_s = 0.02(1 + X)$. Hence, we write.

$$L = \frac{4\pi cG}{0.02(1 + X)} (1 - \beta)M \qquad (4.69)$$

There is a small correction term involving β that Mitalas and Falk attach to Eq. (4.69), but since it is of order one we shall ignore it here. Now, in order to gauge the effects of a changing composition—the evolution of a star—the energy

---

[4]Attention is also drawn to the first part of the research paper: On the Theory of Evolution of Completely Mixed Stars with Mass Loss, by Richard Stothers [*Monthly Notices of the Royal Astronomical Society* (1966), 131, 253–362].

generation term is expressed in quite basic terms. The luminosity is related to the rate at which the hydrogen mass fraction is consumed (as a result of hydrogen fusion reactions) dX/dt, multiplied by Q, the energy generated per reaction per kilogram of matter, with $Q = 0.007\,c^2$, where c is the speed of light, and finally Q is multiplied by the mass of the star (which sets the number of kilograms of hydrogen available for fusion reactions). Accordingly, we write

$$\frac{dX}{dt} = -\frac{L}{QM} \tag{4.70}$$

where the minus sign in Eq. (4.70) indicates that the hydrogen mass fraction is decreasing over time. At this stage, the effects of mass loss from the star can be introduced, and this is written as:

$$\frac{dM}{dt} = -N\left(\frac{L}{c^2}\right) \tag{4.71}$$

where N is a constant. This mass loss formulation was developed by Leon Lucy and Phillip Solomon in 1970, and it generally holds true for massive star winds (although there are many complicating factors and other mass loss laws have been derived; see Sect. 5.6). Eq. (4.71) makes physical sense since the bracketed term on the right-hand side corresponds to the radiation pressure force, and N is accordingly interpreted as the cumulative effect of the radiation pressure acting upon the various atoms and molecules located within a star's outer layers. Conveniently, Eqs. (4.70) and (4.71) can be combined to yield a mass–hydrogen mass fraction relationship:

$$M(X) = M_0 \exp\left(-\frac{NQ}{c^2}(X_0 - X)\right) \tag{4.72}$$

where the 0 subscript indicates an initial value. Eq. (4.72) thus indicates an exponential decrease in mass over time (that is, with decreasing X) under the Lucy and Solomon mass loss law.

(☺) **Exercise 4.26**   Verify the derivation of Eq. (4.72). *Hint:* Divide (4.71) by (4.72) and then integrate setting the initial mass to be $M(X = X_0) = M_0$.

The main sequence lifetime of a massive star undergoing mass loss can be estimated by assuming a typical and constant value for $<\beta>$ during the hydrogen fusion stage. In this manner, Eq. (4.69) can be combined with Eq. (4.70) to give:

$$\frac{dX}{dt} = -\frac{1}{Q}\left(\frac{L}{M}\right) = -\left(\frac{4\pi c G(1 - <\beta>)}{0.02\,Q}\right)\left(\frac{1}{1 + X}\right)$$

which can be integrated to give a time–hydrogen mass fraction relationship:

$$t(X) = \tau_{ms}\left(1 - \frac{X(2+X)}{X_0(2+X_0)}\right) \qquad (4.73)$$

where τ_{ms} is the main sequence lifetime

$$\tau_{ms} = \frac{QX_0(2+X_0)}{400\pi c G(1-<\beta>)} = 8 \times 10^6 \frac{X_0(2+X_0)}{(1-<\beta>)} \quad (yrs) \qquad (4.74)$$

The $(1-<\beta>)$ term approaches unity as the mass of a star increases; so too does t_{ms} decrease with increasing mass.

(☺) **Exercise 4.27** (1) What is the main sequence lifetime of a star that always radiates at its Eddington luminosity (take the opacity to be that of electron scattering)? (2) What do you note about this lifetime? *Hint:* Start with Eq. (4.21) and substitute for the M/L term from Eq. (4.46).

The effect of mass loss is not only to reduce the mass as a function of time—since X is decreasing from X_0 to 0 during the hydrogen fusion stage—but also to reduce the luminosity. This can be seen by rewriting Eq. (4.69) under the conditions of mass loss and constructing the ratio

$$\frac{L(\dot{M} \neq 0, X)}{L(\dot{M} = 0, X)} = \frac{1-<\beta(\dot{M} \neq 0)>}{1-<\beta(\dot{M} = 0)>}\left(\frac{M(X)}{M_0}\right) \qquad (4.75)$$

Since $M(X) < M_0$ and the $<\beta(\dot{M} \neq 0)>$ term is always smaller than $<\beta(\dot{M} = 0)>$ (because the star averaged mean molecular weight $<\mu>$ is always higher in the mass losing star), the luminosity for a given X value will always be smaller in the mass loss situation compared to that in the constant mass case (Fig. 4.11, and recall Fig. 4.10). For similar reasons, mass loss will also extend the main sequence lifetime

Fig. 4.11 The effect of mass loss on the main sequence evolution of a massive star (schematic); see P3 for detailed numerical calculations. The thin dotted lines link models with the same core hydrogen mass fraction X_1. (Credit: Author)

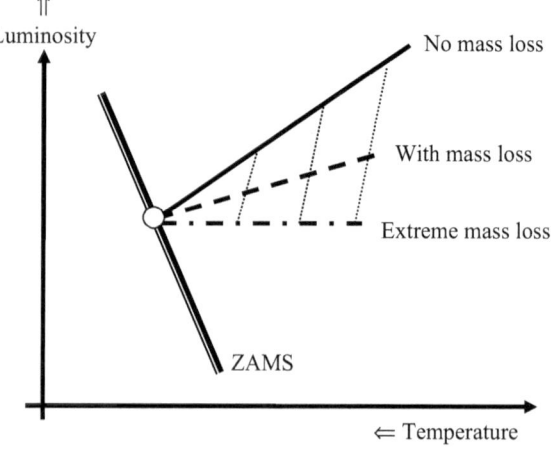

of a star. Not only this, but by evolving to a lower mass, a mass-losing star will evolve with a lower central and surface temperature than its constant mass counterpart for the same value of the core hydrogen mass fraction.

A critical mass loss rate, above which the luminosity of a star will no longer increase as it ages (the extreme—horizontal—mass loss track in Fig. 4.11), can be derived from the mass-luminosity equation (Eq. (4.69)) by setting the derivative $dL/dt = 0$. In this manner, the condition for which the effect of mass loss on the luminosity cancels that due to nuclear evolution is determined. From Eq. (4.69), therefore,

$$\frac{dL}{dt} = 0 = K \left[\frac{\dot{M}}{(1+X)} - \frac{M(dX/dt)}{(1+X)^2} \right]$$

where K absorbs all the various constant terms. Substituting from Eq. (4.7) now gives the critical mass loss rate for evolution at constant luminosity as

$$\dot{M}_{crit} = -\frac{X_0}{(1+X)} \frac{M(0)}{t_{nuc}} \tag{4.76}$$

where $t_{nuc} = Q X_0 M(0) / L(0)$ is the nuclear timescale for a non-mass-losing star of initial mass $M(0)$, luminosity $L(0)$ and hydrogen mass fraction X_0.

(☺) **Exercise 4.28** What is the critical mass loss rate for a 30 $M_\odot$ star? Take the initial hydrogen mass fraction to be $X_0 = 0.71$, and estimate t_{nuc} from Eq. (4.22) with the appropriate mass-luminosity law taken from Eq. (1.24).

While it is not entirely clear from Eq. (4.64) (until the numbers are revealed), a massive, fully mixed star will evolve a smaller radius and a higher surface temperature as it ages (indeed, fully mixed stars don't have a red giant phase; see Sect. 5.8). The increase in surface temperature is primarily driven by the reduction in the X mass fraction and the reduced β terms, the variation in these terms overriding the mass loss-driven reduction in the central temperature. This result is demonstrated in paper P2 and illustrated in Fig. 4.10. This situation goes some way in explaining the reason for the Humphreys-Davidson exclusion zone in the upper right-hand corner of the HR diagram—one of the key characteristics of the LBV stars being their extreme amounts of mass loss (Fig. 4.12). During their quiescent phase, LBVs reside in the so-called S Doradius instability strip, and while described as being quiescent, such stars are still losing mass at a very high rate. During an outburst, when a singularly large amount of matter is lost in a very short time, an LBV will temporarily move into a location above the Humphreys-Davison boundary. It is additionally observed that many massive and LBV stars show an anomalous nitrogen, carbon and oxygen-rich composition in their outer layers. This anomaly comes about due to the mass loss peeling away the outer hydrogen-rich layers to reveal nuclear-processed material that was once in the energy-generating convective core. The exposure of nuclear processed material can be predicted according to the

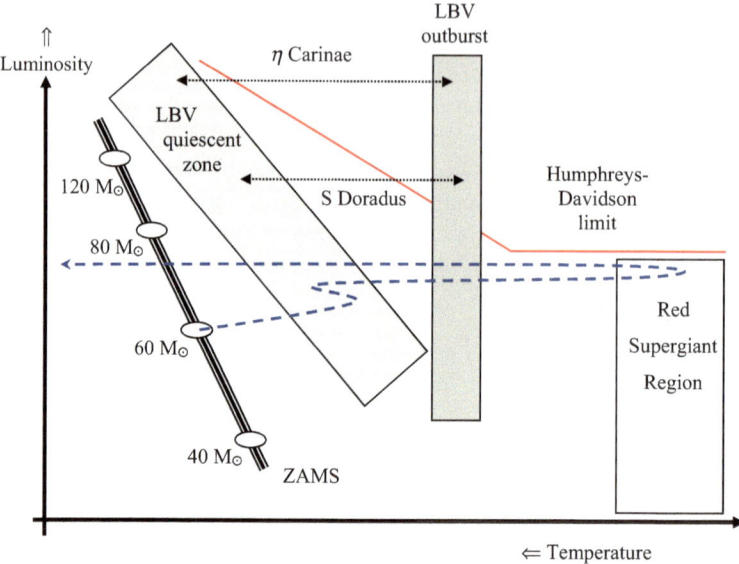

Fig. 4.12 The Humphreys-Davidson boundary (schematic). As a consequence of extreme mass loss, massive stars do not readily enter into the low-temperature red giant region, but evolve instead to higher temperatures. The dashed tracks labeled η Carinae and S Doradus illustrate the outburst position shifts of an LBV. The Humphreys-Davidson limit at low temperatures delineates the maximum luminosity region for the spectral type M, luminosity class Iab supergiants. A schematic evolutionary track for a 60 $M_\odot$ star is shown (dashed curve) as it evolves from the ZAMS in o the red supergiant region and then, during its final stages, as a result of high mass loss, moves towards higher surface temperatures (undergoing supernova disruption not as a red supergiant but a blue supergiant / Wolf-Rayet star). (Credit: Author)

condition: $M(X_{exp}) = q_0 \, M_0$, where q_0 is the initial convective core size (recall Eq. (4.62)). Accordingly, using Eq. (4.72) we have:

$$X_{exp} = X_0 + \frac{143}{N} \ln(q_0) \tag{4.77}$$

where the 143 term is the inverse of $Q / c^2 = 0.007$. The natural logarithm term in Eq. (4.77) is always going to be negative and less than one, and accordingly for exposure the mass loss parameter N must always be greater than several hundred. If $q_0 = 0.5$, and $N = 250$ (a typically deduced value for mass losing stars), for example, exposure of the core will take place at $X_{exp} = 0.3$ (assuming $X_0 = 0.7$), and Eq. (4.73) further indicates that exposure will occur when the star is 64% of the way through its main sequence phase.

   In the last three sections, we have struggled to keep pace with the increasing input physics, although massive stars are found to be reasonably amenable to analytic study. Perhaps the aphorism *the deeper you dig, the muddier you get* now holds true. The analytic story must now essentially come to an end. To go further, with the inclusion of detailed expressions for the energy generation, the opacity, changing chemical composition and mass loss, recourse must be made to detailed numerical methods.

4.7 Answers to Exercises

Following are the answers to all exercises in this chapter. The exercises for which no answer is provided are those in which the student is asked to check a result or complete the algebra steps between two equations given in the text.

Exercise 4.6 The convective turnover time is $T_{conv} \approx 232$ days (taking $V_{conv} = 5$ m/s), and the rotation period to convective turnover time ratio is $P_{rot} / T_{conv} \approx 0.4$.

Exercise 4.7 The characteristic mean-free path length is $l \approx 0.06$ mm.

Exercise 4.9 (iii) The luminosity is $L = 1.3 \times 10^{27}$ Watts $= 3.4$ $L_\odot$.

Exercise 4.10 The core radius fraction is r(fusion zone)$/R_\odot = 0.36$.

Exercise 4.12 The number of conversions required is of order 10^{38} per second, and this amounts to a miniscule fraction of order 10^{-19} of the total number of protons in the Sun ($N_{protons} = 10^{57}$).

Exercise 4.13 The amount of matter converted into energy since the Sun formed is 6.2×10^{26} kg. *Comment:* only a negligible amount of the Sun's mass (0.03%) has been converted into energy since it formed.

Exercise 4.14 If 5% of Sun's mass was in the form of hydrogen, then this constitutes some 6×10^{55} hydrogen atoms. If all these hydrogen atoms combined to form helium, then a total of (6×10^{55} / 4) $\times$ 4×10^{-12} Joules worth of energy would be generated. Given the Sun's current luminosity, this energy supply would last for some 1.5×10^{17} seconds (~5 billion years).

Exercise 4.15 In this case, the energy store is simple $M_\odot c^2$, and the lifetime will be of order 1.5×10^{13} years.

Exercise 4.16 The mass-luminosity law (Fig. 1.11A) indicates that the luminosity increases as the mass to some power $\alpha > 1$. Hence the (M / L) term in Eq. (4.21) will always be a decreasing function of increasing mass.

Exercise 4.18 Assuming the star is not convective within its interior, there are four differential equations. There is one independent variable, r. There are four dependent variables, P, T, M and L. There are four boundary conditions (two at the center and two at the surface). In order to obtain a complete solution to the equations of stellar structure, the only reaming variables to specify are the composition terms X_i, $i = 1$, 2, 3. . . ., and the total mass M_{star} that needs to be realized at $r = R_{star}$.

Exercise 4.19   The answer is option (2).

Exercise 4.27   With $X_0 = 0.7$, $t_{ms} \approx t_{ms}/10 \approx 260$ thousand years. The timescale is independent of the mass.

Exercise 4.28   The nuclear timescale is $t_{nuc} \approx 2 \times 10^6$ years, and the critical mass loss rate is $(dm/dt)_{crit} \approx -6 \times 10^{-6} M_\odot/\text{year}$.

Chapter 5
The Ravages of Time

An outline of star formation was presented in Chap. 2. The key idea employed was that of gravitational collapse, with an extended gas cloud contracting to form a central protostar. Rotation of the initial gas cloud was also considered, and this results in the formation of an accretion disk about the protostar. Once all of the available cloud material has been incorporated into the protostar, it will continue to contract under gravity until the central temperature and density reach levels at which hydrogen fusion reactions can begin. Once the fusion reactions start within its central regions, the protostar will stop contracting and a *bona fide* main sequence star will have formed (as described in Chap. 4).

In Chap. 2, it was additionally reasoned that the mass domain of star production falls within the range from about 0.1 $M_\odot$ to 100 $M_\odot$. The lower mass limit is determined by a degeneracy onset condition halting gravitational collapse prior to the initiation of hydrogen fusion reactions within the central core. The upper mass limit is set according to a star becoming unstable once its interior is predominantly supported by radiation pressure. While these physical conditions determine the mass range of stardom, they do not directly explain the observed mass distribution of stars—the number of stars that are formed within a specific mass range. At a fundamental level, the number versus mass distribution for the stars produced from a particular molecular cloud must be related to the cloud collapse and fragmentation process, but exactly how the process works is not as yet fully understood. What does seem to be clear, however, is that nature likes to make many more low-mass stars than massive ones, and that the mass distribution function is apparently the same (or nearly so) for all star formation regions within our galaxy and galaxies beyond our own. Exactly why this universality applies is not currently understood.

The initial mass function (IMF) $\xi(M)$ is defined so that the number of stars dN that form in the mass range between M and $M + dM$ is $dN = \xi(M)dM$. The total number of stars in the mass range M_1 to $M_2 > M_1$ is accordingly given by the integral

© Springer Nature Switzerland AG 2019   155
M. Beech, *Introducing the Stars*, Undergraduate Lecture Notes in Physics,
https://doi.org/10.1007/978-3-030-11704-7_5

$$N = \int_{M_1}^{M_2} \xi(M)\, dM \qquad (5.1)$$

while the total amount of matter incorporated within the stars in the range M_1 to $M_2 > M_1$ is the integral

$$M_{total} = \int_{M_1}^{M_2} M\, \xi(M)\, dM \qquad (5.2)$$

It is a nontrivial exercise to determine $\xi(M)$, but various mass-domain-related power law expressions appear to provide a good general description. For stars in the mass range $0.4 - 10\,M_\odot$, the so-called Salpeter power law applies. First described in 1955 by Australian astronomer Edwin Salpeter, this law gives $\xi(M) = \xi_0 M^{-2.35}$, where ξ_0 is a constant determined by the local star density. Modifications to the form of the IMF, however, have been introduced at very low masses such that in the mass range $0.1 - 0.5\,M_\odot$, the power law changes to $\xi(M) = \xi_0 M^{-1.3}$. Objects formed with masses smaller than $0.1\,M_\odot$ do not become stars but instead become brown dwarfs and possible free-floating planets; we do not consider them here. The changeover in the slope of the IMF at lower masses appears to be universal, occurring for all star formation regions that have been observed, but the physical reason for the turnover is unclear. It no doubt relates to the processes that determine the way in which a collapsing gas cloud undergoes hierarchical internal fragment.

A schematic IMF is shown in Fig. 5.1. Across the top of the diagram, the various labels indicate a series of domains in which different mechanisms for the final mass distribution are thought to operate. Between about 1 and about $0.5\,M_\odot$, it is a cloud fragmentation mechanism that determines the IMF, while in the range 0.1 to about $0.5\,M_\odot$, it is cloud fragmentation in conjunction with such effects as the ejection of a protostar from its accreting cloud, and/or truncated accretion that determines the

Fig. 5.1 The schematic variation of the initial mass function against star mass. The α-values correspond to an IMF power law M^α. (Credit: Author)

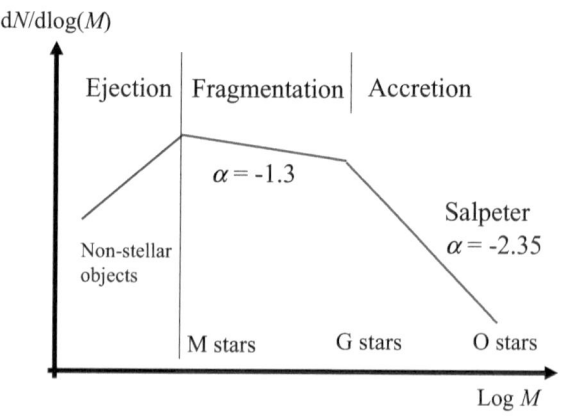

Table 5.1 The relative number of stars and the total mass of stars formed in various mass ranges (in these calculations it is assumed that ξ_0 is constant)

Mass range ($M_\odot$)	0.1 to 0.5	1 to 5	10 to 30	50 to 100
Number	2.5471	0.6564	0.0256	0.0023
Total mass	0.5944	1.2305	0.4074	0.1565

IMF. For masses greater than about $1\,M_\odot$, stars simply grow by accretion, and the final mass is related to the amount of material available to build up the specific stars. Across the bottom of Fig. 5.1, the approximate spectral type of the resultant stars (as a function of mass and number) is indicated, showing that the majority of stars that form are of a low-mass, low-luminosity M spectral type. In contrast, the most massive, high luminosity, O spectral type stars are only rarely formed. Table 5.1 provides a breakdown of relative numbers of stars formed according to a selection of mass ranges, as well as an annotation of the relative total mass of stars formed within the same mass ranges. The two sets of numbers shown in Table 5.1 are derived by completing the integrals shown in Eqs. (5.1, 5.2) respectively. Given the form of the power law for the IMF, it is evident (as seen in Table 5.1) that low-mass stars will dominate with respect to overall number and their total mass distribution. Massive stars may well dominate with respect to their energy output and luminosity, but they are short-lived and only rarely formed objects.

(☺) **Exercise 5.1** Verify the numbers in Table 5.1 by completing the integrals (5.1, 5.2) using the appropriate power laws for the IMF. *Example:* In the mass range $50 - 100\,M_\odot$ the integrals are:

$$N = \int_{50}^{100} \xi_0 M^{-2.35} dM = -\frac{\xi_0}{1.35}\left[100^{-1.35} - 50^{-1.35}\right] = (0.0023)\xi_0$$

$$M_{total} = \int_{M_1}^{M_2} M\xi_0\left(M^{-2.35}\right) dM = -\frac{\xi_0}{0.35}\left[100^{-0.35} - 50^{-0.35}\right] = (0.1565)\xi_0$$

The apparent universality of the IMF makes it useful as a star cluster diagnostic. If a specific cluster, for example, has a turnoff point mass of M_{TP} (recall Fig. 1.13), then the number of main sequence stars N_{MS} will be,

$$N_{MS} = \int_{M_{min}}^{M_{TP}} \xi(M)dM$$

where $M_{min} = 0.1\,M_\odot$ is the minimum stellar mass. The number of white dwarfs that should reside in the cluster, however, will be determined by those stars that have an initial mass smaller than $8\,M_\odot$ (see Sect. 5.4 below for why this specific upper limit applies)

$$N_{WD} = \int_{M_{TP}}^{8 Msun} \xi(M)\,dM$$

Furthermore, the ratio of white dwarf stars to main sequence stars within a given cluster will be of order

$$\frac{N_{WD}}{N_{MS}} = \frac{M_{TP}^{-1.35} - 0.06}{M_{min}^{-1.35} - M_{TP}^{-1.35}} \tag{5.3}$$

where the 0.06 term in the equation is $8^{-1.35}$. In deriving Eq. (5.3), it has been assumed that the Salpeter power law applies all the way to M_{min}. A plot of Eq. (5.3) is shown in Fig. 5.2, and it is seen that, as expected, the number of white dwarfs increases as the cluster ages (that is, as M_{TP} decreases) and that the fraction of white dwarfs is typically only a few percent of the total number of main sequence stars.

Fig. 5.2 The ratio of white dwarfs to main sequence stars for clusters of various ages described according to the turnoff mass M_{TP} (recall Fig. 1.13 and Table 1.3). The red dots correspond to turn-off masses at 2, 3, 5, & 7 Gyr. (Credit: Author)

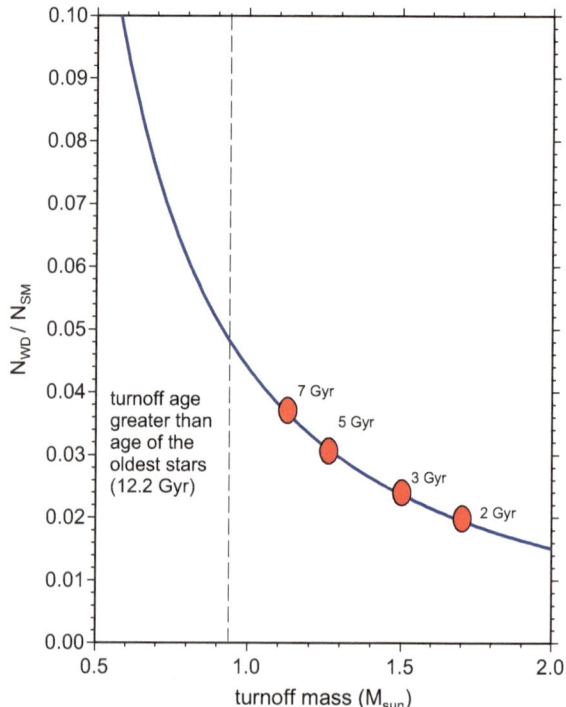

(☺) **Exercise 5.2** It is estimated that the oldest globular clusters have ages of order 12.2 billion years. (1) What is the turnoff mass for such clusters? (2) What is the fraction of white dwarfs to main sequence stars that such clusters will display? *Hints:* For (1) use Eq. (4.22) with $\alpha = 4.0$ and for (2) use Fig. 5.2.

The amount of material ejected from the stars within a given cluster can be estimated by accounting for the mass loss from stars over their lifetime. If M_{eject} is the fraction of material ejected by a star during its lifetime, then

$$M_{eject} = 1 \qquad\qquad \textit{if } M > 8\,M_{\odot}$$

$$M_{eject} = (M - M_{WD})/M \quad \textit{if } M_{TP} < M < 8\,M_{\odot}$$

$$M_{eject} = 0 \qquad\qquad \textit{if } M < M_{TP}$$

where it is assumed (for simplicity) that the entire mass of a star is disrupted in a supernova event and that stars with an initial mass smaller than $8\,M_{\odot}$ evolve into white dwarfs with a characteristic mass of $M_{WD} = 0.5\,M_{\odot}$. Accordingly, the mass of material returned M_{return} to their surroundings by the stars within a given cluster will be:

$$M_{return} = \int_{M_{min}}^{M_{max}} M\,\xi(M)M_{eject}dM = \int_{M_{TP}}^{8} (M - M_{WD})\xi(M)dM + \int_{8}^{M_{max}} M\,\xi(M)dM$$

$$(5.4)$$

where $M_{max} \approx 100\,M_{\odot}$ is the maximum stellar mass. In constructing Eq. (5.4), it is has been assumed that the turnoff mass is smaller than $8\,M_{\odot}$, and accordingly that the cluster is at least 20 million years old (which corresponds to the main sequence lifetime of an $8\,M_{\odot}$ star).

(☺) **Exercise 5.3** Evaluate integral (5.4) for a range of turnoff masses ($M_{TP} = 6$, 3, 1.5 $M_{\odot}$) and determine the mass fraction $M_{return} / M_{initial}$ in each case. Take $M_{max} = 100\,M_{\odot}$, $M_{min} = 0.1\,M_{\odot}$, and $M_{WD} = 0.5\,M_{\odot}$, and assume that the Salpeter power law for the IMF holds over the entire stellar mass range. *Notes:*

$$M_{initial} = \int_{M_{min}}^{M_{max}} M\xi(M)dM$$

In calculations such as the one for $M_{initial}$, the outcome is quite sensitive to the assumed value of M_{min}—convince yourself that this is the case.

5.1   Approaching the Main Sequence

Figure 5.3 shows a schematic evolutionary track in the HR diagram for a protostar as it approaches the main sequence. At this stage, the protostar is gaining energy through gravitational contraction (recall Chap. 2). Initially, as the protostar contracts, it grows in luminosity and its surface temperature increases. Once the protostar encounters the Hayashi boundary, however, it becomes convective throughout its interior. Further evolution then takes the protostar down the Hayashi boundary in the HR diagram until a radiative core develops. At this stage, the protostar evolves to the left in the HR diagram with a somewhat increasing luminosity and towards increasing surface temperatures. This part of the protostar's journey towards stardom is often called the Henyey track after Louis Henyey (University of California, Berkeley) who, along with Robert LeLevier and Richard Levée, first described this pre-main sequence evolutionary phase in the 1950s. Henyey and coworkers pioneered the use of electronic computers in solving the equations of stellar structure (introducing the so-called Henyey relaxation method for solving the equations). The Henyey track is characterized by a relatively slow contraction towards the main sequence, on a Kelvin-Helmholtz timescale, with the protostar in quasi-static equilibrium evolving to higher surface temperatures and luminosity.

The Hayashi boundary, named after Japanese astronomer Chushiro Hayashi, effectively divides the HR diagram into two zones. To the right of the boundary (often called the forbidden zone), a protostar is unable to attain hydrostatic equilibrium and must undergo gravitational collapse (recall Fig. 2.4 and exercise 2.9). To the left of the boundary, hydrostatic equilibrium is possible. A protostar evolving along (and down) the Hayashi boundary is fully convective. A protostar evolves down the Hayashi boundary until it develops a radiative core and thereafter follows a Henyey track, in near-hydrostatic equilibrium, towards the main sequence. Essentially, the condition found by Hayashi builds upon the requirement that there is a minimum surface temperature T_H for which a star can attain hydrostatic equilibrium. For temperatures lower than T_H, the outer layers of a star become optically thin

Fig. 5.3 The evolutionary path of a protostar in the HR diagram as it moves towards the main sequence and the attainment of stardom (recall also Fig. 2.4). (Credit: Author)

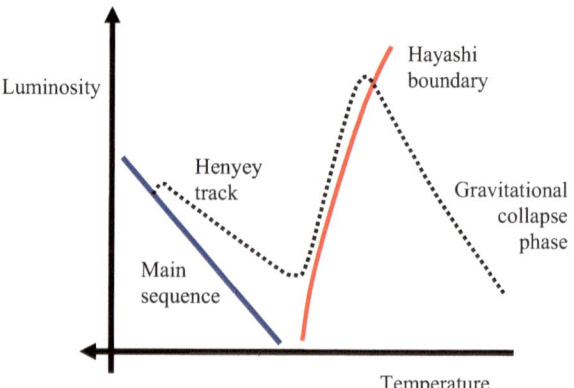

(because the opacity is small), and this violates the condition that the photosphere (and hence the radius of a star) are determined at an optical depth of order one (recall exercise 4.3).

With this notion of a minimum temperature, it becomes clear why a protostar must eventually become fully convective. Since a newly forming protostar is large in physical extent, the Stefan-Boltzmann law dictates that its total luminosity L must also be large (recall Fig. 2.4), and characteristically the luminosity will exceed that which can be carried radiatively—$L > L_{rad}$ (recall Sect. 4.1)—and accordingly the protostar will be convectively unstable with a surface temperature of order $T_H \sim$ 2500–3500 K. Further contraction at a constant temperature of T_H sees the luminosity decrease, and the star evolves down the Hayashi boundary until $L \sim L_{rad}$, at which point a radiative core develops and the protostar picks up its appropriate Henyey track and moves leftward in the HR diagram at near-constant luminosity, but with increasing surface and central temperatures.

Hayashi realized the importance of convection in protostar evolution (and the significance of the forbidden zone) in a series of papers published in 1961. That the Hayashi boundary must be a near-vertical line in the HR diagram can be illustrated by considering the situation (already described in exercise 4.3) where the outer atmosphere of a star has a negligible mass compared to its interior body. Also, we shall adopt the $n = 1.5$ polytropic model as being appropriate to that for a fully convective star. Accordingly, from Eq. (3.40) and 3.46) we have that in the interior

$$P = K\rho^{5/3} \text{ and } K = CM^{1/2}R^{3/2}, \text{ where C is a constant.}$$

The polytropic (convective) interior must eventually match up to the radiative photosphere at some radius $r = R$, and the ability of the photosphere to radiate the energy flux at this radius and remain in hydrostatic equilibrium requires that

$$\frac{dP}{dr} = -\left(\frac{GM}{R^2}\right)\rho$$

where the bracketed term is now taken as being a constant—indeed, it is the surface gravity term $g = GM/R^2$. The pressure variation from R to where it vanishes (effectively at $r = $ infinity) is now described according to the integral equation

$$P_R = g \int_R^\infty \rho\, dr \tag{5.5}$$

The temperature at radius $r = R$ is by definition the effective temperature of the star T_e, and this must satisfy the Stefan-Boltzmann relationship $L = 4\pi R^2 T_e{}^4$. The optical depth τ of the photosphere (again by definition) is taken as being of order one, and accordingly,

$$\tau = \int_R^\infty \kappa \rho \, dr \approx \bar{\kappa} \int_R^\infty \rho \, dr = 1 \qquad (5.6)$$

where $\bar{\kappa}$ is the opacity averaged over the photosphere. Expressing the opacity in terms of the density and temperature as a power law with $\bar{\kappa} = \kappa_0 \rho T_e^b$, where κ_0 and b is are constants, and substituting for this in Eq. (5.6) along with a substitution from Eq. (5.5), we find

$$P_R = \left(\frac{g}{\kappa_0}\right)\frac{1}{\rho T^b} \qquad (5.7)$$

There is one more equation available at this stage: the perfect gas law (assuming negligible radiation pressure), which gives, $P_R = (R/\mu)\rho_R T_e$ We now have four equations that described L, P_R, ρ_R and T_e, and accordingly a relationship between L, T_e and M can be extracted (this involves a lengthy bit of intricate algebra—if you feel adventurous, give it a go). The end result is that.

$$\log(L) = (12 + 2b)\log(T_e) - 4\log(M) + \text{constant} \qquad (5.8a)$$

For a given mass, Eq. (5.8a) provides a description for the locus of the Hayashi boundary in terms of L and T_e in the HR diagram. The relationship derived indicates that there is some mass sensitivity (the $\log(M)$ term), but for a typical value of $b = 4$ for the opacity term, the relationship between luminosity and effective temperature is such that $L \sim T_e^{20}$ (effectively describing a vertical line in the HR diagram). The L-T_e relationship for the Hayashi line is significantly steeper than that describing the locus of the main sequence (recall Sect. 1.3).

Once the $L \sim L_{\text{rad}}$ condition is realized, a star leaves the Hayashi boundary and picks up its appropriate Henyey track. During this phase of its evolution, the temperature gradient within the protostar's interior is described by Eq. (4.5), and accordingly we write

$$L = \frac{16\pi ac R^2 T^3}{3\kappa\rho}\frac{dT}{dr}$$

Setting $dT/dr \approx T_c/R$, approximating the central temperature as $T_c \sim M/R$, writing the density as $\rho \sim M/R^3$ and adopting a Kramers opacity law with $\kappa \sim \rho \, T^{-3.5}$, we find upon substitution that a mass-luminosity-radius relationship results, with

$$L \, R^{0.5} = M^{5.5} \times \text{constant}$$

Henyey and coworkers, using a detailed opacity table rather than a simple Kramers law, found a luminosity-mass relationship of the form $L \, R^{0.78} = \text{constant}$.

Using the Stefan-Boltzmann relationship to eliminate the radius R term, the Henyey track is described according to the relationship

$$\text{Log } (L) = (4/5) \text{ Log}(T_e) + (22/5) \text{ Log}(M) + \text{constant} \qquad (5.8b)$$

Equation (5.8b), which complements Eq. (5.8a), indicates that Henyey tracks form a sequence of parallel lines within the HR diagram and that these lines are systematically shifted to higher luminosity with increasing mass M.

5.2   The Main Sequence

Slowly evolving along its appropriate (mass-dependent) Henyey track, a protostar's path in the HR diagram eventually intercepts that of the main sequence locus, and it is at this point that fusion reactions involving the conversion of hydrogen into helium are initiated. The protostar is now a star, and it has settled into the longest-lived phase of its existence, evolving on a nuclear timescale (recall Eq. (4.21)).

The main sequence is the backbone of the HR diagram. The main sequence (recall from Sect. 1.3) stretches from the high-luminosity, high-temperature O spectral type stars in its upper left-hand corner to the low-luminosity, low surface-temperature M spectral type (red dwarfs) in its lower right-hand corner. It cuts a serpentine swath across the diagonal of the diagram, and it is where stars are mostly plotted (recall Fig. 1.9). There is over a 90% chance that if you pick a star at random within the galaxy, it will be a main sequence star. The number of stars in a given luminosity-temperature region of the main sequence is determined by the dwell-time in that region (which relates to the initial mass and the appropriate nuclear timescale; recall Eq. (4.22)) and, for a given cohort of newly formed stars, the initial mass function. Massive stars, on the main sequence or elsewhere, are rare because they evolve much more rapidly than low-mass stars, and because nature makes very few of them. This latter result follows from the strong negative power in the mass term displayed by the Salpeter IMF (recall Fig. 5.1). As revealed in Table 1.1, there are on average about a million times more low-mass M spectral type stars than massive O type stars per unit volume of space within the solar neighborhood.

Along the main sequence, the luminosity varies from over a million times greater than that of the Sun to of order one millionth that of the Sun's luminosity. The mass range varies from about $1/10^{th}$ that of the Sun to perhaps as high as 150 times more massive. The temperature range across the main sequence varies from about 2500 K to about 30,000 K (of order ½ that of the Sun's surface temperature to some five times greater). The radii of main sequence stars vary from about 1/10th that of the Sun for the lowest mass M dwarf stars to about 20 times larger than the Sun for the most massive O spectral type stars. The mass-luminosity and mass-radius relationships for main sequence stars are shown in Fig. 1.11.

The internal structure of the main sequence shows a distinct variation with mass—as illustrated earlier in Fig. 4.2—with the lowest mass stars (masses smaller

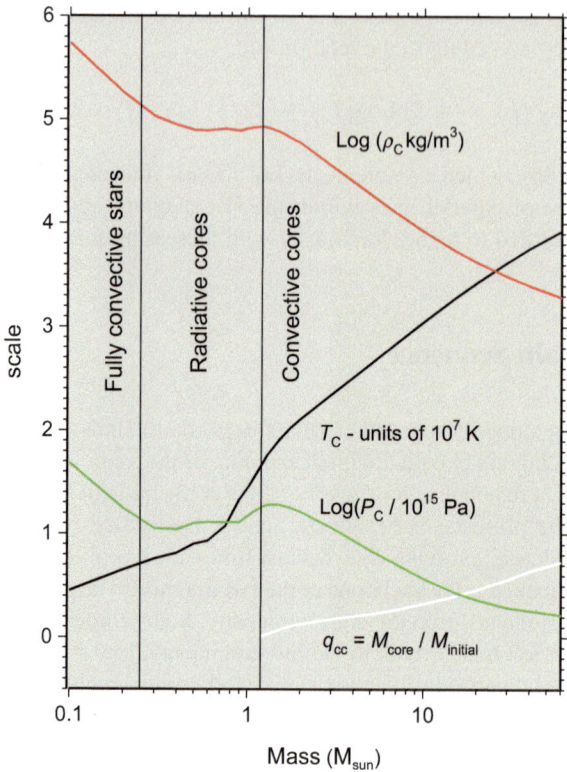

Fig. 5.4 The central temperature (in units of 10^7 Kelvin), the logarithm of the central density (kg/m^3) and the logarithm of the central pressure (in units of 10^{15} Pascal) for main sequence stars plotted against initial mass. Also shown is the variation in the convective core mass fraction $q = M_{core} / M_{initial}$. The vertical lines at 0.25 M$_\odot$ and 1.2 M$_\odot$ correspond to those regions that are fully convective $M < 0.25$ M$_\odot$, stars with radiative cores and convective envelopes and stars with convective cores and radiative envelops. Data from Table 2.2 in Hansen and Kawaler, *Stellar Interiors* (Springer, New York, 1994). (Credit: Author)

than ~ 0.25 M$_\odot$) being fully convective. The upper and lower main sequence stars are further divided according to their internal structure, with upper main sequence stars having convective cores and radiative envelopes. In contrast, the lower main sequence stars have radiative cores and convective envelopes. The changeover from upper to lower main sequence structure occurs at a mass slightly more massive than that of the Sun and is largely due to the onset and then dominance (in terms of energy generation) of the CNO fusion cycle over that of the PP chain (recall Sect. 4.3)—the former *running* at much higher temperatures (and being much more temperature sensitive) than the latter. The temperature and density variation within main sequence stars is illustrated in Fig. 3.7, with the more massive main sequence stars having higher central temperatures. The variation of the central temperature, central density and central pressure as a function of initial mass on the main sequence is shown in Fig. 5.4.

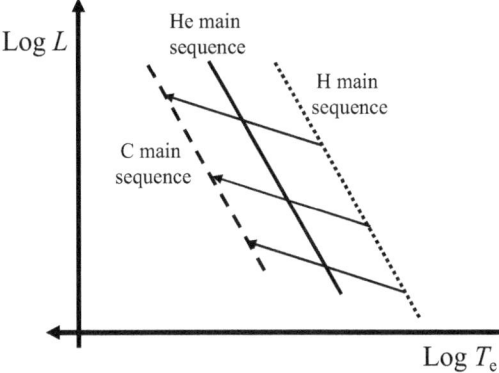

Fig. 5.5 Schematic diagram for fully mixed stars on main sequences with different homogeneous chemical compositions. As the composition from hydrogen (H) to helium (He) to carbon (C) changes, the locus of the main sequence is shifted to successively higher temperatures. The arrowed lines show the schematic variation of the luminosity and temperature at a fixed mass. (Credit: Author)

Table 5.2 Variation of the luminosity, radius, surface and central temperature for a zero-age main sequence solar mass model

Z	$L/L_{\odot}$	$R/R_{\odot}$	T_e (K)	T_c / 10^6 (K)
0.03	0.616	0.880	5456.3	13.35
0.02	0.700	0.886	5615.6	13.37
0.01	0.910	0.913	5903.2	13.68
0.004	1.220	0.940	6262.6	14.19
0.001	1.534	0.911	6736.9	14.65

The location of a star on the main sequence in the HR diagram, in accordance with the Vogt-Russell theorem (recall Sect. 4.4 and Fig. 4.7), will vary according to composition. The trend in variation is qualitatively determined by the mean molecular weight μ (recall Eq. (3.2) and Table 1.2) and is illustrated in Fig. 5.5. Detailed numerical calculations indicate that the luminosity and temperature of a star of given mass and composed entirely of helium are higher and hotter than the same mass star composed entirely of hydrogen; similarly, a star composed entirely of carbon will have a higher luminosity and be hotter than a similar mass star composed entirely of helium.

Exercise 5.4 Go to the *mad star* webpage at http://www.astro.wisc.edu/~towsend/static.php?ref=ez-web and generate a series of solar mass ($M = 1\,M_{\odot}$), zero-age main sequence models with varying values for the metallicity parameter $0.001 \leq Z \leq 0.03$. (1) Construct a table of data (e.g. as given below) and/or plot the data points in an HR diagram. (2) Determine fitting formulae for the variation of the luminosity with log Z and the variation of surface temperature with log Z. The results should look something like those in Table 5.2 below.

Simple linear fitting formula for the luminosity and surface temperatures given in Table 5.2 reveal that

$$(L/L_\odot) = -0.2059 - 0.5581 \log(Z)$$
$$(T/T_\odot) = 0.7325 - 0.1429 \log(Z)$$

where $T_\odot = 5778$ K is the present-day Sun's effective temperature. The luminosity along with the surface and central temperatures systematically increase with decreasing metallicity. This latter effect comes about (seemingly in contrast to Fig. 5.5) since as the metallicity is reduced, so the composition being adopted in the numerical models becomes increasingly helium rich. Fred Adams and Gregory Laughlin (University of Michigan) have speculated upon a new kind of very low-mass "frozen star" that may be produced once the metallicity of the interstellar medium exceeds $Z \approx 0.05$. Objects formed out of such a metal rich medium with masses as small as $0.05\ M_\odot$ (much smaller than the present minimum star mass limit) will be able to initiate low-level hydrogen fusion reactions within their cores, but they will have surface temperatures of order 300 K—the freezing point of water. For further details on the future of stars and star formation see Adams and Laughlin's *The Five Ages of the Universe* (Touchstone Books, New York, 1999).

5.3   Rotation and Magnetic Fields

In the discussion that followed a protostar's evolution from the Hayashi boundary to its main sequence location, no specific mention was made of either rotation or magnetic fields. This may seem surprising since these effects were singled out as being of fundamental importance with respect to the onset of star formation and disk formation (recall Chap. 2). In general (but certainly not always), the effects of rotation upon stellar structure are small. Some measure of its effect can be gauged by considering the ratio of the centrifugal force F_{cen}, arising from rotation, to that of gravitational attraction F_{grav}. Accordingly, for a star of mass M, radius R and equatorial rotation velocity V_{Rot}, we have:

$$\frac{F_{cen}}{F_{grav}} = \frac{V_{Rot}^2/R}{GM/R^2} = \frac{RV_{Rot}^2}{GM} \tag{5.9}$$

Substituting numbers for the Sun yields $F_{cen}\ /\ F_{grav} \sim 2 \times 10^{-5}$, and it is clear from this evaluation that the Sun is spinning well below the critical rate where $F_{cen}\ /\ F_{grav} \sim 1$, and the centrifugal and gravitational forces are nearly equal.

(☺) **Exercise 5.5** Given that the Sun has an equatorial rotational period of 24.47 days, (1) show that the equatorial rotation velocity is $V_{Rot} \approx 2$ km/s, and (2) verify the result given above for $F_{cen}\ /\ F_{grav}$.

The ratio expressed in Eq. (5.9) must be much smaller than one throughout the interior of a star if rotation is to have no significant affect upon its structure. One area where the effects of rotation may become significant, however, is in the region of the nuclear core. In this case rotation could provide a significant fraction of the pressure support, and accordingly the temperature gradient need not be so high there (that is, high enough to service the pressure gradient required to support the weight of overlying layers). In the standard model of the Sun (and recall exercise 4.9), the entire luminosity is generated within the inner 32% or the radius (which encompasses the inner 66% of the mass). For the Sun, therefore, if the centrifugal force is to be no more than 1% of the gravitational force, then

$$\frac{F_{cen}}{F_{grav}} = V_{Rot}^2 \frac{R_{core}}{GM_{core}} = 0.01$$

and accordingly, the maximum core rotation velocity is $V_{rot} = 63$ km/s. In other words, the Sun's spin-rotation period would need to increase to about 0.81 days at the core boundary—a factor of 30 increase over the present equatorial spin-rotation period of 24.47 days. Detailed models of the Sun's interior (constrained by helioseismology[1]) indicate that at the core boundary, the rotation period is about 29 days, indicating that $F_{cen} / F_{grav} \sim 10^{-6}$. Accordingly, rotation is not (generally) taken as an important modifier of the internal structure of Sun-like main sequence stars.

The observed rotation velocities for main sequence stars vary significantly (these stars, recall, are the luminosity class V stars; see Sect. 1.3 and Fig. 1.10), attaining a maximum of order 200 km/s at spectral type B and falling to about 25 km/s at spectral type F (Fig. 5.6). In general, it is found that evolved giant stars of luminosity class III and IV rotate at lower velocities than their main sequence (luminosity class V) counterparts. This is an evolutionary effect associated with the conservation of angular momentum and the increase in stellar radius during the evolution from luminosity class V to luminosity class IV and III.

The angular momentum of a star can be written as $J = I\omega$, where I is the moment of inertia and $\omega = V_{rot} / R$ is the equatorial angular velocity. The moment of inertia is given by the volume integral

$$I = \int_V \rho(r) r^2 \, dV \tag{5.10}$$

and this integral can be evaluated under the polytropic approximation $\rho = \rho_c \theta^n$, where ρ_c is the central density and n is the polytropic index (recall Sect. 3.3). Accordingly, $I = R^5 \rho_c k$, where k is given by the integral equation

---

[1] This method uses observations of the Sun's surface oscillation modes (caused by trapped sound waves) to constrain the variation of the density, pressure and temperature of its deep interior.

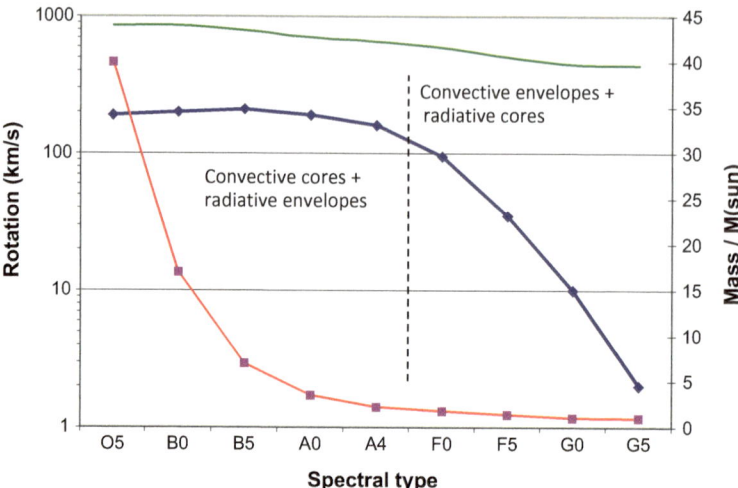

Fig. 5.6 The typical rotation velocity of main sequence stars against spectral type (middle curve). The upper curve indicates the maximum rotation velocity for a given stellar mass (lower curve). The vertical line between the A4 and F0 spectral type stars indicates a transition boundary across which the energy transport mechanism in the outer envelope of a star changes from being radiative (stars hotter than A4) to convective (stars cooler than A4). (Credit: Author)

$$k = \left(\frac{8\pi}{3}\right)\frac{1}{\xi_1^5}\int_0^{\xi_1} \vartheta^n \xi^4 d\xi \tag{5.11}$$

If the initial rotational velocity, central density and radius of a main sequence star are V_{ms}, $\rho_{c,ms}$ and R_{ms}, and those of its evolved, giant configuration are V_g, $\rho_{c,g} > \rho_{c,ms}$ and $R_g > R_{ms}$, then conservation of angular momentum dictates $I_{ms} V_{ms} / R_{ms} = I_g V_g / R_g$, which can be written with the aid of Eq. (5.10) as

$$V_g = V_{ms}\left(\frac{R_{ms}}{R_g}\right)^4\left(\frac{\rho_{c,ms}}{\rho_{c,g}}\right) \tag{5.12}$$

Since (assuming the polytropic index n is constant) all the terms on the right-hand side of Eq. (5.12) are smaller than one, $V_g < V_{ms}$: the velocity of rotation slows as the star evolves.

(☺) **Exercise 5.6** Show that the moment of inertia for a uniform density sphere of radius R and mass M is $I = (2/5) M R^2$. **Hints:** Solve Eq. (5.11) for the $n = 0$ polytrope, and use the definition: $I = R^5 \rho k$.

(☺) **Exercise 5.7** Using the computer program developed in exercise 3.22, find solutions to the integral in Eq. (5.11) when the polytropic index is $n = 1.5$ and $n = 3$.

Magnetic fields, if strong enough, can also provide additional pressure support within the interior of a star. In this case, the fractional increase provided will be of order

$$\varepsilon = \frac{B(r)^2/2\mu_0}{P(r)} = 2\mathrm{x}10^{-11}B(r)^2\left[\frac{P(0)}{P(r)}\right]$$

where $B(r)$ is the magnetic field strength, in Tesla, at radius r, $\mu_0 = 4\pi \times 10^{-7}$ Henries per meter is the permeability of free space and $P(r)$ is the pressure at position r. The central pressure for a standard solar model is of order $P(0) \approx 2 \times 10^{16}$Pa, while that at the nuclear core boundary is $P(r = 0.32\,\mathrm{R}_\odot) \approx 10^{15}$ Pa. Accordingly, ε will be of order 1% if the magnetic field at the nuclear core boundary is of order $B(r = 0.32\,\mathrm{R}_\odot) \approx 5 \times 10^{-4}$ Tesla. At the surface of the Sun, the magnetic field strength is typically of order 10^{-4} Tesla, and current models suggest that the magnetic field strength in the deep interior (below the tacholine[2]) will be smaller than 10^{-5} Tesla. Accordingly, as with rotation, it seems that in general it is quite safe to ignore the effects of any magnetic fields when considering the internal structure of Sun-like main sequence stars.

Having just argued that neither rotation nor magnetic fields are of great importance in determining main sequence structure, a qualification is immediately required. Rotation and magnetic fields are indeed of critical importance in determining the observed characteristics of stars. The appearance of sunspots and flares in the Sun's outer layers, for example, are all related to magnetic field activity and rotation. The point, however, is that these are essentially outer layer *surface* phenomena, rather than deep interior phenomenon. The study of solar activity and the development of an understanding its 22 year magnetic cycle is of vital importance to human society, since such factors directly influence space weather (important for the safe maintenance of space platforms) and drought cycles on Earth.

Not only are rotation and magnetic fields the key mechanisms behind solar flare activity and the solar wind, but it is also the case that a combination of rotation, magnetic fields and mass loss were of fundamental importance in carrying angular momentum away from the young Sun. This has a direct bearing on the formation of planets within the solar nebular and by default our very existence. Such complex matters are not considered here.

In general, mass loss, rotation and magnetic fields do result in the gradual spin-down of stellar rotation periods, and in the early 1970s Andrew Skumanich (at the High Altitude Observatory in Boulder, Colorado) found an empirical law indicating that the rotation velocity of main sequence, Sun-like stars decreases as the inverse square root of their age. A recent calibration of the Skumanich law by Eric Mamajek

---

[2]The tacholine is the transition boundary between the Sun's radiative interior and its convective envelope that occurs at a radius of about $r = 0.7\,\mathrm{R}_\odot$. It marks the transition region whereby the solid-body-like rotation of the interior changes to the differential rotation exhibited in the Sun's outer layers.

(University of Rochester) reveals that the spin period P in days and the system age T_9 in billions of years correlate as: $\text{Log}[P(\text{days})] = 1.07 + 0.5\text{Log}[T_9]$. That such a spin-down (increasing period) with age relationship should exist is entirely reasonable and related to the conservation of angular momentum. Given the angular momentum J is constant, and $J = I\omega = IV_{rot}/R$, and taking $I \approx k\,MR^2$ (recall exercise 5.6), where k is a constant, $J = k\,M\,V_{rot}R = \text{constant}$. Accordingly, since a main sequence star increases in size as it ages (see next section), so V_{rot} must decrease in order that angular momentum is conserved. This picture is somewhat complicated by the fact that Sun-like stars also have relatively strong surface magnetic fields and undergo mass loss; this results in a so-called magnetic braking effect that further slows the rotational velocity.

5.4   Slow Change

While evolution on the main sequence is on a slow nuclear timescale, this does not mean that no change is taking place. Indeed, all three of the intrinsic stellar properties of luminosity, temperature and radius change during the hydrogen fusion stage. An estimate of the change in the luminosity of Sun-like stars can be attained by considering the homology relationships derived in Sect. 4.4. Eq. (4.26) indicates that for those stars in radiative equilibrium subject to a Kramers' opacity law, the luminosity L, mass M and radius R of a star are related as:

$$L = K\mu^{7.5}M^{5.5}R^{-0.5} \tag{5.13}$$

where K is a constant and μ (now extracted from the K_5 term in Eq. (4.26)) is the mean molecular weight. Let us rewrite this equation by eliminating the radius R in terms of the mass and density ρ. Accordingly,

$$L = K\mu^{7.5}M^{5.5}\left(\frac{M}{\rho}\right)^{-1/6} = K\mu^{7.5}M^{5.333}\rho^{0.167}$$

During the main sequence phase, assuming no mass loss takes place, the mass term M remains constant. The characteristic density term will vary as the star evolves, but here we allow a physicist's approximation dodge and note that while the density value may change, the small exponent of 0.167 to which the density is raised will largely offset any changes that occur. Accordingly, this term will also remain essentially constant during the main sequence phase[3] In this manner, the time dependency of the luminosity change over the main sequence phase can be written as

---

[3]This rather heavy-handed approach echoes (somewhat out of context) Eddington's famous comment, 'the mathematics is not there till we put it there'.

$$L(t) = L(0) \left[\frac{\mu(t)}{\mu(0)} \right]^{7.5} \tag{5.14}$$

where the (0) label indicates an initial value. The luminosity change during a star's main sequence is largely determined, therefore, by the change in its mean molecular weight. Provided $\mu(t) > \mu(0)$, the luminosity must increase as a star ages, and the change can be fairly substantive since the relevant power involved is a high 7.5. If we further assume that a star is composed entirely of hydrogen and helium, then Eq. (3.2) gives the mean molecular weight as

$$\mu = \frac{4}{3 + 5X(t)} \tag{5.15}$$

where $X(t)$ is the time varying hydrogen abundance. Initially $X(0) = 0.75$ and $\mu(0) = 0.59$, and if all of the hydrogen within a star could be mixed into the central core, then at hydrogen exhaustion $\mu(t \ @ \ X = 0) = 0.8$, and accordingly the luminosity might increase by as much as a factor of 10. This does not actually happen, as we have previously discussed in Sect. 4.3, because stars do not in general mix all of the available hydrogen into the central nuclear burning regions. We can still describe the change in luminosity as a function of time by looking at the time derivative of the mean molecular weight. From (5.15) we have

$$\frac{d\mu(t)}{dt} = -\frac{5}{4} \mu^2(t) \frac{dX}{dt} \tag{5.16}$$

and, as we saw in Sect. 4.3, we can rewrite the time derivative of the hydrogen abundance in terms of the luminosity and mass, with

$$\frac{dX}{dt} = -\frac{L(t)}{QM} \tag{5.17}$$

where $Q = 0.007c^2$ is the energy released per kilogram of hydrogen converted into helium. If we now differentiate Eq. (5.14) we obtain

$$\frac{dL(t)}{dt} = L(0) \frac{15}{2} \frac{1}{\mu(t)} \left[\frac{\mu(t)}{\mu(0)} \right]^{13/2} \frac{d\mu(t)}{dt} \tag{5.18}$$

and finally, combining Eqs. (5.16) and (5.17) and substituting the result into Eq. (5.18), eliminating the $\mu(t)$ term using Eq. (5.14), and writing $l = L(t) / L(0)$, we find

$$\frac{dl}{dt} = A \, l^{32/15} \tag{5.19}$$

where $A = \frac{75}{8} [\mu(0)L(0)/QM]$ is a constant dependent upon the initial conditions.

(☺) **Exercise 5.8** Verify the derivation of (5.19). This is a large piece of paper exercise.

Equation (5.19) can now be integrated to determine the time variation of the luminosity during the main sequence phase. That is,

$$L(t) = L(0)\left[1 - \frac{17}{15}A\,t\right]^{-15/17} \tag{5.20}$$

(☺) **Exercise 5.9** (1) Verify the derivation of (5.20). This is a large piece of paper exercise. (2) Determine $L(0)$ in Eq. (5.20) by using the Sun's current luminosity and current age: $L(t) = 1\,\mathrm{L}_\odot$ and $t_\odot = 4.56$ billion years. (3) Construct a graph of $L(t)$ versus t (in units of 10^9-years) for the Sun. Compare your results against Fig. 5.7 below. Eq. (5.20) does in fact provide a reasonably good description to the evolution of a Sun-like star during most of its main sequence phase, underestimating the luminosity at 10 Gyr by about 20%.

Not only does the luminosity of a star increase during its main sequence phase, but so too does its radius—just like ageing professors, stars grow bigger around the waist as they age. The increase in the radius is approximately exponential, with

Fig. 5.7 The time evolution of the Sun's luminosity according to Eq. (5.20). Also shown are the results for a detailed solar mass model obtained via the *mad star* webpage (recall exercise 4.23). (Credit: Author)

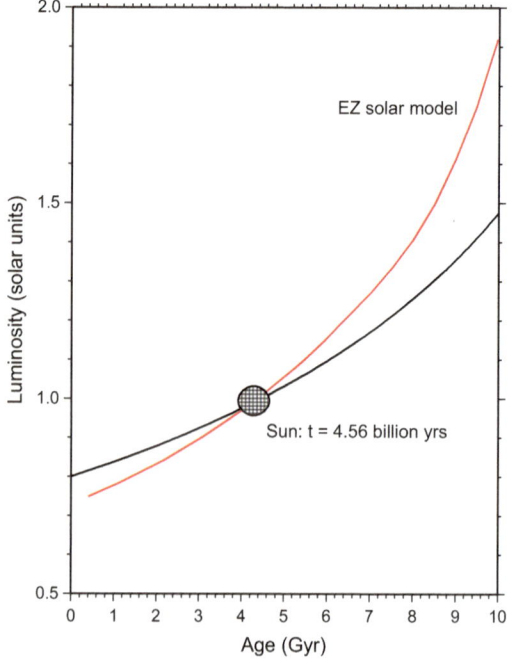

$$R(t) = R(0)\exp\left[\phi \frac{t}{t_{MS}}\right] \tag{5.21}$$

where $\varphi > 0$ and in the case of the Sun, $\varphi \approx \frac{1}{2}$. This increase in size is only modest, however, in comparison to that experienced during the red giant (recall Figs. 1.9 and 1.10) and core helium burning phases.

5.5 Tapping the Pulse

The luminosity, radius and surface temperature of a star gradually change during the main sequence phase. These quantities can also be modulated by pulsation. Figure 1.15 indicates that the instability strip cuts through the mains sequence at spectral types falling between F (on the cool side) and A (at the hotter), this region encompassing the so-called delta-Scuti variables and the rapidly oscillating Ap stars.

The characteristic pulsation period Π can be expressed in terms of the sound travel time across the diameter of a star: $\Pi = 2\,R\,/\,V_s$, where $V_s = \sqrt{\gamma P/\rho}$ is the sound speed and where $\gamma = 5/3$ for a monatomic gas, P is the pressure and ρ is the density. The variation in the pressure, especially in the outer layers where the mass is essentially constant and equal to the total mass M, can be gauged according to the constant density model (recall Sect. 4.2.1). Indeed, in the outer layers of a star the equation of hydrostatic equilibrium gives

$$\frac{dP}{dr} = -\frac{GM\rho}{r^2} = -\frac{4\pi}{3}G\rho^2 r$$

which, when integrated to the surface at $r = R$, where $P(R) = 0$, yields the pressure variation as

$$P(r) = \frac{2\pi}{3}G\rho^2\left(R^2 - r^2\right) \tag{5.22}$$

combining Eq. (5.22) with the expression for the pulsation period Π further yields,

$$\Pi = 2\int_0^R \left(\frac{2\pi}{3}\gamma G\rho\left(R^2 - r^2\right)\right)^{-1/2} dr = \left(\frac{18}{5\,\pi G\rho}\right)^{1/2}\int_0^R \frac{dr}{\sqrt{R^2 - r^2}} \tag{5.23}$$

the integral in Eq. (5.23) may be solved for by substitution. Setting $r = R\sin\theta$, we obtain

$$\Pi = \left(\frac{18}{5\,\pi\,G\rho}\right)^{1/2} \int\limits_{0}^{\pi/2} d\vartheta = \sqrt{\frac{9\,\pi}{10\,G\,\rho}} \tag{5.24}$$

Equation (5.24) indicates, as we would expect, that the pulsation period is of order the dynamic period of the star (as expressed in Eq. (3.9)). Substituting for constant terms, Eq. (5.24) becomes

$$\Pi(\text{days}) = 0.06\sqrt{\frac{\rho_{Sun}}{\rho}} \tag{5.25}$$

where $\rho_{Sun} = 1410$ kg/m^3 is the Sun's average density (Eq. (5.25) above is the derivation for the previously stated Eq. (1.26)).

(☺) **Exercise 5.10** A simplified model of a pulsating star is one in which the entire mass M of the star is envisaged as being concentrated at the center and the surface layer is a thin shell of mass $m << M$ located at a distance r from the center. This is the so-called one-zone model. The star is additionally imagined to a have a uniform pressure P within its interior. Outside of the shell, the pressure is taken as being zero. Introducing the small perturbations δR and δP such $R = R_0 + \delta R$ and $P = P_0 + \delta P$, where the 0 subscript indicates an equilibrium value, show that to first order approximation the shell must undergo simple harmonic motion. Assume that the expansion and contraction of the shell is adiabatic, such that $PV'' = $ constant, where V is the volume of the shell and γ is the ratio of specific heats.

Outline of Solution This is a classic problem, and the idea is to start with Newton's second law of motion—this is analogous to the development used in Eq. (3.7) earlier. Newton's second law indicates that if there is an imbalance between the pressure force at the base of the shell and gravity, then the resulting acceleration of the shell will be

$$m\frac{d^2R}{dt^2} = 4\pi R^2 P - G\frac{mM}{R^2}$$

Now, substitute for the radius and pressure in terms of the perturbed quantities $R = R_0 + \delta R$ and $P = P_0 + \delta P$, and remember that for the equilibrium (0 subscript) terms, we have that $4\pi R_0^2 P_0 = GmM/R_0^2$ and that as it is a linearized approximation only, those quantities involving δR and δP are kept—terms like $(\delta R)^2$ and $(\delta R)(\delta P)$ are ignored. See additional hints below. This manipulation will result in the following equation

$$m\frac{d^2(\delta R)}{dt^2} = 2G\frac{mM}{R_0^3}(\delta R) + 8\pi R_0 P_0(\delta R) + 4\pi R_0^2(\delta P)$$

Now, using the adiabatic condition, $PV^{\gamma} = $ constant, with the volume being given by $4\pi R^3/3$, so $PR^{3\gamma} = $ constant, and accordingly, $\delta P/P_0 = -3\gamma\, \delta R/R_0$. Eliminating δP from the linearized equation, the end result is a second-order differential (wave) equation in δR, such that:

$$\frac{d^2(\delta R)}{dt^2} = -(3\gamma - 4)G\frac{M}{R_0^3}(\delta R)$$

Provided $\gamma > 4/3$, the wave equation has a solution in terms of simple harmonic motion, with $\delta R = A\sin(\omega t)$, where A is the pulsation amplitude and ω is the angular pulsation frequency given by

$$\omega^2 = (3\gamma - 4)\,G\frac{M}{R_0^3} = \frac{4\pi}{3}(3\gamma - 4)\,G\rho_0$$

The pulsation period will be $\Pi = 2\pi/\omega$, and this is fully compatible with the result derived for the constant density model and given in Eq. (5.25).

Additional hints This is a big sheet of paper question, but the algebra is relatively straightforward. Use the binomial theorem to obtain the linearized approximation to $(R_0 + \delta R)^{-2} \approx (1 - 2\delta R/R_0)/R_0^2$. *Comment:* Notice that the amplitude A cancels out in this approximation, and accordingly the one-zone linearized model makes no prediction as to the amplitude of pulsation. To get at this number, a more detailed (non-linearized) calculation is required.

5.6   Mass Loss

The qualitative effect of mass loss upon the evolution of massive stars was described in Sect. 4.5.4. At some level, all stars lose mass into space from their outer regions. Most of the time, at least for low-mass main sequence stars, the amount of material lost through a stellar wind is generally insignificant and has no profound effect upon stellar evolution—the Sun's mass loss rate, recall, is some $10^{-14} M_{\odot}$/year. This situation, however, is not the case in later post-main sequence phases, where the mass loss can be substantial. Indeed, it is mass loss that is responsible for determining the end fate of intermediate mass stars.

That the mass loss rate is likely to depend upon the luminosity of a star makes physical sense, since each photon leaving the photosphere carries away momentum corresponding to its energy $E = hf$, where f is the frequency and h is Planck's constant, divided by the speed of light c. The total momentum of all the photons emitted per second by a star will accordingly be its luminosity divided by the speed of light: L/c. This photon momentum can couple with the gas and dust situated in a star's outer layers, thereby producing a mass outflow, which is expressed in terms of the mass loss rate and the terminal velocity V_∞, giving,

$$\frac{L}{c} = \alpha \dot{M} \; V_\infty \tag{5.26}$$

where α is an efficiency constant $0 < \alpha < 1$ and characteristically V_∞ is of order 10 to 100 km/s. The full picture is more complicated than just described, but the argument is a fundamental one that establishes the basic idea. Various empirical mass loss laws have been introduced over the years, and they all tend to have slightly different modification factors and a range of dependencies upon physical stellar parameters such as mass, radius and luminosity. One commonly used empirical mass loss formulation is that developed by Dieter Reimers (University of Hamburg) in the mid- to late-1970s. This formula gives the mass loss rate as

$$\dot{M} = -\eta \frac{LR}{M} \tag{5.27}$$

where η is a constant. Problematic with the use of Reimers formula, however, is deciding what exactly the value of the constant η should be. Typically, when solar units are used for the mass, radius and luminosity, then $\eta = 2.5 \times 10^{-13}$ and the mass loss rate is in solar masses per year. Indeed, as will be discussed in Sect. 5.7 below, the future existence of the Earth when the Sun passes through its so-called asymptotic giant phase is highly sensitive to the value of η. Specifically, a series of solar models constructed by I-Juliana Sackman, Arnold Boothroyd and Kathleen Kraemer found that the Sun expanded to a size larger than Earth's orbit (to a radius greater than $1\,AU \approx 213\,R_\odot$) if the mass loss parameter was $\eta = 1.6 \times 10^{-13}$. If the mass loss parameter was taken to be $\eta = 2.5 \times 10^{-13}$, however, the Sun did not expand to a radius that engulfed the Earth in its orbit. The future existence of the Earth in the later stages of the Sun's evolution hangs upon a knife-edge with a factor of just 1.5 in the value of η deciding whether it will survive or be vaporized.

(☺) **Exercise 5.11** Rewrite Reimers' mass loss law for main sequence stars in terms of the luminosity only. Use the mass-luminosity law (Eq. (1.24c)) and the mass-radius law (Eq. (1.25b)).

(☺) **Exercise 5.12** Use Reimers' mass loss formula to estimate how long it will take for the Sun during its advanced red giant phase to be reduced to half of its present mass. Take the luminosity to be constant and equal to 5000 $L_\odot$ and assume a temperature of 3000 K. *Hints:* (1) First calculate the radius R from the given luminosity and temperature—use the Stefan-Boltzmann Eq. (1.14). (2) Taking L, η and R to be constant (and remember M, L and R are in solar units and that the time is in years), Reimers' Eq. (5.27) can be directly integrated to yield

$$\int_1^{0.5} m\, dm = -(\eta L R) \int_0^t dt$$

and accordingly,

$$t = \frac{3}{8\,\eta\,L\,R}$$

With $\eta = 2.5 \times 10^{-13}$, the latter equation provides an expression for the time over which the Sun will lose half of its mass.

The effect of pulsation on mass loss can be gauged by taking the time differential of Eq. (5.25). Using the dot-notation to signify a time derivative, we have

$$\dot{\Pi} = \left(\frac{\Pi}{2}\right)\left(3\frac{\dot{R}}{R} - \frac{\dot{M}}{M}\right) \tag{5.28}$$

As the rate of mass loss is $\dot{M} < 0$, so $\dot{\Pi} > 0$, indicating that the period will increase provided $\dot{R} \geq 0$. This latter condition is generally satisfied—even taking $\dot{R} = 0$ provides a lower limit to the positive change in the pulsation period. Combining Eqs. (5.25) and (5.27) now yields

$$\dot{M} = -\eta \left(\frac{T}{T_{sun}}\right)^4 \left(\frac{\Pi}{0.06}\right)^2 \tag{5.29}$$

where we have substituted for the temperature T of the star from the Stefan-Boltzmann law. Since the temperature of a variable star does not vary significantly over its pulsation cycle, Eq. (5.29) indicates that the mass loss rate should be proportional to the square of the pulsation period. Indeed, the Sun will eventually enter as a red giant into a Mira variable phase (during its advanced helium fusion stage), and at this time, as a bloated pulsational variable, it will lose nearly half of its present mass (see Fig. 1.15 and Sect. 5.7 below).

We will see later (Sect. 5.9) that there is a maximum mass (the so-called Chandrasekhar limiting mass: $M_{CH} = 1.44\ M_\odot$) above which the gravitational collapse of an evolved stellar core (after the exhaustion of helium as a 3α fusion reaction fuel) is inevitable. Below this maximum, however, the core of a star can evolve into a state in which electron degeneracy dominates the pressure support and an equilibrium white dwarf structure comes about. The maximum core mass condition determines whether a star will end its existence as a stable, everlasting[4] white

---

[4]Nothing, of course, lasts forever, but in principle a white dwarf can evolve to a temperature of absolute zero within its interior and still remain stable. The ultimate end fate for an aged white dwarf will be disruption and consumption by a massive black hole—a process that will take place in the very deep future of the universe.

dwarf, or utterly destroy itself as a supernova. That mass loss is involved in the process of forming white dwarfs is evidenced by the fact that some such objects have masses as small as $0.2\,M_\odot$. Using Eq. (4.22), the main sequence lifetime of a star with this mass is about 112 billion years, which is some eight times larger than the estimated age of the universe. Clearly, to be stripped down to the lowest masses observed for white dwarfs requires the removal of substantial amounts of matter, principally from the outer envelope.

In order to see what mass range of stars might eventually produce white dwarfs, let us consider a star of initial mass M_0 that loses mass from its surface at a rate proportional to its luminosity L: $\dot{M} = -\alpha L$, where $\alpha \sim 10^{-14}$ kg/J is a constant[5]. As the star evolves, fusion reactions will convert some of the core mass into energy (recall Sect. 4.3) at a rate Q (Joules / kg), and accordingly the core mass M_c at time t will be given by

$$M_c(t) = \frac{L}{Q}t \tag{5.30}$$

Now, the envelope of the star loses mass from both its inner boundary where it adjoins the core and at its surface, and accordingly

$$\dot{M_e} = -\dot{M_c} + \dot{M} = -\frac{L}{Q} - \alpha L = -L\left(\frac{1}{Q} + \alpha\right) \tag{5.31}$$

We can integrate Eq. (5.31) using the boundary condition that $M_e = M_0$ at time $t = 0$. The variation in the envelope mass is therefore

$$M_e(t) = M_0 - L\left(\frac{1}{Q} - \alpha\right)t \tag{5.32}$$

The core mass at the time that the envelope mass is fully lost can be obtained by eliminating t between Eqs. (5.30) and (5.32), and this step yields

$$M_c = \frac{M_0}{1 + \alpha Q} \tag{5.33}$$

Setting the core mass to be no more than the Chandrasekhar limiting mass $(M_c = M_{CH})$, we have from Eq. (5.33) that

$$M_0 < M_{CH}(1 + \alpha Q) \tag{5.34}$$

---

[5]This characteristic value is based upon the observation that during a star's most dramatic mass loss phase (as a bloated supergiant), the mass loss rate is of order $10^{-7}\,M_\odot$/yr and the luminosity is of order 5000 $L_\odot$.

The question now is what are the characteristic values of α and Q? In order to convert a kilogram of solar material into carbon and oxygen, the required binding energy per nucleon (the energy needed to split a nucleus into its component neutrons and protons) will be of order 7.68 MeV and 7.98 MeV respectively—say, 7.85 MeV (these are the sorts of numbers that we look up on the web when needed). Given that there are of order $N = 1/m_P \approx 6 \times 10^{26}$ nucleons per kilogram of solar material, the total energy released per kilogram by fusion reactions (Q) will be of order $7.85 \times 10^6 \times 6 \times 10^{26}$ eV / kg $= 7.5 \times 10^{14}$ Joules /kg. Accordingly, with $\alpha \sim 10^{-14}$ kg/J, Eq. (5.34) indicates that M_0 could be as high as ~12 $M_\odot$. Detailed numerical studies, using a more realistic mass loss formulation (i.e. that due to Reimers') find that the upper initial mass limit for producing a white dwarf is more like $M_0 \leq 8\,M_\odot$.

(☺) **Exercise 5.13** (1) How long will it take the Sun to shed, given its present (solar wind) mass loss rate of $10^{-14}\,M_\odot$/year, an amount of mass equivalent to that of the Earth? (2) What is the mass flux (kg/s/m^2) at the Earth's orbit due to the solar wind? *Hints:* For (1), take the Earth's mass to be 6×10^{24} kg. For (2), take the Earth's orbital radius to be 1.5×10^{11} m.

Mass loss is generally thought of in terms of material being lost into space from the surface of a star. However, more extreme possibilities can be imagined and may exist in the universe (somewhere). The more extreme situation in mind here is that in which matter is lost to a star by falling out of all causal contact with our universe. This is the situation for matter falling into a black hole. The mass distribution of black holes is unknown at the present time. Observations indicate that supermassive black holes exist at the centers of galaxies—our own Milky Way galaxy has a central black hole with a mass of some four million solar masses (see exercise 5.14). Observations at the Laser Interferometer Gravitational-Wave Observatory (LIGO) have also found evidence for the existence of black holes in the mass range from 10 to 30 $M_\odot$. Black holes with substellar masses are suspected to exist, as first discussed by Stephen Hawking in the 1970s, but to date no clear evidence for any such objects has been found.

(☺) **Exercise 5.14** Figure 5.8 (left) shows a small 2×2 arc second region centered on the compact radio source Sagittarius A* at the center of our galaxy. To the right is shown the orbit of star S2 over a 10 year interval of data collection. The orbit of star S2 is elliptical about the central galactic black hole, with an orbital semi-major axis of 5.602 light days and an orbital period of 16.0518 years. Determine the mass of the black hole located within Sgr A*. *Hint:* Use Kepler's third law (Eq. (1.15)), watch your units (convert light days to AU) and note that the mass of the central black hole greatly exceeds that of the star S2.

The proto-typical stellar mass black hole system is that identified as Cygnus X-1. First detected as a strong X-ray source in the mid-1960s, it is now known that the emission is from a binary system containing a massive ~30 $M_\odot$ O spectral type star (HDE 226868) and a 15 $M_\odot$ black hole. The two objects orbit the system's center of mass every 5.6 days and are located some 1.9 kpc from the solar system. The X-ray

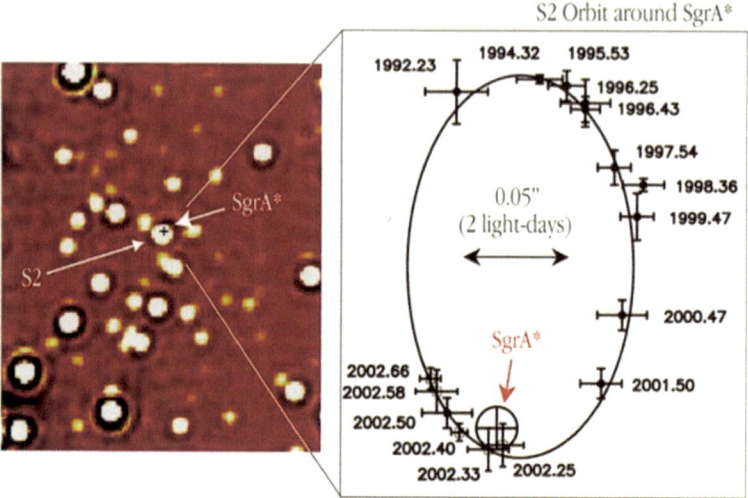

Fig. 5.8 (Left) infrared image of the region surrounding strong radio source Sgr A* at the galactic center. (Right) The orbit of star S2 about Sgr A*. Image courtesy of ESO

emission is produced in a small accretion disk that has formed about the black hole; the disk being fed by matter lost from the O star companion. The estimated mass loss rate from HDE 226868 is some $2.5 \times 10^{-6}\, M_\odot/yr$.

(☺) **Exercise 5.15** Use Kepler's third law, as given in Eq. (1.15), to determine the distance between HDE 226868 and its black hole companion. Compare this separation to that of the planets in the inner Solar System.

Only a handful of binary systems containing black holes have been identified. The closest such system is V616 Monoceros, located about 1 kpc away, which contains an estimated $7\, M_\odot$ mass black hole. After V616 Mon we encounter Cygnus X-1, and then V404 Cygnus at about 2.4 kpc distance. How many isolated black holes there might be in this same region of space is unknown at the present time. One way that an isolated black hole might be identified, however, is through its direct interaction with a star.

Astrophysicists Kip Thorne and Anna Zytkow considered in 1977 the situation in which an isolated neutron star collides with and is captured by a red giant star. Such a collision and capture are only likely to take place within the crowded core region of a globular cluster, or within a close binary system. Nonetheless, should such a capture take place, material accreted from the red giant core onto the neutron star will undergo nuclear fusion reactions and drive the red giant to become a red supergiant. Not only this, but it may additionally be the case that the neutron star accretes enough material to collapse into a black hole. In this case, material falling across the black hole event horizon is lost from our universe—although its gravitational influence is still felt by the surrounding star. Matter falling towards the black hole will become compressed and heated, and sufficient energy might be released to power the parent star. Ultimately, however, the black hole will grow and effectively consume the entire star from the inside out. It was suggested in the 1970s that the solar neutrino problem (as it was then known; recall exercise 4.16) might be solved by invoking the existence of a small black hole at the Sun's center. Such a black hole would enable the Sun to *run* at a lower central temperature, and this would in turn reduce the number of fusion reaction neutrinos generated. This explanation is no longer required, since it is now known that the solar neutrino problem was in fact due to a misunderstanding about neutrino properties rather than a misunderstanding about the internal structure of stars.

(☺) **Exercise 5.16**   Imagine that a small black hole of initial mass $M_{BH,0}$ has come to rest at the center of the Sun. Assuming that the accretion rate ($\Delta M / \Delta t$) of material onto the black hole is such that the accretion luminosity L_{acc} is at the maximum possible Eddington luminosity L_{Edd}, (1) show that the black hole mass increases exponentially. (2) Assuming an efficiency factor f for the conversion of the accretion energy into radiative energy, determine how long will it be before the Sun is fully consumed, given $M_{BH,0} = 10^{12}$ kg and $f = 0.1$. *Hints:* Recall Sect. 1.1, Eq. (1.4) and Sect. 4.5.1. If a quantity ΔM of matter falls into the black hole in time Δt, then Einstein's mass energy-equivalence formula $E = M c^2$ indicates that the energy liberated will be ΔE, and $L_{acc} = f \Delta E / \Delta t = f (\Delta M / \Delta t) c^2$.

(☺) **Exercise 5.17: Term Paper Topic**   A lawsuit was brought against CERN prior to initiating operations of the Large Hadron Collider (LHC) in 2008. It was suggested in the legal brief that miniature black holes might be created within the collider's beams and that such black holes might sink to the center of the Earth and accumulate, ultimately growing to a size that would destroy the Earth. Look into the circumstances surrounding this particular court case [Hawaii District Court case 1:2008cv00136] and develop counterarguments to show that the concerns were completely unfounded.

5.7   Stellar Evolution and the Future Sun

There is no set of first-principle analytic equations that describe stellar evolution.[6] In order to unweave the life history of a star, one has to make recourse to detailed numerical methods and advanced computational techniques. For all this, in the modern era of superfast computers, the numerical solution process is rapid, reliable and robust, and many codes are available for public use (recall exercise 5.4). Figure 5.9 shows a set of five stellar evolution histories in the HR diagram for

---

[6]This being said, analytic fitting formula can be obtained from the results of detailed stellar models. By far the most useful and comprehensive set of such analytic formulae is that found in the research paper by P. Eggleton, M. Fitchett and C. Tout. The distribution of visual binaries with two bright components. The *Astrophysical Journal*, 347, 998–1011 (1989) (and associated addenda).

Fig. 5.9 Evolutionary tracks in the HR diagram for stars of initial mass 0.1, 1, 5, 10 and 50 $M_\odot$. The Sun's evolutionary track is shown in red, while that of the 0.1 $M_\odot$ star is shown in blue. The dashed lines from the 1 and 5$M_\odot$ tracks indicate the evolution towards the planetary nebula phase, where the cores eventually evolve onto their appropriate white dwarf cooling track. Model data from http://www.astro.wisc.edu/~towsend/static.php?ref=ez-web. (Credit: Author)

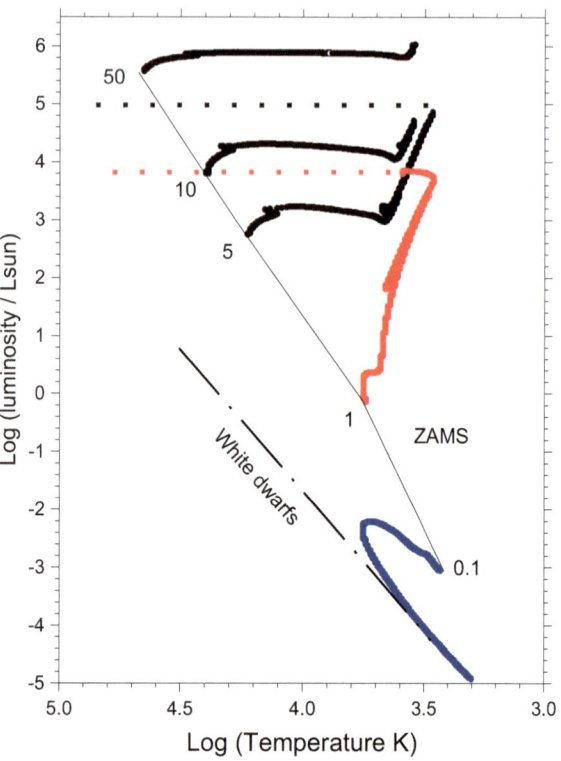

stars of initial mass 0.1, 1, 5, 10 and 50 $M_\odot$. Each track has been computed until something catastrophic happens (that is, some physical situation is encountered that the computer code is not designed to handle). This means the attainment of a pre-supernovae state in the case of the 10 and 50 $M_\odot$ models, and the onset of deep and rapid pulsation cycles in the case of the 1 and 5 $M_\odot$ models. Each calculation begins with a main sequence configuration and then follows the journey of the star through the HR diagram as it ages. The main sequence life time is given by Eq. (4.22) and accordingly, the 50 $M_\odot$ star has evolved off the main sequence in a time equivalent to just 3/1000ths of the 5 $M_\odot$ star; or just 6/100,000ths of the main sequence lifetime of the Sun.

As described in Chap. 4, stars on the mains sequence generate their internal energy through hydrogen fusion reactions. Stars are mostly made of hydrogen (well of order 75% by mass) so the main sequence phase is by far the longest-lived phase in a star's life. As the hydrogen in the central core is used up, however, the star gradually increases in size and luminosity (but its surface temperature decreases). During this phase, the changes in a star's physical characteristics are relatively small (recall Sect. 5.2). With hydrogen exhaustion, however, more dramatic changes take place. For stars like the Sun, the hydrogen-exhausted core will be become isothermal, and a hydrogen-fusing shell source will develop at its outer boundary. At this time, the star begins to evolve into a red giant, with the core contracting and growing in temperature and density and the envelope expanding. For stars with initial masses in the range from 0.8 to 2 $M_\odot$, the onset of helium fusion reactions is rapid and dramatic, earning it the name 'helium flash'. In the case of the Sun, this phase will see its luminosity briefly exceed several thousand $L_\odot$. As fusion reactions involving helium fusion settle down, the star adopts a lower luminosity. In the case of the Sun, this will be of order 50 $L_\odot$.

Stars with initial masses greater than $2M_\odot$ evolve towards a red giant configuration without undergoing a helium flash, but those stars less massive than about $10\,M_\odot$ will undergo blue-loop excursions (evolution towards higher surface temperatures and smaller radii) during their core helium burning phase. At these times, such stars will enter into the instability strip (recall Figs. 1.13 and 1.15), becoming RR Lyrae variables in the lower mass case and Cepheid variables in the higher mass case. With the exhaustion of helium within a star's central core, if its initial mass was less than about $8 - 10\,M_\odot$, it once again evolves towards a higher luminosity, lower surface temperature and larger radius, and enters into its so-called asymptotic giant branch (AGB) phase. This phase is characterized by the onset of deep breathing cycles in which extensive convection zones develop in the outer envelope, and the luminosity varies in a complex manner as a helium burning shell (situated above the now-dormant core) develops and interacts with a slightly further out hydrogen burning shell. At this stage, such stars will be identified as Mira variables (recall Fig. 1.15). A star in its AGB and post-AGB phase undergoes substantial mass loss via powerful stellar winds, and the envelope eventually develops into an extended planetary nebula, with the central core becoming a white dwarf.

Stars more massive than $8 - 10\,M_\odot$ evolve in a much more direct manner in comparison to their lower mass companions. Such stars can evolve within their cores temperatures and densities at which advanced fusion reactions (beyond that of the helium triple-α reactions) can take place, and indeed, fusion reactions all the way to the generation of iron occur. The development of an iron core, however, marks the end of energy generation via fusion reactions, and massive stars eventually (and very rapidly) undergo supernova disruption (see Sect. 5.8 below for details). While intermediate and massive stars undergo evolution towards the red giant region in the HR diagram, the very lowest mass stars show no such tendency. Indeed, these stars initially evolve slightly small radii and higher surface temperatures. These very low-mass stars are essentially fully mixed, and the key mechanisms responsible for red giant formation do not come into operation. With the exhaustion of hydrogen, the lowest mass stars evolve directly towards the white dwarf cooling region in the HR diagram.

Figure 5.10 shows a summary of the general results derived from numerous stellar evolution calculations. Provided that the initial mass is greater than $0.08\,M_\odot$, a *bona fida* star can form. It will settle onto the main sequence, converting hydrogen into helium via the PP chain if its mass is smaller than about 1.1 $M_\odot$, and via CNO cycle reactions if its mass is larger than about 1.1 $M_\odot$ (recall Sect. 4.3; note the kink in the ZAMS in Fig. 5.9). Objects less massive than 0.08 $M_\odot$ are not destined to become stars, but rather objects identified as brown dwarfs. Such objects

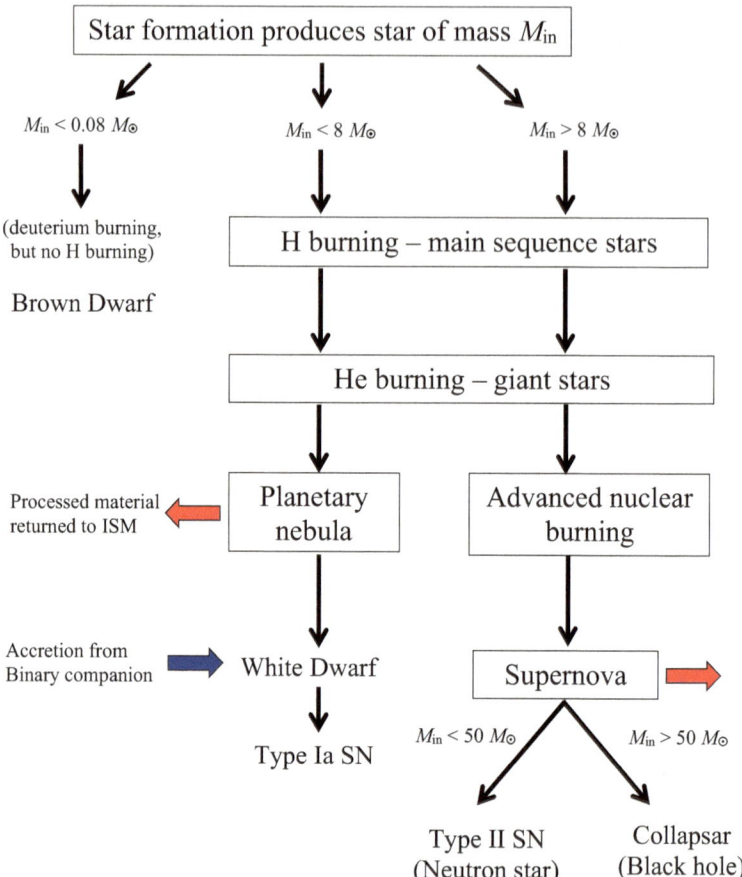

Fig. 5.10 A schematic flowchart indicating the nuclear fusion reactions accessible to and end states attained for stars of increasing initial mass. Red arrows indicate stages at which matter is lost from a star via a stellar wind or supernova disruption, while the blue arrow indicates accretion from a binary companion. (Credit: Author)

may undergo a brief phase in which deuterium fusion takes place, but they do not initiate long-lived hydrogen fusion reactions within their cores. Deuterium fusion proceeds according to the reaction $D + P \Rightarrow {}^{3}He + \gamma$ and liberates some $Q_D = 5.494$ MeV of energy per reaction. The deuterium-burning lifetime t_D follows in the same manner as that used for the main sequence lifetime (Eq. (4.21)), and for a mass M and luminosity L, the lifetime is

$$t_D = q\left(\frac{XM}{m_p}\right)\left[\frac{D}{H}\right]\left(\frac{Q_D}{L}\right) \qquad (5.35)$$

where [D/H] is the deuterium to hydrogen ratio by number and q is the mass fraction of the brown dwarf over which deuterium is consumed. The initial luminosity of a 0.08 $M_\odot$ brown dwarf is as high as 0.1 $L_\odot$, and accordingly the deuterium-burning lifetime is of order a few million years.

(☺) **Exercise 5.18** Given a cosmic deuterium-to-hydrogen ratio by number of [D/H] $= 2.5 \times 10^{-5}$, verify that $t_D \approx 10^6$ years when $M = 0.08$ $M_\odot$ and $L = 0.1\, L_\odot$. Since brown dwarfs are fully convective when first formed, take $q = 1$; also, take the hydrogen mass fraction to be X $= 0.7$. **Comment:** the cosmic deuterium-to-hydrogen ratio is determined by primordial nucleosynthesis and the conditions that prevailed just a few minutes after the universe came into existence. Accordingly, at that time, some 25 deuterium atoms were created per one million hydrogen atoms. The D/H ratio for the gas giant planets within our Solar System is the same as the cosmic ratio, but intriguingly, that of Earth's oceans is slightly lower. This latter observation relates to the origins of Earth's water supply through impacts from asteroids and cometary nuclei.

One of the key points of the detailed numerical studies has been the identification of a critical initial mass at about 8 $M_\odot$. This critical mass determines the end phase of the star. For stars of initial mass less than about 8 $M_\odot$, the end phase will be marked by extensive mass loss (leading to the formation of a planetary nebula) with the core eventual evolving into a white dwarf. For stars initially more massive than 8 $M_\odot$ the final end phase will be that of catastrophic supernova disruption (see Sect. 5.8 below).

Figure 5.11 shows the time variation of the Sun's radius and luminosity, and these guide us in determining the future for life on Earth. The answer is that there is no very long-term future. There are two key issues here: (1) the attainment of a temperature at which a runaway greenhouse effect comes into play and (2) the growth of the Sun to such an extent that it literally envelopes the Earth within its orbit and thereafter consumes it. The first issue will signal the end of the biosphere, since it results in the loss of Earth's oceans, while the second issue will signal the end of the Earth itself. While the former condition will definitely come about in roughly 2 to 3 billion years, when the Sun is some 10 to 15% more luminous than at present, the latter fate may or may not occur, with the outcome being a sensitive function of the Sun's actual mass loss during its red giant and asymptotic giant branch phases of evolution.

(☺) **Exercise 5.19** Return to exercise 5.9 and Eq. (5.20). Show that at the present epoch, the Sun's luminosity is increasing by about 7.5% per billion years. **Hint:** Evaluate the differential $d(L/L_\odot)/dt$ at $t = 4.5 Gyr$ (the age of the Solar System).

The core helium-burning lifetime of a star is much shorter than that of its main sequence lifetime. This comes about primarily because there is a relatively small total amount of helium compared to hydrogen, and the triple-α reaction only provides about $1/4^{th}$ of the amount of energy per reaction provided by hydrogen fusion ($Q_{3\alpha} = 7.275$ MeV versus $Q_{PP} = 27.732$ MeV). Additionally, the luminosity L_{He}

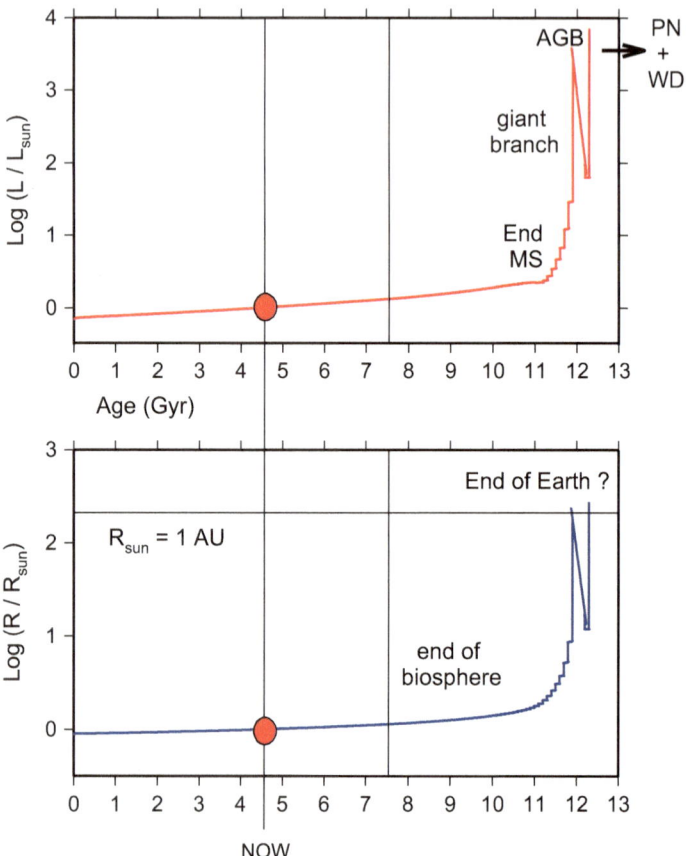

Fig. 5.11 Time evolution of the luminosity and radius of the Sun. Model data from http://www. astro.wisc.edu/~towsend/static.php?ref=ez-web. (Credit: Author)

during the core helium-burning phase is much higher than that adopted by the same mass star on the main sequence. The helium burning lifetime t_{He} is accordingly

$$t_{He} = \left(\frac{M_{core}/M}{m_{He}}\right)\frac{Q_{3\alpha}}{L_{He}} \qquad (5.36)$$

where M_{core} is the mass of the helium burning core and m_{He} is the mass of the helium atom nucleus ($m_{He} \approx 4m_p$). For the Sun, the core mass will be of order $M_{core} \approx 0.5\,M_\odot$ and the luminosity will be of order $L_{He} \approx 45\,L_\odot$. Accordingly, the helium-burning lifetime will be $t_{He} \sim 10^{16}$ s, or 320 million years (about 3% or its main sequence lifetime).

It is after the exhaustion of helium within the core of the Sun that the fate of the Earth will be decided. As the helium-burning shell interacts with the hydrogen-burning shell above it, a series of stronger and stronger thermal pulses begins to develop, and as the Sun ascends to the top of its AGB it will take on the appearance of a Mira variable. At this time, the Sun's outer envelope will become dynamically unstable, with an increasingly strong stellar wind reducing its total mass, eventually by nearly a factor of two. The outer envelope will grow to surround the Sun as an extended gas shell, and the core will rapidly evolve towards a white dwarf configuration. The post-AGB evolution of the Sun is very quick (amounting to a few hundred thousand years) and at very high luminosity. Indeed, the detailed computer models indicate that the post-AGB luminosity of a star is determined according to the mass of its carbon-oxygen core, with $L/L_\odot = 6 \times 10^4 (M_{core} - 0.5)$, where M_{core} is expressed in solar mass units. For the Sun, this formula indicates: $L_{Post-AGB} = 3000$ $L_\odot$. The Sun will attain its maximum size while located at the tip of red giant branch (or possibly at the tip of its AGB), and it is this maximum radius that is sensitive to the actual mass loss rate that it experiences.

In a series of detailed model calculations,[7] K. P. Schröder (Universidad de Guanajuato, Mexico) and R. C. Smith (University of Sussex, England) determined that the maximum radius obtained by the Sun amounts to some 256 $R_\odot$ or about 1.2 AU. With this amount of expansion, the Earth will be consumed by the future Sun in some 7.6 billion years from the present. This being said, the analysis by Schröder and Smith is based upon a number of assumptions and approximations and upon a mass loss algorithm that may or may not be correct for the future Sun. The mass loss from the Sun may just result in the Earth surviving a fiery end. Once again, the conservation of orbital angular momentum comes into play—recall the caption to Fig. (2.5). The Earth's orbital angular momentum is $\Lambda_E = M_E\, a\, V_E$, where M_E is the Earth's mass, V_E the Earth's orbital velocity and a is its orbital radius. Assuming the Earth orbits the Sun in a circular fashion, and recalling exercise 3.23, its orbital velocity will be $V_E^2 = GM_{Sun}/a$. Given, however, that the Sun's mass is changing during its red giant and asymptotic giant branch phases, in order to conserve angular momentum (that is, to keep $\Lambda_E = $ constant), the Earth's orbital radius will have to increase. Accordingly, at any specific time t, the product $M_{Sun}(t)\, a(t) = $ constant. Schröder and Smith find that at the time of maximum radius, the Sun will have lost about 1/3rd of its current mass, and accordingly at that time the Earth's orbital radius will have increased to $a = 1$ AU $[M_\odot/2M_\odot/3] = 1.5$ AU. In this case, the outward expansion of the Earth's orbit due to the loss of mass from the Sun carries it to a region that is not enveloped by the Sun's bloated giant body. In some sense it is a game of cat and mouse that will play out with respect to the Earth's ultimate survival or consumption by the Sun. While the Earth may survive the Sun's red giant and AGB phases, planets Mercury and Venus will not, and they are doomed to a fiery destruction.

(☺) **Exercise 5.20**   Derive the equation $M_{Sun}(t)\, a(t) = $ constant. *Hint:* Start with the equation $\Lambda_E = M_E\, a\, V_E = $ constant, then substitute for V_E, and assume that the Earth's mass remains constant. *Comment:* During the Sun's post-AGB phase, the

[7]See Schröder and Smith, Distant future of the Sun and Earth revisited, *Monthly Notices of the Royal Astronomical Society*, 386, 155–163 (2008).

Earth, given that it actually survives destruction, will nonetheless be heated to such an extent that the outer kilometer (or so) of its crust will be lost through sublimation. Not only this, but the strong solar wind during the post-AGB phase will remove, in short order, the Earth's entire atmosphere—there is no escaping the conclusion that all life on Earth will be fully snuffed out once the Sun enters its later stages of evolution. *Aside:* It has been suggested that one good location to look for the signs of extraterrestrial civilizations and/or extraterrestrial mega-structures (such as large space-based colonies, massive spaceflight activity and terraforming) is around those stars at or near the end of their main sequence phase, such civilizations being forced to leave their home worlds because of the impending red giant expansion of their previously nurturing main sequence parent suns.

The post-AGB phase of the Sun carries it to the planetary nebula region of the HR diagram. At this time it will have a surface temperature of order 30,000 K, a radius of order $2\,R_\odot$ and a luminosity of some $3000\,L_\odot$. The flux of ultraviolet photons from the Sun at this time will be sufficient to ionize a large volume of space, and indeed, the remnants of its dispersed outer envelope will now take on the appearance of a planetary nebula. This short-lived phase soon sees the Sun evolve towards a white dwarf cooling configuration. Keeping now a constant radius of some $0.01\,R_\odot$, the Sun will begin to cool off and become increasingly less luminous.

With no access to internal energy generation via fusion reactions, a white dwarf slowly cools off over time. Electron conduction is so efficient within the interiors of white dwarfs, however, that they are effectively isothermal, with the temperature only dropping significantly in their thin, non-degenerate outer layers. The internal thermal energy E_{th} of a white dwarf, for a pure helium composition, is determined as

$$E_{th} = \frac{3}{2} N k T = \frac{3}{2} \left(\frac{M}{m_p} \right) \left(\frac{1}{2} + \frac{1}{4} \right) kT = \frac{9}{8} \left(\frac{M}{m_p} \right) kT \qquad (5.37)$$

where k is the Boltzmann constant, N is the total number of particles, T is the temperature and M is the mass. The second bracket term in Eq. (5.37) comes about since the number of nuclei in a white dwarf's interior will be $M / (4m_p)$ and the number of electrons will be $M / (2m_p)$. The energy radiated into space at the surface of a white dwarf is derived from the internal thermal energy, and an upper limit to the cooling time can be derived by assuming a constant temperature T for the interior and equating the thermal energy loss per unit time dE_{th} / dt to the blackbody luminosity L. Accordingly, using the Stefan-Boltzmann relation (Eq. (1.14)), we have for a white dwarf of radius R,

$$L = 4\pi R^2 \sigma T^4 \approx \frac{dE_{th}}{dt} = \left(\frac{9}{8} \right) \left(\frac{M}{m_p} \right) k \frac{dT}{dt} \qquad (5.38)$$

Equation (5.38) can now be rearranged to provide a characteristic cooling time t_{cool} to a temperature T, since

$$\int \frac{dT}{T^4} \approx \frac{1}{3\,T^3} = \frac{32\,\pi\,\sigma}{9\,k}R^2 \left(\frac{m_p}{M}\right)\int dt \qquad (5.39)$$

and this further yields the result that

$$t_{cool} = 5.6x10^5 \left(\frac{M/M_{sun}}{(R/R_{sun})^2}\right)\left(\frac{T}{10^3}\right)^{-3} \text{years} \qquad (5.40)$$

The (cooling) time to acquire a surface temperature of 10^3 K for a 0.5 $M_\odot$ white dwarf with a radius of some 0.01 $R_\odot$ is accordingly of order 3 billion years.

(☺) **Exercise 5.21** Verify the derivation of Eq. (5.40) from (5.39).

As a white dwarf ages, its mass and radius remain fixed, but it gradually cools to lower and lower temperatures until ultimately becoming a black dwarf. Such objects probably do not exist within our universe—yet. This is due to the fact that the cooling time to produce a black dwarf is longer than the present age of the universe.

(☺) **Exercise 5.22** Imagine that an isolated white dwarf has evolved all the way into a fully cooled off black dwarf and accordingly come into thermal equilibrium with its surroundings. What surface temperature will it have? Will it be absolute zero, or something else? *Hint:* Think about (and investigate) the cosmic microwave background.

5.8 Becoming Giants

The wonderfully descriptive word 'giganterithrotropism' was introduced by Peter Eggleton and John Faulkner in a conference paper published in 1981. This term describes the phenomenon or tendency of stars to grow seemingly ever larger in size as they evolve off the main sequence to form red giants (recall Figs. 1.9 and 1.13). This tendency was evident from the earliest detailed computer simulations and from the observations relating to the HR diagram, but the reason behind it is entirely unclear. It is unclear in that it is hidden in the details of the equations and the complexity of their solution.

Eggleton and Faulkner, however, were more interested in finding the physical reasons for why stars might be expected to become red giants, rather than just accepting it as an outcome of the detailed numerical models. To this end, they first identified a number of reasons that, while commonly identified, do not actually explain why stars become red giants. Among the inappropriate reasons they listed were the development of convective envelopes and/or the development of a degenerate core. It is the case that red giants do have such features, but their coming about does not explain the dramatic increase in size. Eggleton and Faulkner also identified another common fallacy for the formation of red giants in terms of the envelope absorbing the energy released by the contracting core. The point was, as Eggleton and Faulkner noted, that there is absolutely no reason why a star's envelope should "know" where the energy flowing through it should specifically be coming from. The key processes involved in Giganterithrotropism are the development of a molecular weight gradient between the core and the envelope, and the changeover from central to shell-burning fusion reactions. The former process is the same as that described earlier in Sect. 5.2 and Eq. (5.16). It is perhaps surprising that the molecular weight gradient should enter into consideration and have such a dramatic effect since it changes by a factor of about 2, while the radius changes by a factor of order 100 to 1000. Nonetheless, Eggleton and Faulkner argued that the consequence of the ignition of a shell source and the development of a molecular weight gradient conspire to push the polytropic index within a star's interior towards $n = 5$. To see that such a mechanism might well be at play, we can combine the logarithmic differential of the polytropic pressure law (Eq. (3.44)) and that of the perfect gas law (Eq. (3.1)). The differential of the former gives

$$\frac{d\log\rho}{d\log P} = \frac{n}{n+1},$$ (5.41)

while the differential of the latter yields

$$1 = \frac{d\log\rho}{d\log P} - \frac{d\log\mu}{d\log P} + \frac{d\log T}{d\log P}$$ (5.42)

and these two equations combine to give

$$\frac{n}{n+1} = 1 + \nabla_\mu - \nabla$$ (5.43)

where we have introduced the so-called del-notation with $\nabla_\mu = d\log\mu/d\log P$ and $\nabla = d\log T/d\log P$. Eggleton and Faulkner pointed out that the ∇ term tends towards a limiting value of 1/4 moving inwards from the surface of a star, and accordingly only a relatively modest increase in the ∇_μ term, contributing a factor of order 1/12 to the sum, can push the polytropic index as given in Eq. (5.43) towards its critical value of 5/6, corresponding to a polytropic index of $n = 5$. Why is this important for the red giant story? The answer is seen in Table 3.1, where the characteristics of an $n = 5$ polytrope are revealed. The table shows that they have a finite mass but an infinite radius, and a small (technically zero) average density: these are the approximate characteristics of a red giant.

Eggleton and coworkers further developed their ideas on red giant formation by constructing a multiple-component polytropic model.[8] Specifically, they developed

---

[8]See, P. Eggleton, J. Faulkner and R. Cannon. A small contribution to the giant problem. *Monthly Notices of the Royal Astronomical Society*, 298, 831–834 (1998).

an analytic model in which the core of the red giant is described by an $n = 5$ polytrope, and to this they attached an $n = 1$ polytropic envelope. The solution equation for the $n = 1$ polytrope was introduced in Sect. 3.3 and has the form

$$\theta_1 = \frac{\sin \xi}{\xi} \tag{5.44}$$

with the *surface* being located at $\xi_1 \Rightarrow \pi$. The solution for the $n = 5$ polytrope is

$$\theta_5 = \frac{1}{\sqrt{1 + \xi^2/3}} \tag{5.45}$$

and in this case, the *surface* is located at $\xi_1 = \infty$ (recall Table 3.1). The $n = 5$ polytrope has the interesting (strange) property that although formally $\xi_1 = \infty$, the mass contained within such a configuration is finite.

(☺) **Exercise 5.23** The mass contained within a polytropic sphere of radius ξ is given by the integral

$$M(\xi) = 4\pi \alpha^3 \rho_c \int_0^\xi \xi^2 \theta^n d\xi$$

Show that even as $\xi \Rightarrow \infty$, the mass remains finite when $n = 5$.

In the double polytrope model, the run of the pressure and density are described according to the equations: $P = P_c \theta_5^6$ and $\rho = \rho_c \theta_5^5$, where P_c and ρ_c are the central pressure and density. In the envelope, the pressure and density variations are given by the equations $P = P_c \theta_c^4 \theta_1^2$ and $\rho = (\mu_e/\mu_c)\rho_c \theta_c^4 \theta_1$, where θ_c is the value of θ_5 at the core-envelope boundary, and where μ_e and μ_c are the mean molecular weighs corresponding to the core and the envelope. Since the mean molecular weights of the core and envelope are different, with $\mu_c > \mu_e$, there must be a discontinuity in the density at the core-envelope boundary in order that the pressure and temperature remain continuous.

(☺) **Exercise 5.24** Verify the formula for the envelope pressure and density variation. *Hint:* Begin with the perfect gas equation and then substitute for the pressure, density and temperature in terms of the appropriate polytropic variable θ_5 or θ_1.

By varying the location of the boundary, various red giant configurations can be studied. One of the interesting characteristics of the model that Eggeleton, Faulkner and Cannon found was that once $\mu_c \geq 3\,\mu_e$, the core mass fraction cannot exceed a value of $2/\pi$. This finding has a parallel with another classic result established by Mario Schönberg and Subrahmanyan Chandrasekhar in 1942. This particular limit establishes the conditions under which an isothermal core embedded within an extended outer envelope must begin to contract. Usefully, the Schönberg-Chandra-sekhar limit can be investigated via a double polytropic model[9] similar in construc-tion to that for the Bonner-Ebert sphere as described in Sect. 3.4; in this case, a central isothermal core is constructed under the approximation of a polytrope of index $n = \infty$, and at some specified boundary the core is *attached* to an envelope modeled according to a polytrope of index $n = 1$. In the Schönberg-Chandrasekhar case, the maximum core mass fraction before core contraction must begin is given by the condition: $(M_{core}/M)_{max} = 0.37 \ (\mu_e/\mu_c)^2$.

(☺) **Exercise 5.25: Term Paper Topic** Find a copy of the research paper by Eggleton, Faulkner and Cannon, and work through their analysis. Specifically look at their development of the equation describing the core-mass fraction. *Comment:* There is an extended body of literature concerning why and how stars become red giants, and not everyone agrees with the analysis presented by Eggleton, Faulkner and Cannon. Delve into some this literature and produce a term paper addressing the topic of why stars become red giants.

Before moving on to consider the end phases of stellar evolution, it is worth returning for one more look at Fig. 5.9, and noting that the evolutionary track for the 0.1 $M_\odot$ star does not actually show any tendency towards becoming a red giant. This is predominantly due to the fact that very low-mass stars remain nearly fully convective throughout their entire evolution, and accordingly never develop a distinct mean molecular weight difference across a core-envelope divide. The same reason lies behind the evolutionary tracks for a fully mixed 30 M star shown in Fig. 4.10.

5.9 Becoming Dwarfs

The existence of the first white dwarf was known even before it was actually seen and before anyone knew what it was. This is the history of Sirius B, the diminutive companion to Sirius A (the brightest star in the night sky at the present epoch). That the star Sirius must in fact be part of a binary system was demonstrated by Friedrich Bessel in 1844. Bessel found that the proper motion path of Sirius was not a straight line on the sky but rather a sinusoidal curve. The only reasonable explanation for this observation was that Sirius (technically now Sirius A) had an unseen companion (Sirius B) and that it was in periodic motion about the system's barycenter. Contin-ued observations by other astronomers soon pinned down the system parameters, revealing a close pairing of the two stars with an orbital period of 50.13 years. At their closest, Sirius A and Sirius B are just 3.0 AU apart, while at their greatest

---

[9]See, M. Beech, The Schönberg- Chandrasekhar limit: a polytropic approximation. *Astrophysics and Space Science*, 147, 219–227 (1988).

separation they are 11.9 AU distant. The entire Sirius AB binary would comfortably fit interior to the orbit of Uranus within our Solar System.

The mass of Sirius A is about twice that of the Sun, while Sirius B has about the same mass as the Sun. Being close enough to have an accurately measured parallax, giving a system distance of 2.64 pc, the absolute luminosity of each component can be determined, and it is found that while Sirius A is some 25 times more luminous than the Sun (giving it an A0 spectral type; recall Table 1.1), Sirius B in contrast has a luminosity of about 1/37th that of the Sun.

The first observer to see Sirius B was Alvan Clark with the newly constructed 18.5-inch refractor at Dearbourn Observatory in 1882. The first spectrum of Sirius B was made by Walter Adams at Mount Wilson Observatory in 1915. The observations by Adams were particularly surprising in that the spectrum he obtained indicated that Sirius B was a hot star and not a cool one as had been expected. This was significant since the low luminosity and high temperature betrayed by Sirius B implied a very small radius—indeed, it is about the same size as the Earth, being some 12,000 km across. While the small size of Sirius B was a surprise, the implied density for the object was even more surprising: some 10^8 kg/m^3, one-hundred-million times greater than the bulk density of the Sun. At first, this result for Sirius B was considered absurd, but in the early 1920s Arthur Eddington argued that since the atoms within stars would be completely ionized within their interiors, high density configurations might well come about. Eddington then noted that this implied a fundamental problem: If the material in a star such as Sirius B behaved like an ideal gas, then as long as it was hot it must radiate energy into space. Eventually, however, the star will have radiated so much energy into space that it will have less energy than a similar mass body made of ordinary atoms at an ordinary density with a temperature corresponding to absolute zero. The point, Eddington noted in his influential text *The Internal Constitution of the Stars* (published in 1926), was, 'When the star cools down and regains the normal density ordinarily associated with solids, it must expand and do work against gravity. The star will need energy in order to cool…Imagine a body continually losing heat but with out sufficient energy to grow cold'.

Physicist Ralph Fowler resolved Eddington's apparent paradox about the energy content of such objects as Sirius B by noting in 1926 that the pressure support provided by a degenerate gas is independent of the gas temperature, and accordingly a degenerate gas can still support a star-like object against collapse even if it has cooled to a temperature close to absolute zero. Fowler built his arguments around the then-newly published work by Enrico Fermi concerning the quantum physics of electron degeneracy. Indeed, Fowler found that provided the electrons were non-relativistic, the pressure varied according to the density as

$$P = \frac{(3\pi^2)^{2/3}}{5} \left(\frac{\hbar^2}{m_e}\right) \left(\frac{\rho}{\mu_e m_p}\right)^{5/3} \tag{5.46}$$

where $\mu_e = 2 / (1 + X)$ is the mean molecular weight per electron. Importantly, Fowler's formula indicates that objects such as Sirius B (now identified as a white dwarf[10]) will behave like polytropes of index $n = 1.5$. Fowler's non-relativistic equation of state was expanded to include relativistic conditions by the German-Estonian physicist Wilhelm Anderson in 1929, and importantly the determinacy of the pressure support changes to

$$P = \frac{(3\pi^2)^{1/3}}{4} (\hbar c) \left(\frac{\rho}{\mu_e m_p} \right)^{4/3} \tag{5.47}$$

where now the 4/3rd power in the density term indicates that a white dwarf supported by relativistic degenerate electron gas will behave like a polytrope of index $n = 3$. It is this turnover in the density power law from 5/3 in the non-relativistic case to 4/3 in the relativistic case that implicates a limiting mass M_{lim} for white dwarf structures. We have in fact seen this limiting condition in Sect. 3.3 and specifically within Eq. (3.50). When $n = 1.5$, Eq. (3.46) indicates that the radius will vary as the inverse 1/3rd power of the mass: $R \sim M^{-1/3}$. What this indicates is that more massive white dwarfs are smaller than less massive ones—the exact reverse of objects made out of ordinary matter for which $R \sim M^{1/3}$. By adding mass to a white dwarf, it becomes smaller and more compact, thereby increasing in density. Ultimately, as the mass increases, the electrons will become relativistic and accordingly the behavior approaches that expected of an $n = 3$ polytrope. Eq. (3.46) indicates that the radius dependency drops out and that there is a unique (limiting) mass for such structures—the mass depending upon the constant K only—that in turn is defined by the constants entering into Eq. (5.47). The sequence of possible stable, degenerate white dwarf models ends at the critical mass determined by K; objects more massive than this limit cannot be held in hydrostatic equilibrium by degenerate electron pressure (Fig. 5.12).

The first estimate for the limiting mass of a white dwarf was made by Edmund Stoner (a student of Ralph Fowler) in 1930, and it was found that $M_{lim} = 1.1 \, M_\odot$. This was soon followed by another estimate by Subrahmanyan Chandrasekhar in 1931, who found $M_{lim} = 0.91 \, M_\odot$. Renowned Russian physicist Lev Landau, working independently of the Cambridge University group of Stoner, Fowler and Chandrasekhar, also determined a critical mass limit for white dwarf structures, finding in 1932 that $M_{lim} = 1.5 \, M_\odot$. The definitive derivation of the limiting mass condition for relativistic degenerate white dwarfs, however, was published by Chandrasekhar in 1934, when he showed that $M_{lim} = M_{CH} = 1.46 \, M_\odot$. This limiting mass dictates that any structure with a mass greater than M_{CH} must collapse without limit—in other words, it must find a new equilibrium condition (supported by something other than electrons), find a new steady energy generation source or become a black hole.[11]

---

[10]The term white dwarf was apparently coined by Dutch astronomer Willem Luyten in 1922.

[11]The term black hole was coined in the 1960s and made particularly popular by physicist John Wheeler.

Fig. 5.12 A selection of white dwarf stars in the mass-radius diagram. The lines indicate the mass-radius relationships corresponding to $n = 1.5$ and $n = 3$ polytropes. No white dwarf will appear to the right of the vertical line ($n = 3$ polytrope), since such objects would have a mass greater than the Chandrasekhar limiting mass. (Credit: Author)

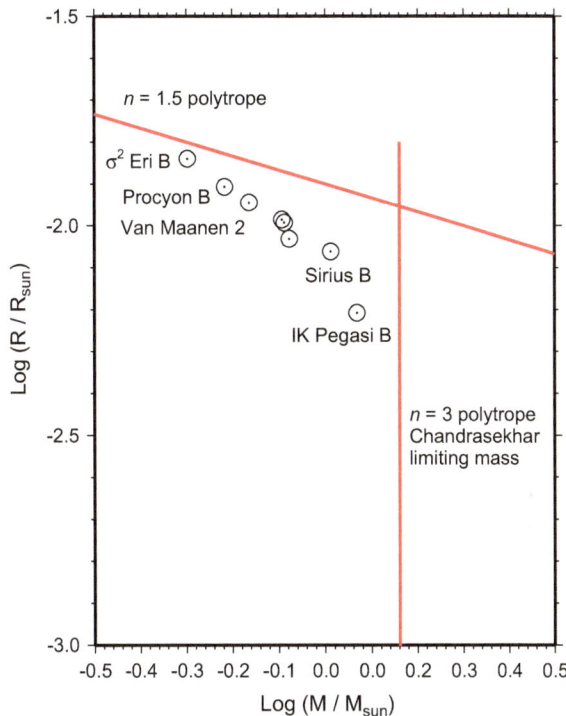

The actual expression for the Chandrasekhar limiting mass is (interestingly) largely dependent upon fundamental constants, with

$$M_{CH} \approx \left[\frac{1}{m_p^2}\left(\frac{\hbar c}{G}\right)^{3/2}\right] \times \left[\frac{M_3}{2\mu_e^2}(3\pi)^{3/2}\right] \approx \frac{7}{2}m_p\,\alpha_g^{-3/2} \qquad (5.48)$$

where $M_3 = -\xi_1^2(d\theta/d\xi)_{\xi_1} = 2.02$ is a constant (recall Eq. 3.46 and Table 3.1). The second square bracket on the right-hand side of Eq. (5.48) is dependent upon the assumed composition through the μ_e (the mean molecular weight per electron) term, but characteristically evaluates to about 3.5 when $\mu_e = 2$ (for a pure helium composition).

Recalling the fundamental constants described in Sect. 2.3, the gravitational fine-structure constant α_G has a value of some 5.9×10^{-39}, and accordingly the Chandrasekhar limit is akin to the process of trying to compress more than $\sim 10^{58}$ protons into a region not much larger than the Earth. At this limit something has to give, and what gives is the electron degeneracy pressure support, with gravity winning free range to crush the material into an even smaller volume of space. Gravity's win, however, is short lived, and the white dwarf is now primed for total destruction. As the degeneracy at the center of the white dwarf is lifted, the collapse raises the temperature to that at which carbon fusion reactions can begin ($T > 500$

million Kelvin and $\rho > 10^9$ kg/m^3). Although a new energy source has been ignited at the center of the white dwarf, its pressure support is largely provided by the still-degenerate outer layers, and this inhibits the ability of the white dwarf to expand and cool—that is, to attain some form of hydrostatic equilibrium with respect to its newfound supply of fusion-generated energy. Accordingly, the central temperature increases dramatically and without constraint, initiating a thermal runaway with the white dwarf being fully destroyed through a process called carbon deflagration. This rapid and violent disruption sequence is identified with the precursor phase of a Type Ia supernova (see Sect. 5.10). Since such supernova are produced by similar mass white dwarfs—white dwarfs with a mass very close to the Chandrasekhar limit—they are believed to be good standard candles, and since the energy released in such events are truly enormous (with a peak luminosity of several billion L$_\odot$), Type Ia supernova are particularly important cosmological distance indicators.

While Chandrasekhar's 1934 paper is now considered a classic work, it was not well received by Eddington, who argued that while Chandrasekhar's calculations were correct, the theory of relativistic degeneracy that they were based upon was incorrect. At the time that he made it, Eddington's argument was not entirely unreasonable, but he was definitely on the wrong side of the dispute. Interestingly, Eddington actually made a correct prediction (and then abandoned it) at a meeting of the Royal Astronomical Society on January 11, 1935, arguing that '[as] the star has to go on radiating and radiating and contracting and contracting until, I suppose, it gets down to a few km. radius, when gravity becomes strong enough to hold in the radiation, and the star can at last find peace'. This is Eddington essentially arguing that the final collapse to a black hole state is inevitable. Unfortunately, Eddington continued in his critique of Chandrasekhar's work, commenting, 'This was almost a *reductio ad absurdum* of the relativistic degeneracy formula'. With this statement, Eddington loses the historical battle.

Around the time that Eddington wanted to avoid the possibility of stars finally collapsing into black holes, Lev Landau also felt that the identification of a limiting mass was a consequence of applying the theory of relativistic degeneracy incorrectly, arguing in 1932 that stars 'do not show any such ridiculous tendencies [and] we must conclude that all stars heavier than 1.5 M$_\odot$ certainly possess regions in which the laws of quantum mechanics (and therefore quantum statistics) are violated'. Even Albert Einstein as late as 1939 argued that black holes (gravitational singularities) could never occur in nature and specifically that they could not form via the gravitational collapse of stars. Indeed, Einstein wrote, 'Schwarzschild singularities do not exist in physical reality—matter cannot be concentrated arbitrarily'. In the same year that Einstein argued that black holes cannot exist, Robert Oppenheimer, along with coworkers Robert Serber, George Volkoff and Richard Tolman, were laying the general relativistic foundations of stellar collapse and confirming the existence of a limiting mass.[12] Indeed, even the 'great heroes of science' can get their

---

[12]In this case, the limiting mass applied to neutron stars—these objects are supported, up to another limiting mass, by the pressure of degenerate neutrons.

arguments wrong, and perhaps the most important lesson to take away from the story recounting the troubled origins of the white dwarf limiting mass is that, in spite of the flamboyant posturing and nay-saying of dominant and respected voices, scientists working as an unbiased collective can eventually find the right answers and the correct paths.[13]

While the internal density and the compaction of white dwarfs are high (with $\rho\sim10^8$ kg/m^3, and $R\sim0.01$ R$_\odot$) there is no specific requirement to add general relativistic corrections to the equations that describe their internal structure. This cannot be said of neutron stars, where the densities and compaction are even higher than those found in white dwarfs, reaching of order 10^{17} kg/m^3 and radii as small as 10 km respectively. The neutron was first identified as a component of the atomic nucleus by James Chadwick in 1932, and almost immediately it was speculated that neutron stars—stars stabilized against gravitational collapse by degenerate neutron pressure—might exist. Lev Landau in his remarkable paper published in the journal *Physikalische Zeitschrift der Sowjetunion*, in 1932 argued for the possible existence of stars that resembled giant atomic nuclei (which is essentially what a neutron star is). Likewise, Walter Baade and Fritz Zwicky in 1934 suggested that neutron stars might form during supernova explosions (see Sects. 5.10 and 6.2). This was a brilliant leap of theoretical deduction, but the Baade-Zwicky idea lay dormant for over 30 years, only being given substance in the late-1960s.

The key observational breakthrough was the detection of radio pulsars by Jocelyn Bell and coworkers at Cambridge University in 1967. The idea that a highly magnetized, rapidly spinning neutron star might be at the heart of the pulsar phenomenon was developed independently by Franco Pacini and Thomas Gold. The model invoked is that of the oblique rotator in which the magnetic field axis and the spin axis of the neutron star are not aligned. Accordingly, a form of lighthouse effect comes into effect with the magnetic poles, where the electromagnetic radiation is actually generated, being swept around the sky. The first pulsar discovered by Bell and coworkers (now designated PSR B1919 + 21) has a period of 1.3373 s, but other much more rapidly rotating pulsar systems were soon detected. The association between pulsars and supernovae remnants was also soon established, verifying the ideas of Baade and Zwicky. The classic example is the pulsar at the core of the Crab Nebula. This particular supernova event was actually observed and recorded as a naked-eye 'guest-star' in 1054, but the pulsar observations revealed a remnant neutron star spinning 30 times per second (period = 33.5 milliseconds).

One of the first questions to be answered upon the initial discovery of pulsars was: Could the variation be due to the rapid pulsation of a white dwarf? The instability strip (recall Fig. 1.15) does cut through the white dwarf region, resulting in the appearance of the ZZ Ceti variables. These systems can show rapid brightness variations with periods as short as a few tens of seconds. Accordingly, using

---

[13]Author Arthur C. Clarke has more famously written that 'when a distinguished but elderly scientist states that something is possible, he is almost certainly right. When he states that something is impossible, he is very probably wrong'.

Eq. (5.25), the implied density for the pulsator is some $\rho \sim 4 \times 10^7$ kg/m^3 (when $\Pi = 30$ s), which is certainly characteristic of that of a white dwarf. To obtain pulsation periods as short as 1s or less, the implied density is $\rho > 4 \times 10^{10}$ kg/m^3, which is more characteristic of a neutron star density than that of a white dwarf. Finding millisecond pulsars pushed the argument to the point where it was accepted that the pulsar phenomenon could not be due to pulsating white dwarfs. It is also the case that the phenomenon is not due to the pulsation of neutron stars either. Rather, it is a variation driven by the spin of the neutron star. This new argument further begs the question: Can a neutron star spin at the rate required and not fly apart? To answer this, Eq. (5.9) indicates that an object will begin to fly apart once $F_{cen} \sim F_{grav}$, and this implicates a maximum spin velocity of

$$V_{max} = \sqrt{GM/R} \qquad (5.49)$$

This in turn provides a maximum spin period of $P_{max} = 2\pi R / V_{max}$, where R is the radius of the neutron star. With characteristic values of 1.3 M$_\odot$ and 10 km for the mass and radius of a neutron star respectively, $P_{max} = 5 \times 10^{-4}$ seconds. The fastest spinning pulsar detected (designated PSR J1748-2446ad) is powered by a neutron star spinning some 716 times per second, giving $P = 1.4 \times 10^{-3}$ s $\sim 3 \times P_{max}$. Accordingly, it would appear that neutron stars can spin fast enough, without destruction, to drive the pulsational modulation.

(☺) **Exercise 5.26: A Pause for Thought** Physics and astronomy often push the imagination to contemplate extreme circumstances and highly bizarre objects, but we should never become too blasé about the end results. Think one more time about the explanation for a pulsar: it is an object with the mass of the Sun, compressed into a sphere just 20 kilometers across, spinning (in some cases) many hundreds of times per second. If such an object does not fill you with awe and dread and wonder, then perhaps it is time to close this book and change your subject of study.

The energy associated with a pulsar is ultimately related to the neutron star's kinetic energy of rotation. This energy is determined according to $E_{Rot} = \frac{1}{2} I \omega^2$, where I is the moment of inertia and ω is the neutron star's angular velocity ($\omega = V_{rot} / R = 2\pi / P$, where R is the neutron star radius and P is the spin period). Taking $I = (2/5) M R^2$, where M is the mass of the neutron star (recall exercise 5.6) the conversion of rotational energy into radiative energy can be gauged via the conservation of energy, with

$$\frac{dE_{rad}}{dt} + \frac{dE_{Rot}}{dt} = 0 \qquad (5.50)$$

At this stage we do not specifically consider the mechanism by which the rotational energy is being converted into radiative energy, but simply write $L = dE_{rad} / dt$. Accordingly,

$$\frac{dE_{Rot}}{dt} = \frac{1}{2}\frac{d}{dt}\left(I\omega^2\right) = -\frac{8\pi^2}{5}\left(MR^2\right)\left(\frac{1}{P^3}\frac{dP}{dt}\right) = -L \qquad (5.51)$$

(☺) **Exercise 5.27** Verify the derivation of Eq. (5.51).

Equation (5.51) provides a link between the luminosity L—associated now with the radiative energy output of the surrounding supernova remnant—and the spin-down rate of the central neutron star. As the neutron star ages and loses energy to the surrounding nebula, its rotation rate slows down and the nebula becomes less luminous. For the Crab nebula, the integrated energy output is observed to be some 3×10^{31} Watts (about 75, 000 $L_\odot$), and the estimated mass and radius of the central neutron star are 1.3 $M_\odot$ and 10 km, respectively. With these numbers, Eq. (5.51) indicates that the spin-down rate for the Crab pulsar should be some $dP/dt \approx 3 \times 10^{-13}$ seconds per second (or 26 nanoseconds per day).

The spin-down rate can actually be used to provide an estimate for a pulsar's age τ (and that corresponding to its associated supernova remnant). The characteristic age τ is defined as

$$\tau = \frac{P}{2(dP/dt)} \qquad (5.52)$$

The factor of 2 in Eq. (5.52) is a little odd looking but relates to the idea that the spin-down rate is determined by the so-called braking index formula, with $\dot{P} = kP^{(2-n)}$, where k is a constant and n is the braking index determined by the magnetic field structure of the neutron star. Typically, $n = 3$ is assumed, and this corresponds to the magnetic field having a simple dipole (bar magnet-like) structure.

For the Crab pulsar, the observed values are $P = 0.0335$ seconds and $dP/dt = 4.22 \times 10^{-13}$ s/s, giving $\tau = 7.9 \times 10^{10}$ s or 1258 years. This is a little on the long side, since, recall, the originating supernova was actually observed in the year 1054, some 964 years ago. In presenting Eq. (5.52), the concept of pulsar magnetic fields was introduced, and indeed, it is the rotation of the magnetic field anchored to the neutron star that drives the pulsar emission mechanism and that ultimately powers the emission from the rest of the surrounding nebula. That neutron stars should have very strong magnetic fields is determined by the conservation of magnetic flux. In this case, the conserved quantity is AB, where A is the characteristic area through which a magnetic field of field strength B threads its magnetic field lines. The typical main sequence star that might produce a neutron star during its end phase supernova disruption will have an O or B spectral type (recall Table 1.1). Such stars have a characteristic radius of 5 $R_\odot$ and observed magnetic field strengths of order 10^{-2} Tesla and larger. Taking $A \sim R^2$, this implies a neutron star magnetic field strength of $B_{NS}\sim(5R_\odot/10$ km$)^2 \times 10^{-2} \approx 10^9$ Tesla. For comparison, the Sun's typical magnetic field strength is about 10^{-4} Tesla (that of the Earth's magnetic field is about 5×10^{-5} Tesla).

   With such intense magnetic fields at their surface, equally intense electric fields must also exist at the surface of a neutron star. Indeed, the electric field is so strong that it will pull electrons out of a neutron star's solid outer crust. These electrons will be accelerated along the rotating magnetic field lines, emitting synchrotron radiation in a tight beam along the magnetic field axis. As the neutron star rotates, this lighthouse beam of synchrotron emission is brought into and then out of the observer's line of sight, and it is this rotational modulation that produces the characteristic pulsar signal.

   In terms of their internal structure, neutron stars are decidedly complicated and include zones in which the neutrons behave more like a (zero viscosity) superfluid liquid than a gas, and in which, at the very center, a quark-gluon plasma may appear.[14] And, while neutron degeneracy is the principle pressure support mechanism, the equation of state no longer follows a simple polytropic power-law dependency. Accordingly, the limiting mass of neutron stars is not well constrained, but it is thought to lie between 2 and 4 $M_\odot$.

   Beyond the limiting mass for a stable neutron star, there is but one end phase: collapse to form a black hole. Remarkably, given the incredible complexity of neutron star structure, black holes are the simplest of structures. In order to describe a black hole, just two parameters are needed: its mass and its spin period. Of course, it is somewhat flippant to say that black holes are the simplest of structures, but effectively a black hole partitions three-dimensional space into two domains, with the inside and outside bounded by a smooth two-dimensional surface called the event horizon. The extent of the event horizon is determined by the Schwarzschild radius, which it turn depends upon just the mass of the black hole (and the theory of general relativity). It is the event horizon that transforms black holes into the ultimate Vegas: what happens inside a black hole stays inside of a black hole, forever.[15] Outside of the event horizon, the black hole can be taken as a sphere with a specific radius R_{EH} and mass M. Just inside of the event horizon, the physics is initially no different to that operating just outside of it, except that now the escape velocity for any corporal object is greater than that of the speed of light. Deeper below the event horizon and closer to the center, however, behaviors begin to become extreme, and gravity fully rules the roost. At the very center of a black hole lurks the mysterious singularity, a highly compact region of spacetime within which physics as we presently understand it holds no domain. What happens at the singularity is literally anyone's guess at the present time.

   The classic black hole radius[16] at which the escape velocity (recall Eq. (1.3)) equals that of the speed of light gives a minimum sized sphere (of radius R_{min}) that a

---

[14]This is a hypothetical state of matter that is thought to come about at extremely high densities and temperatures. In this state the quarks and gluons, which are normally confined inside atomic nuclei, can move about (at least for a very short time) as free particles.

[15]As to whether "forever" is really for all eternity or just a very, very long time is still open to debate.

[16]The classical radius for a black hole is identical to the more formally—and correctly—derived (via general relativity) Schwarzschild radius.

given amount of material (of mass M) must be squeezed into in order to form a black hole (this is the radius of the event horizon):

$$R_{min} = R_{EH} = \frac{2GM}{c^2} \tag{5.53}$$

In the case of a 1 solar mass object, $R_{EH} \approx 3$ km, just three times smaller than the radius of a neutron star. So, while neutron stars are perfectly stable (provided that they are not a member of a close binary system in which mass transfer can take place), they are only marginally larger in size than the event horizon of an equivalent mass black hole. Indeed, if the central core of a collapsing star is squeezed just a little more than that required to make a neutron star, then a black hole will be the final end state of a massive star's core. The initial mass limit beyond which the end state of a massive star's core is to become a black hole is thought to be above 50 $M_\odot$ (recall Fig. 5.10).

While the singularity is hidden from direct observation, black holes can nonetheless be detected by observing the behavior of matter at or close to their event horizon. Black holes can accrete matter directly and in the process form an accretion disc with an inner radius of R_{EV}. Material in such a disk will spin more rapidly as it approaches the inner radius and as discussed in Sect. 2.3 (and illustrated in Figs. 2.6 and 2.7) may become heated and highly luminous. Some idea of the size of the event horizon can in fact be gauged according to timescale of rapid flickering t_{flick} in the accretion luminosity, since the material at the inner edge of the accretion disk will be traveling close to the speed of light. Accordingly, at the inner edge of the disk, which has a radius R_{EV}, the rotation period of material will be of order $t_{flick} = 2\pi R_{EV}/c$. Once the event horizon radius is known, the mass of the black hole can be found via Eq. (5.53).

(☺) **Exercise 5.28** Gunther Witzel (Universität zu Köln) and coworkers have found evidence for a 24 min variation in the infrared brightness of Sagittarius A*—the location of the black hole at the center of our galaxy—although shorter timescale variations of between 10 and 30 min have also been reported. (1) Taking $t_{flick} = 20$ min, determine the size and mass of the black hole at the center of our galaxy. (2) Compare your mass result with that derived in exercise 5.14 and discuss the two estimates. Think hard about the uncertainties associated with each method of determination. **Data source:** G. Witzel et al., Source-intrinsic near-infrared properties of SGR A*: total intensity measurements. *Astrophysical Journal*, 203, article id.18, 36 pp (2012).

5.10   Supernovae

Stars only exist because of the transformation of atomic elements. To begin with, hydrogen is converted into helium on the main sequence. Red giant stars take the process one step further and convert helium into carbon. With the exhaustion of helium various evolutionary pathways become possible, and which way a star then goes is dependent upon the initial mass and the mass loss history (recall Fig. 5.10). Stars with an initial mass less than about eight times that of the Sun enter into a planetary nebula stage, with the central core becoming a white dwarf. Stars initially more massive than eight times that of the Sun follow a different path, entering into a phase in which advanced fusion reactions involving carbon can take place. These advanced fusion reactions work their way up step by step to the production of iron. Surrounding the iron core is a series of shells of different composition, the compo- sition relating to earlier fusion reaction stages. A so-called onion shell model develops: the core is iron rich and has a size similar to that of the Earth (about $1/100^{th}$ $R_\odot$) and a mass of about 1 $M_\odot$. Around the iron core is a silicon-rich shell, and around this is an oxygen-rich shell, then a carbon-rich shell, then a helium-rich shell, and around this is an outermost shell that is rich in hydrogen. Fusion stops at iron for the straightforward reason that further transformations require the input of energy rather than the generation of energy, and indeed, it is the exothermic property of fusion reactions beyond iron that eventually results in stellar death.

(☺) **Exercise 5.29**   The binding energy for ^{56}Fe is 8.8 MeV per nucleon. How much energy is released per kilogram of matter by the sequence of fusion reactions taking hydrogen to iron?

Once an iron core has formed, it begins to contract. This happens since there are no fusion reactions taking place within the core. Eventually, the electrons in the core become degenerate. Once the iron core mass exceeds about 1.46 $M_\odot$, the Chandra- sekhar limiting mass for iron, the core begins to collapse rapidly and its temperature begins to rise dramatically. It is the unrestrained temperature increase that leads to structural catastrophe through photodisintegration, in which ^{56}Fe $\Rightarrow$ 13 ^{4}He + 4n − 100 MeV. The 100 MeV robs the core of energy, and the collapse turns into a freefall. The temperature and compression at the center increase evermore: the high energy gamma ray photons are now energetic enough to break helium nuclei into protons and neutrons, further robbing the core of energy—some − 6 MeV per nucleon in this case. Eventually, the compression in the core becomes so high that the protons and electrons interact to produce neutrons. This not only absorbs more energy from the core but also reduces the number of particles, further reducing the pressure support and intensifying the core collapse. Ultimately, the neutron-rich core becomes degenerate (the neutrons now being the degenerate component). The core density at this stage is of order 10^{18} kg/m^3, and provided the core mass is not more massive than 2–3 $M_\odot$, the collapse is halted. At this time, the neutron-rich core is perhaps 40 km across. Using Eq. (3.9), the dynamical collapse time for the iron core is of $t_{coll} \sim (G \rho)^{-1/2} \sim 10^{-3}$ s (about ten times quicker than the blink of an eye).

While the collapse of the core comes to a halt (provided it does not exceed the limiting mass for a neutron star), the envelope must react to the vast amount of energy released through core collapse. An estimate of the energy liberated can be deduced from Eq. (1.7), giving the release of gravitational energy as:

$$\Delta_{grav} = -\frac{3}{5} GM_{core}^2 \left(\frac{1}{R_{core}} - \frac{1}{R_{NS}} \right) \approx \frac{3}{5} \frac{GM_{core}^2}{R_{NS}} \sim 1 \times 10^{46} \text{ Joules} \qquad (5.54)$$

The amount of energy absorbed by nuclear processes in the core will be of order

$$\Delta_E = 100 \text{ Mev } (M_{\text{core}}/M_{\text{Fe}}) + 6\text{Mev } (M_{\text{core}}/(13M_{\text{He}})) \sim 6 \times 10^{44} \text{Joules}$$

which is about $1/10^{\text{th}}$ of the gravitational energy released. The leftover energy results in the outward acceleration of the material located outside of the core prior to collapse. Remarkably, what drives the outward expansion of the envelope and the eventual supernova destruction of the star is the copiously produced but normally ghost-like neutrinos. As the protons within the collapsing core are converted to neutrons, a neutrino is produced, and accordingly some 10^{57} neutrinos must be produced (recall the core has an initial mass of about 1 $M_\odot$). These neutrinos can easily carry away the 10^{46} Joules of energy generated during core collapse, and while neutrinos typically interact only rarely with matter, the density, neutrino flux and neutrino opacity are so high in the central regions of the star that a significant fraction of the neutrino energy is absorbed by the envelope. Given a typical core mass of 1 $M_\odot$, a pre-supernova mass of 10 $M_\odot$ and a characteristic supernova expansion velocity of 5000 km/s, the kinetic energy of expansion will be of order

$$\Delta_{\text{ke}} = \frac{1}{2} (M - M_{\text{core}}) V_{\text{exp}}^2 \sim 2 \times 10^{44} \text{Joules}$$

which is, again, easily accommodated for within the total energy budget, and is appropriately of order the gravitational binding energy Δ_{bind} of the envelope:

$$\Delta_{\text{bind}} \sim G M(M - M_{\text{core}})/R_{\text{core}} \sim 4 \times 10^{45} \text{Joules}.$$

All in all, while about 1% of the available energy goes into the kinetic energy of the expanding envelope, only about 0.1% is carried away by photons. These will characteristically be in the X-ray part of the electromagnetic spectrum.

(☺) **Exercise 5.30** How long would it take the Sun at its current luminosity to radiate into space the same amount of energy produced during core collapse?

(☺) **Exercise 5.31** (1) What is the luminosity L_ν of the core collapse carried off by the neutrinos? (2) Given that there are some 10^{11} galaxies in the universe, and that each galaxy contains 10^{11} Sun-like stars, how does the neutrino luminosity compare to the total optical luminosity of the universe?

(☺) **Exercise 5.32** What is the expected neutrino flux (neutrinos/m^2) at Earth from a core collapse supernova located at a distance of $D = 50,000$ pc?

(☺) **Exercise 5.33** The most extensively studied supernova of recent times is that observed in the Large Magellanic Cloud (a satellite galaxy to the Milky Way at a distance of 50 kpc) in February 1987 (this event is now identified as SN1987A). The Kamiokande-II neutrino detection facility in Japan detected a total of 12 neutrinos from SN1987A. Given that the Kamiokande II detector has a cross-section area of about 250 m^2, approximately how many neutrinos from SN1987A passed through the detector?

(☺) **Exercise 5.34** Given a characteristic neutrino cross section of interaction of a miniscule 10^{-47} m^2, and that the interstellar medium has a density of 10^{-17} kg/m^3, what is the typical column length L along which one neutrino interaction might take place?

The core collapse scenario outlined above essentially holds true for those stars with initial masses between $8 < M_{initial} / M_\odot < 60$. For stars initially more massive that 60 M$_\odot$, it turns out that the higher internal temperatures favor the production of oxygen over carbon through the reaction ^{12}C + ^{4}He $\Rightarrow$ ^{16}O. Additionally, at temperatures in excess of a billion Kelvin, pair production can take place whereby a high energy photon spontaneously produces an electron-positron pair ($\gamma \Rightarrow e^+ + e^-$), driving dynamical instability. The initial mass determines whether the photo-disintegration or the pair-instability comes about, and whether the core collapses to form a neutron star or a black hole; additionally, the mass loss history of the star determines the eventual observed characteristics of the supernova light curve (brightness versus time diagram).

Fig. 5.13 Supernovae classification scheme. (Credit: Author)

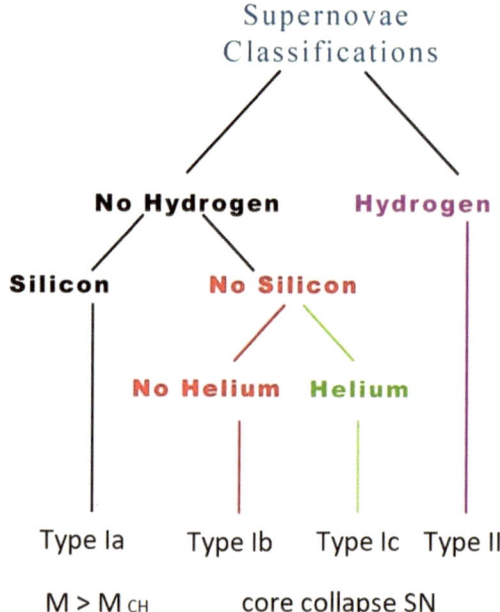

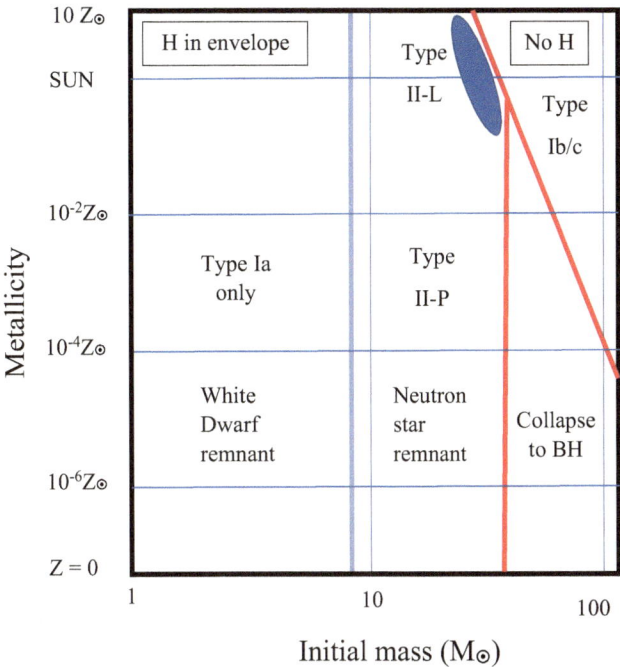

Fig. 5.14 Supernovae mass-metallicity diagram. (Credit: Author)

The classification scheme for supernovae is based upon the observed shape of the light curve and the chemical composition of the ejecta. Figure 5.13 shows a basic classification scheme for the various supernovae types. What determines the appearance of hydrogen (or no hydrogen) and helium (or no helium) lines in the spectrum is mass loss. Within the onion skin model, as mass loss proceeds, the outer shells—first the hydrogen-rich shell, and then the helium rich shell—are removed. If hydrogen lines are present in the ejected material, a Type II supernovae designation is given. If there are no hydrogen lines, a Type Ib designation is given. If there are no hydrogen or helium lines present, then a Type Ic designation is given. If the light curve shows a steady linear decline, an L designation is given. If the light curve shows a temporary plateau (constant luminosity) feature, a P designation is given. Accordingly, a supernova might be classified according to the observational data as a Type I L/b or a Type II-P, and so on. The Type Ia label is reserved for those supernovae involving white dwarf thermonuclear deflagration, which will be discussed later.

Detailed numerical models reveal that not only are initial mass and the mass loss history of a star important in determining the final supernovae phase, but so too is the initial composition. Figure 5.14 gives some idea of the complexity of the situation. For initial masses smaller than 8 $M_\odot$, only Type Ia supernovae can occur. Above an initial mass of 8 $M_\odot$, various channels and end states occur according to metallicity and mass loss effects. Above an initial mass of about 150–200 $M_\odot$, the only end phase is that of black hole production. At lower masses, a range of end states supporting the formation of a neutron star are possible.

The diagonal red line in Fig. 5.14 indicates the effect of mass loss prior to core collapse onset, and this separates out the Type II and Type Ib/c supernovae. The light curve itself is dominated by the initial rapid rise to maximum brightness, followed by a gradual decline. The initial peak brightness is associated with the breakout of the neutrino-accelerated shockwave, but thereafter the brightness is dominated by the decay of radioactive elements. The radionuclides are produced via neutron capture events, and one of the predominant species produced is that of nickel-56. This particular radionuclide decays to cobalt-56, with a half-life time of 6.1 days. During the decay process, it emits a 1238 KeV gamma ray photon. The cobalt-56 further decays into the stable nucleus of iron-56 with a half-life time of 77.1 days. This final decay is associated with the emission of an 847 KeV gamma ray photon. Since the number of parent radionuclides decays exponentially, the brightness of the supernova, if plotted against a logarithmic time scale, will be a straight line. This characteristic decline is particularly prominent in Type Ia supernovae, since most of the progenitor white dwarf is converted into nickel-56 via explosive (runaway) nucleosynthesis. Aluminum-26 is another radioactive element produced during supernovae outbursts, and this element has a half-life time of 720,000 years. This particular radionuclide played an important role in early Solar System evolution. The detection of its daughter product, magnesium-26, within meteorites indicates that the Sun's birth cluster must have contained at least a few large supernovae-producing stars (see Sect. 6.2).

In addition to the classification scheme shown in Fig. 5.13, the various supernovae types are additionally distinguished according to where they are observed to occur within a galaxy. Type II Ib/c supernovae are only found in spiral galaxies, and particularly in those regions undergoing active star formation. This makes sense in that such supernovae are associated with the most massive, short-lived stars. Type Ia supernovae, in contrast, are observed predominantly in elliptical galaxies where active star formation regions are conspicuous by their absolute absence. Recall that such supernovae are associated with the explosion of a carbon-oxygen rich white dwarf star with a mass close to the Chandrasekhar limit. It is the fact that the white dwarf progenitors in a Type Ia supernovae must be close to the upper stability mass of 1.4 $M_\odot$ that indicates such supernovae must come about through some form of mass accretion—either as a consequence of mass exchange with a main sequence or evolved binary companion, or through coalescence with another white dwarf (also within a binary system). The buildup of mass or spiral-in to produce a white dwarf with a mass close to its stability limit takes time, and this is why Type Ia supernovae are found in regions that are not associated with active star formation.

It is the fact that Type Ia supernovae progenitors must all have essentially the same mass at the time of their destruction (the Chandrasekhar limiting mass) that makes them valuable standard candles—objects for which a constant peak luminosity can be determined (this turns out to be of order 10^{36} J/s or $2.5 \times 10^9 L_\odot$). Since Type Ia supernovae are so bright, they have proven particularly useful in the determination of the scale of the universe. The available data relating to supernovae observed in distant galaxies indicates that the galactic production rate of supernovae is about 1 per 30 – 50-year time interval. Accordingly, there should be of order 2 –

3 supernovae per century within the Milky Way galaxy. The last supernova located within the Milky Way galaxy that was visible to the naked eye occurred over 400 years ago in 1604 (this was Kepler's supernova, Type Ia). Remarkably, Kepler's supernova was observed just 32 years after Tycho Brahe's supernova (again of Type Ia), which occurred in 1572.

Exercise 5.35   The energy flux associated with a 6th magnitude star, the faintest star that the eye can see, is about 10^{-10} Watts/m^2. At what maximum distance (in light years) might a human eye thus detect a Type Ia supernova?

5.11   Answers to Exercises

Following are the answers to all exercises in this chapter. The exercises for which no answer is provided are those in which the student is asked to check a result or complete the algebra steps between two equations given in the text.

Exercise 5.2   The turnoff mass is $M_{TP} \approx 0.94$ M$_\odot$, and $N_{WD}/N_{MS} \approx 0.05$—that is, about 5% of the original cluster members have evolved into white dwarfs.

Exercise 5.3 $M_{initial} = 5.8263$ ξ_0, and, $M_{return} / M_{initial} = 0.3579, 0.4210, 0.4904$ for $M_{TP} = 6$, 3 & 1.5 M$_\odot$ respectively. As would be expected the older a given cluster (the smaller the value of M_{TP}), the more matter will have been ejected and returned to the ISM from the constituent stars.

Exercise 5.9   (2)The initial Sun luminosity is $L(0) = 0.80$ L$_\odot$.

Exercise 5.11   The relationship is $M = - \eta L^{31/35}$, which is nearly a linear dependency in the luminosity, as argued for in the development of Eq. (5.26).

Exercise 5.12   The radius is $R = 261$ R$_\odot$, and the time for the Sun to lose half of its mass is $t = 1.15$ million years.

Exercise 5.13   (1) $t \approx 300$ million years. (2) The mass flux is $\sim 2 \times 10^{-15}$ kg/s/m^2.

Exercise 5.14   The central black hole mass is $M_{BH} = 6.6 \times 10^{36}$ kg, or 3.3×10^6 M$_\odot$.

Exercise 5.15   The separation is $a = 2.6 \times 10^{10}$ m, or 0.17 AU. This separation is about half the size of the orbit of Mercury about our Sun.

Exercise 5.16   (1) The accretion rate of matter onto the black hole will be:

$$\frac{\Delta M_{BH}}{\Delta t} = (1-f)\frac{\Delta M}{\Delta t} = (1-f)\frac{L_{Edd}}{f\,c^2} = \frac{M_{BH}}{\tau}$$

which integrates to give: $M_{BH} = M_{BH,0}$ exp$[t/\tau]$, where

$$\tau = \frac{0.02\,(1+X)\,c}{4\pi G}\left(\frac{f}{1-f}\right) = 3.4\times 10^8 \frac{f}{1-f}\ \text{years}$$

(2) In this case, $M_{\text{BH}}(t) = 1\,M_\odot$, and the consumption time will be $t \approx 1.6$ billion years.

Exercise 5.28 (1) $R_{\text{EVA}*} = 5.7 \times 10^{10}$ m $\approx 82\,R_\odot$, and $M_{A*} = 3.8 \times 10^{37}$ kg ≈ 19 million solar masses.

Exercise 5.29 The total number of nucleons per kilogram is $N = 1/m_P \approx 6 \times 10^{26}$. Each nucleon releases $\varepsilon = 8.8$ MeV in the formation of iron, so the total energy released is $\varepsilon\,N = 5.3 \times 10^{33}$ eV $= 8.4 \times 10^{14}$ Joules. *Notes:* The chemical energy available in 1 kg of oil is 4.2×10^7 Joules. One kilogram of hard coal (SKE) provides 2.9×10^7 Joules of energy.

Exercise 5.30 We have $t{\sim}\Delta_{\text{grav}}/L_\odot = 2.5 \times 10^{19}$ seconds $\approx 8 \times 10^{11}$ years—174 times longer than it has currently existed.

Exercise 5.31 The neutrino luminosity is $L_\nu{\sim}\Delta_{\text{grav}}/t_{\text{coll}}{\sim}10^{49}$ J/s $= 3 \times 10^{22}L_\odot$, and $L_{\text{univ}}{\sim}10^{22}L_\odot$.

Exercise 5.32 The flux will be $10^{57}/(4\pi D^2) \approx 3 \times 10^{13}$ neutrinos/m^2.

Exercise 5.33 The neutrino flux f_ν at Earth is that found in exercise (5.32), and accordingly the number of SN1987A neutrinos within the detector during the fraction of a second that they passed through the detector will be of order $250\,f_\nu \sim 10^{16}$ neutrinos. *Notes:* The detector is a cylinder of height 16 m and diameter 15.6 m (giving a cross-section area of about 250m^2, and the cylinder contains some 2×10^6 kg of water, giving some 10^{32} target nuclei). The interaction targets are actually the electrons within the water molecules, which will emit a flash of light (Cherenkov radiation) if they interact with a neutrino.

Exercise 5.34 The typical distance that a neutrino will travel before interacting with another particle is its mean free path (the distance L). For a neutrino interaction to be at all likely, it must pass through a volume of space at least equivalent to the volume of a nucleon $V = m_P/\rho$, where in our case m_P is the proton mass and ρ is the density of protons. Given a cross section of interaction σ, so $V = m_P/\rho = L\,\sigma$, which gives, $L = m_P/(\sigma\,\rho) \approx 2 \times 10^{30}$ meters. *Notes:* The distance L just derived is larger than the size of the observable universe. If the neutrino was passing through a column of lead with density 11,000 kg/m^3, then $L \approx 1.5 \times 10^{16}$ meters $= 1.6$ light years.

Exercise 5.35 Using Eq. (1.10), $D \approx 3 \times 10^{22}$ meters $\approx 280{,}000$ lyr (this is about three times the diameter of the Milky Way galaxy, but only 1/10th of the way to the Andromeda Galaxy). *Note:* The calculation just made does not take into account the severe dimming effects of interstellar dust, which greatly limits the ability of astronomers to detect supernovae within the disk of our Milky Way galaxy. Looking out of the galactic plane, however, where there is virtually no interstellar dust obscuration, supernovae can be seen all the way to the edge of the observable universe (out to distances of order tens of giga-parsecs). As of the end of 2018, The Open Supernova Catalog (http://sne.space/) contains details on some 52,422 recorded supernovas.

Chapter 6
Selected Topics and Case Studies

The brief reviews and essays that make up this final chapter are intended to provide some detailed information and reference material on recent research topics in astronomy and astrophysics. It is hoped that the material will be of use with respect to term paper ideas. References are given to full research papers, not so much because their content is readily accessible to the reader (at this stage), but because they will give some idea of how science communication works. Indeed, one of the most important aspects of being (and becoming) an astronomer is to *know* the literature—what are other researchers doing; where have are their arguments weak, where are they strong; where have other researchers gone wrong in the interpretation of the data, where have they made the right choices, etc.

At this stage, however, when reading through the variously referenced research papers, operate in a gestalt manner—that is, don't worry about all the details and minutia, but try to get a feel for the overall content and arguments that are being presented—and then think hard about what those arguments mean. Some of the viewpoints expressed in the essay topics below are deliberately controversial. It is often said that astronomy is the most ancient of sciences, and while this may be so, it is also one of the most modern, dynamic and—on occasion—downright controversial of sciences. The Royal Society of London, founded in 1660, adopted as its motto *Nullius in verba*, which translates to something like 'on the word of no one': it is a good motto, a good general principle for living everyday life, and an especially good way of *doing* science.

6.1 The Atypical Sun

This section is based upon an article I published in the *Journal of the Royal Astronomical Society of Canada* in 2011, and considers the question: Is the Sun a typical star? By being such a common, everyday and familiar sight, the Sun is often overlooked as a *bona fide* object of astronomical interest. There is perhaps a

© Springer Nature Switzerland AG 2019
M. Beech, *Introducing the Stars*, Undergraduate Lecture Notes in Physics,
https://doi.org/10.1007/978-3-030-11704-7_6

historical underpinning for this sentiment, and it should be remembered that it has been barely 150 years since it became demonstrably clear through spectroscopic studies that the Sun is an actual star and (up to a point) vice versa (Tayler 1989; Arny 1990). 136 years ago, for example, Arthur Searle commented in his *Outlines of Astronomy* (1875: p. 14), 'Very little, indeed, is known of the stars'. However, he later asserted (p. 53) that 'observations with the spectroscope have also confirmed the belief previously grounded on the brightness and remoteness of the stars, that they are bodies resembling the Sun'. Charles Young further wrote in his 1899 *A Text-Book of General Astronomy* (p. 184) that 'the Sun is simply a star; a hot, self luminous globe of enormous magnitude. . .although probably of medium size among its stellar compeers'. With this statement, Young confirmed the star-like nature of the Sun and introduced yet another characteristic, stating that the Sun is 'probably only of medium size'. Accordingly, not only are stars like the Sun, but there is also a range of stellar sizes, and by implication temperatures and masses as well. The fact that stars have varying degrees of energy output (luminosity) had already been established about 60 years before Young wrote his text.[1]

Hector Macpherson in his wonderfully named *The Romance of the Modern Astronomy* (1923: p. 185) picked up on Young's point by writing, 'the stars are Suns, This is a very good truth which we must bear in mind'. Macpherson continued (p. 193) to explain that the Sun is a yellow dwarf star. William Benton in his 1921 *Encyclopedia Britannica* entry concerning the Sun also commented upon its size and noted that, 'the Sun is apparently the largest and brightest of the stars visible to the naked eye, but it is actually among the smallest and faintest'. The comments by Macpherson and Benton, while in contrast to those of Young, actually build upon the monumentally important results of Ejnar Hertzsprung and Henry Norris Russell, who circa 1910 independently introduced the idea of dwarf and giant stars along with what has become known as the HR diagram (recall Fig. 1.9). In terms of stars being blackbody radiators (again a theory not actually established in its modern form until the appearance of the pioneering quantum mechanical model of Max Planck in 1900), the size (radius, R), temperature (T) and luminosity (L) are related according to the famous Stefan-Boltzmann law: $L = \text{constant } R^2 T^4$ (recall Eq. (1.14)). That the luminosity is further related to the mass of a star, as we have seen in Chap. 4, was a result established by Arthur Eddington in the early 1920s.

By arranging the stars in the HR diagram (recall Fig. 1.13) it is possible to begin comparing the Sun's physical characteristics against those of other stars in general. Accordingly, Simon Newcomb in his *Astronomy for Everybody* (published 1932) explained (p. 267), albeit rather tentatively, 'What we have learned about the Sun presumably applies in a general way to the stars'. With respect to the HR diagram

---

[1]This was evident as soon as the first (believable) stellar parallax measurements were published in 1838/9. Indeed, since the star Vega (as observed by Friedrich Struve) was found to be some 2.2 times further away than 61 Cygni (as observed by Friedrich Bessel) and yet was 6 magnitudes brighter, it must have a greater intrinsic luminosity.

(p. 274), he noted, 'The dot for the Sun, class[2] G0, is in the middle of the diagram'. With Newcomb's latter comment, we begin to see a new and quite specific picture of the Sun emerge: it is an average, middle-of-the-road sort of star. This point was emphasized by Arthur Eddington in his book *The Nature of the Physical World*, published in 1935. Eddington wrote (p. 164–5), 'Amid this great population [the galaxy] the Sun is a humble unit. It is a very ordinary star about midway in the scale of brilliance... In mass, in surface temperature, in bulk the Sun belongs to a very common class of stars'. To this he later added (in classic Eddingtonian language), 'In the community of stars the Sun corresponds to a respectable middle-class citizen'. Extending Eddington's anthropomorphic scheme of stellar personification, Eugene Parker (2000) described the Sun as being 'a pedestrian star'.

Even with the continued acquisition of data and the development of new astrophysical theories, from the 1930s onwards the idea of an ordinary Sun became entrenched within the minds of astronomers. A few examples, all gleaned from the general literature, of how the Sun has been systematically 'normalized' over the years are given in the list below:

- W. H. McCrea (1950): *Physics of the Sun and Stars.* (p. 58): 'The Sun is a typical star', and (p.106): 'it turns out, indeed, that the Sun is a pretty average star in almost every respect'.
- Deutsch (1962) in *Stars and Galaxies* (Ed. T. Page). (p. 44): 'The Sun is a typical; star, a hot sphere of gas'.
- J. Meadows (1967): *Stellar Evolution.* (p. 109): 'The Sun is so important to us that we tend to think of it as just a typical star. Yet its characteristics are perfectly normal'.
- J. P. Wild (1976): *Focus on the Stars.* (p.73): 'The Sun is located in a nondescript place within a spiral arm of the Galaxy'.
- C. Sagan (1980): *Cosmos.* (p. 243): 'Our ancestors worshiped the Sun, and they were far from foolish. And yet the Sun is an ordinary, even a mediocre star'.
- R. W. Noyes (1982): *The Sun: our star.* (p. 7): 'It turns out that our Sun is very much a run-of-the-mill star... [and it] lies midway along the main sequence'.
- H. Zirin (1988): *Astrophysics of the Sun.* (p. 1): 'Contrary to popular belief, it [the Sun] is a fairly large star; eighth brightest among the 100 nearest stars. Although it falls in the middle of the sequence of spectral classes, most stars are dwarfs of later and smaller types'.
- C. J. Caes (1988): *Studies in Starlight: understanding our universe.* (p. 132): "The Sun, as far as can be detected, is simply an average star. Generally speaking there is nothing especially unusual about its physics. It is of average temperature for a star, of average brightness, and of average age'.
- M. Zeilik and J. Gaustad (1990): *Astronomy: The Cosmic Perspective.* (p.374): 'What is the Sun? Basically, an ordinary star'.

---

[2]The Sun's spectral class is now taken to be G2.

- Stephen Hawking (1995): Radio interview with Ken Campbell: 'The human race is just a chemical scum on a moderate sized planet, orbiting a very average star in the outer suburb of one among a hundred billion galaxies'. Sited in Deutsch (2011)
- G. J. Babu and E. D. Feigelson (1996): *Astrostatistics*. (p. 26): 'Our solar system is located in a rather ordinary part of the Galactic disk.... Our Sun, formed 4.5×10^9 years ago,[3] is a typical star and most of its neighbors are similarly low mass, middle-aged main sequence stars'.
- L. Golub and J. M. Pasachoff (2001): *Nearest star; the surprising science of our Sun*. (preface): 'Our Sun is a fairly ordinary star, a bit brighter than most but not exceptionally so. There are many stars much bigger and brighter, while most stars are smaller and fainter'.
- S. G. Ryan and A. J. Norton (2010): *Stellar Evolution and Nucleosynthesis*. (p. 14): 'The Sun is a typical star'.
- C. Impey (2010): *How it Ends: from you to the universe*. (p. 160): 'The Milky Way, our position in it, and the star we orbit aren't unusual or special'.

The literature survey reveals a convergent trend that is remarkable in both being wrong and entirely misguided. Indeed, it is patently clear that the Sun is an extraordinarily special star. In spite of much historical nay-saying, the Sun's essential characteristics are not average or ordinary, and they do not even correspond with expectation when subjected to an analysis by the Copernican Principle—also called the Principle of Mediocrity.

In its modern form, the Copernican Principle has become something that would have entirely horrified Copernicus. This shift in interpretation aside, the Copernican Principle is generally expressed in a form that asserts the non-favored location and non-special viewpoint of humanity[4] within the universe. That is, we are not privileged or even unique observers of the cosmos. The idea behind the Principle is of general importance in the practice of science and in the field of cosmology appears to be demonstrably true: the universe on the large scale is isotropic and homogeneous, and our general viewing circumstances are no different from those of any other potential observer in it. Be that as it may, the point is that by adopting an unguarded devotion to the Copernican Principle, astronomers have (unintentionally) developed and propagated a set of incorrect conclusions about the place of the Sun in the cosmos. The argument apparently runs along the lines that since, by the Copernican Principle, humanity as observers of the universe are not specially located, the star (the Sun) about which the Earth orbits and from which humans observe cannot

---

[3]We note here that the only average or middling quantity that the Sun apparently has is that of its age, in the sense that in terms of its main sequence lifetime, it is middle-aged. This characteristic, ironically, raises interesting and difficult questions with respect to the apparently rapid appearance of life on Earth.

[4]In contrast to its modern usage, Copernicus would have been appalled by the idea that the Earth, Sun and humanity were not special, and indeed that they were not unique and highly favored entities.

be special, therefore it must be an average sort of stellar object in a non-special, "nondescript" location within the Milky Way galaxy. Such conclusions are demonstrably wrong.

Exercise 6.1: Term Paper Topic Investigate the historical and present day uses and abuses of The Copernican Principle. A good (but technically demanding) place to start is David Deutsch's book, *The Beginning of Infinity: explanations that transformed the world* (Allen Lane, 2011).

In recent years, many authors have suggested that the Sun is a typical or average star because it is located in the middle of the main sequence on the HR diagram. There are numerous problems with this deduction. Firstly, the HR diagram is invariably shown as a plot of logarithmic quantities, and this automatically compresses the range of values to be plotted (see the discussion in Spiegelhalter, Pearson and Short, 2011). If one drew the HR diagram with a linear scale, for example, the Sun would not be located in the (geometrical) middle range of stellar luminosities and/or temperatures. Secondly, if one takes the (naive) middling (arithmetic mean) value of any stellar quantity, then the resultant value is not a quantity typical of the Sun. For example, the range of stellar masses varies from about 0.1 to about 100 $M_\odot$, and the midway arithmetic mean value is accordingly 50.05 $M_\odot$—a quantity that is hardly the mass of the actual Sun. Third, and finally for this discussion, even if a star (the Sun or otherwise) is located at the midway point on the main sequence in the HR diagram, this does not mean that it must be the most likely or most typical star to be encountered within the Milky Way galaxy.

The Sun from our collected literary quotations has been described as 'typical', 'average', 'run-of-the-mill', 'ordinary', 'mediocre' and 'normal'. All such expressions are apparently employed in the sense that if one picked a star at random within the galaxy then it would be a Sun-like star, and/or if one measured a range of values for stellar mass, radius, temperature and luminosity, then their averages would all somehow reduce to intrinsic solar quantities: $1M_\odot$, $1R_\odot$, T ~ 5800 K and $1L_\odot$, respectively. There are clearly a number of problems with such expectations. When the Sun is described as being an 'average' or 'typical' star, it is rarely (in fact never, according to our literature survey) stated with respect to a specific distribution of stars. There are, for example, some very obvious comparisons where the Sun would be an extreme and highly untypical object. To the stars in a globular cluster, for example, the Sun would in comparison be an extremely young star with a very odd chemical composition (a high metal abundance). And yet, to the stars in a newly formed galactic cluster, the Sun would by comparison be a low-mass, low-luminosity, very old star, with a relatively low metal abundance. Even if we make a more sensible comparison between the Sun's properties and those stars that reside in the solar neighborhood, the Sun in no manner has typical stellar characteristics.

Statistically speaking, terms such as 'typical', 'most common' and 'average' will only coincide if the objects being studied have a normal (or Gaussian) distribution.

Table 6.1 Summary of RECONS data

Objects	Systems	O	B	A	F	G	K	M	WD	BD	Planets
369	256	0	0	4	6	20	44	247	20	28	15

The first column indicates the total number of objects (stars as well as white and brown dwarfs) within 10 pc of the Sun, while the second column indicates the number of stellar systems (single, binary, triple, etc). Columns three through nine indicate the number of stars of a given spectral type (the Sun, included in the dataset, is a G spectral type star). Columns 10 and 11 indicate the number of white dwarf (WD) and substellar brown dwarf (BD) objects. The last column indicates the number of planets so far detected. Data from www.recons.org

The most important point about stars is that their mass distribution[5] is decidedly non-normal—it is a strongly peaked negative power law distribution in stellar mass, with the lowest mass stars being by far the most numerous. As we shall see below, the most 'typical' star within the galaxy, and the star most likely to be encountered or 'picked' at random in the solar neighborhood, is an M dwarf main sequence star— one resembling Proxima Centauri with a mass of order 0.1 $M_\odot$.

The most complete catalog of stars located close to the Sun with well-measured physical characteristics is that provided by the Research Consortium On Nearby Stars (RECONS). Table 6.1 is a summary of the RECONS dataset for the stars located within 10 pc of the Sun (see Henry et al., 2006 and www.recons.org/). It is believed that the vast majority of stellar objects within 10 pc of the Solar System are identified within the RECONS catalog (this result is probably not true for the Brown dwarfs, but they do not concern us here), and it is also believed that the dataset is representative of that which might be found in any region of the galaxy at the Sun's galactocentric distance of 8 kpc. A quick glance at the entries in Table 6.1 immediately indicates a predominance of low-mass, low-temperature, small radii K and M spectral type dwarf stars. This is entirely consistent with the form of the initial mass function (IMF) described at the beginning of Chap. 5.

The number of stars of mass M, within the RECONS 10 pc catalog, is described by a power law with the mass function $N(M) = 4.6 / M^{1.20}$. If there were equal numbers of objects at any given stellar mass, then the exponent in the mass function would be zero, but as it stands, of the 320 stars in the 10 pc survey the Sun is among the top 25 most massive. The most massive star within 10 pc of the Sun is Vega, weighing in at 2.135 ± 0.074 $M_\odot$ (Yoon et al., 2010). The modal—most common— mass value in the 10 pc survey falls in the range between 0.1 and 0.15 $M_\odot$, and the median value, for which half of the systems have a greater mass and half have a smaller mass, is 0.35 $M_\odot$. That the latter results are further typical for the rest of the galaxy is revealed by the available data relating to the stellar initial mass function (IMF). While the slope of the IMF varies in a complex manner according to the stellar mass range, the peak number of stars formed is invariably (even universally)

---

[5]The mass distribution is the most important quantity since on the main sequence, where stars spend most of their lifetime, the radius, luminosity and temperature are all determined by the amount of stellar material.

found to fall in the 0.1 to 0.5 $M_\odot$ mass range. (Bastian et al., 2010, and recall exercise 5.1).

Sun-like stars having $M \approx 1$ $M_\odot$ and a G spectral type are found to make up just 6% of the stars within the 10 pc volume. In contrast, the M spectral type stars constitute 77% of the total number. In addition, the modal absolute magnitude for the stars in the 10 pc data set is found to be $M_V \approx +13.5$, a value some 8.5 magnitudes fainter than that of the Sun. Compared to the most typical (that is, ordinary, common, run-of-the-mill, pedestrian, etc.) star in the solar neighborhood, the Sun is nearly ten times more massive, ten times larger, two times hotter and ten thousand times more luminous. The Sun is not a typical star even within its own precinct.

Given that the Sun is not an average, ordinary or typical star within the solar neighborhood, does it qualify as special? The question is not meant to focus on humanity's existence being dependent upon it, as in this sense it is obviously very special and we would not exist without it. Rather, the question refers to its extensive characteristics, such as being a single star, and then a single star with an attendant planetary system, and so on. Again, one can turn to reasonably well-known and well-understood datasets to answer this question. Following Dole (1964) and Adams (2010), the answer to our question can be expressed as a probability. Accordingly, the probability P_{Sun} of finding a star within the galaxy having similar observable characteristics as the Sun can be written in the form of a Drake-like equation[6]:

$$P_{Sun} = 100 \times F_1 \, F_{SB} \, F_Z \, F_P \, F_H \qquad (6.1)$$

The terms entering Eq. (6.1) relate to F_1, the fraction of stars with a mass equal to 1 $M_\odot$; F_{SB} the fraction of solar mass stars that are single as opposed to being members of a binary or multiple system; F_Z, the fraction of stars with a metal abundance corresponding to that of the Sun at the Sun's location within the galaxy; F_P, the fraction of solar mass stars harboring planets; and F_H, the fraction of planet-harboring Sun-like stars in which one (or more) might reside within the habitability zone. All of the terms in Eq. (6.1), in contrast to those in Drake's more famous (perhaps notorious) equation, [6] are reasonably well known. Looking at each quantity in turn, it is evident that $F_1 = 0.06$, corresponding to the fraction of spectral type G stars within the annotated spectral sequence distribution (see e.g. Table 6.1; Mihalas and Binney, 1981). To a good approximation $F_{SB} = 1/3$, with the majority of Sun-like stars being found in binary systems (such as in the case of our nearest neighboring system α-Centauri AB). F_Z is again reasonably well constrained, and the Sun in fact has a relatively high metal abundance, with the survey by Rocha-Pinto and Maciel (1996) indicating that within the solar neighborhood,$= F_Z = 0.25$ for $Z \geq Z_\odot$. Indeed, it should be noted that the composition exhibited by the Sun does not correspond to just any radial location within the Milky Way galaxy—a condition that negates statements that imply the Sun is somehow situated in an

---

[6]The parallel here is to Frank Drake's famous equation for estimating the number of extraterrestrial civilizations within the Milky Way galaxy.

ordinary or nondescript region of the galaxy. The fraction of Sun-like stars supporting large planets is known to vary with the composition of the parent star (and hence galactic location), and following the study by Wyatt et al. (2007), the fraction of Sun-like stars with Jovian planets varies as $F_P = 0.03 \times 10^{Z/Z_\odot}$; which is to suggest that $F_P = 0.3$. And finally, the least well-known quantity is that related to the finding of a planet within the habitability zone. At present, this number may only be constrained via theoretical modeling, and, for example, Gowanlock et al. (2011) have argue that for planet-hosting systems $F_H = 0.05$.

With our various quantities now in place, the following evaluation is found: $P_{Sun} \approx 0.01\%$. In other words, if one picked a star at random within our galaxy, there is a 99.99% chance that it will *not* have the same intrinsic characteristics as our Sun and (basic) Solar System. Clearly, the Sun is not an ordinary star. In addition, the special characteristics associated with the Sun and Solar System apply irrespective of the origins of life on the habitable planet. If we wish to include our own existence in the calculation, then P_{Sun} will be (perhaps according to one's bias) many orders of magnitude smaller. Irrespective of this addition, by any reasonable standard, the Sun and its attendant planets constitute a rare and uncommon type of system within our galaxy.

The main aim of this discussion has been to argue that, counter to common convention, the Sun is a very special star. It is special most obviously from an astrobiological viewpoint for supporting a habitable planet. Less obviously, it is a non-typical star within the galaxy—the most typical or ordinary kind of star within the solar neighborhood (and the larger galaxy) being something more like Proxima Centauri: a low-mass, M dwarf main sequence star. The Sun and Sun-like stars are not now, never were, nor ever be typical objects within the Milky Way galaxy. Not only is the Sun an atypical star, but it is a rare and special type of solar mass star in that it is not located within a binary system, it is located at some 8 kpc from the galactic center (a non-random location that links directly to its metallicity), and that it is the host star of a planetary system of which one planet resides within its current habitability zone. Furthermore, the Sun is actually a relatively young star for its position within the galaxy: most solar mass stars situated within the galactic habitability zone (GHZ), a region located within the annulus having radii between 7 and 9 kpc from the galactic center, are estimated to be at least of order 1 billion years older than the Sun (Lineweaver, Fenner and Gibson, 2004).[7]

In his recent far-ranging book *The Beginning of Infinity: explanations that transform the world* (published 2011), physicist David Deutsch took specific aim at the philosophy underlying the Copernican Principal (and/or the Principal of Mediocrity). He described it as being a paradoxical, parochial and philosophically limiting tool upon which to base any work. Indeed, it is by a blinkered application of this very Principle that has resulted in the incorrect (now long-running) misconception of the Sun—a false deduction based upon the idea that there can be

---

[7]Lineweaver, Fenner and Gibson (2004) also point out that at the present epoch the GHZ is composed of less than 10% of the stars that have ever formed within the Milky Way.

nothing special about the circumstances under which humanity has evolved to become an observer of the large-scale universe. To deny the highly exceptional and unlikely manner in which the Solar System came about, along with its various specific characteristics, and how intelligent life managed to evolve so rapidly on planet Earth, only blinkers and limits our understanding of life's very origins. It should be emphasized that there is no additional value to be associated with the Sun's rare and highly exceptional characteristics: they are, quite simply, what they are observed and measured to be.

6.2 Supernovae and Life on Earth

Supernovae are the bright jewels in the darkness of the universe. They shine with a brightness (luminosity) that outstrips even that of their parent galaxies, and they are the great movers and shakers of the interstellar medium. Indeed, our very existence is predicated on the fact that supernovae can and do occur. This latter condition need not always apply, and some universes could be constructed in which no stars undergo supernova disruption (see Sect. 6.5).

Since the earliest of times, supernovae have inspired the thoughts and actions of sky-watchers. Many ancient records point to the existence of 'guest stars'; it has been made clear from the late-sixteenth century onwards, specifically through the observations of Tycho Brahe and Johannes Kepler, that supernovae are phenomena set within the stellar realm (Clark and Stephenson 1977). In the modern era, supernovae are known to be responsible in part for the generation of cosmic rays and the enhancement of the metal abundance within the interstellar medium.

From an astrobiological perspective, supernovae are essential for producing the critical atomic elements out of which terrestrial planets can form and from which life can ultimately evolve. From an astroengineering perspective, a pre-supernova star situated close to a planet supporting an advanced civilization is a direct incentive to either initiate interstellar migration or begin large-scale engineering projects to minimize the potential threat, such structures possibly being detectable at interstellar distances (see, e.g., Beech 2008). From an anthropocentric perspective, the very origins of the Solar System (and indirectly us) were initiated by a supernova explosion, as is interpreted by the presence of ^{60}Ni within meteorite fragments (Bizzarro et al. 2007). This particular isotope, only produced via supernova nucleosynthesis, is derived from the decay of ^{60}Fe, which has a short (relative to the age of the Solar System) half-life of 2.6 million years. The presence of ^{60}Ni in the meteorite precursor bodies therefore indicates that ^{60}Fe must have been well mixed within the natal solar nebular.

In the strange sense of being both potential midwife and undertaker, nearby supernovae may have extinguished, or at the very least placed under considerable environmental stress, life on prehistoric Earth and possibly on a young (Noachian epoch) Mars. An understanding of the various supernovae types and their formation mechanisms is thus more than just an academic exercise: it is a topic that makes contact with Earth's distant past and will have a direct bearing at various times in the future upon the structure of the biosphere and the lives of our distant descendants. At the present epoch, the closest stellar system to the Sun that will ultimately end its days as a supernova is that of IK Pegasi, a binary star system containing one of the most massive white dwarf stars known (recall Fig. 5.12), located some 46 parsecs away (Beech 2011). While currently the closest supernova progenitor system to us, IK Pegasi affords us absolutely no cause for concern, since its detonation time is set several billion years into the future, at which time the system will be many kiloparsecs distant (recall exercise 1.25).

There are strong selection effects acting against detecting supernovae within the disk of the Milky Way galaxy—dust within the interstellar medium being the primary extinction mechanism at optical wavelengths. Accordingly, only five supernovae have been observed from Earth (that is, achieved naked-eye visibility) during the past 1000 years. Chinese chronicles list 'guest stars' for April 1006 and August 1181, and these are believed to have been Type II supernovae (recall Sect. 5.10, and Fig. 5.13 for the SN classification scheme). The well-known Crab Nebula, associated with SN 1054, was also produced by a Type II detonation, while SN 1572 (Tycho's Nova) and SN 1604 (Kepler's Nova) are both remnants of Type Ia supernovae. The supernova remnant Cassiopeia A (SN 1680—1667?) was possibly observed, without realizing it, by John Flamsteed, who was at the time mapping the appropriate part of the sky, but the full circumstances of the situation remain unclear (Clark and Stephenson 1977). The youngest known galactic supernova remnant is SNR: G1.9 + 0.3, which was formed some 150 years ago (Carlton et al., 2011). Situated at an estimated 8.5 kpc away, the supernova most likely achieved naked-eye visibility and certainly contemporaneous telescopic brightness, and yet it was not actually recorded. Reynolds et al. (2008) suggest that the precursor body to G1.9 + 0.3 was a Type Ia supernova. Thus, in the past millennium, the observational record indicates that perhaps seven supernovae (here including SNR Cas A and G1.9 + 0.3) have attained naked-eye brightness, of which three are believed to be Type Ia and 4 Type II supernovae. Given that the expected supernova rate for the Milky Way galaxy (as deduced from observations of other spiral galaxies) is about 19 per millennium, the historical results suggest that of order 1 in 3 supernovae detonations are actually seen from Earth at naked-eye brightness.

Exercise 6.2: Term Paper Topic Investigate the historical and present-day study of the Crab Nebula.

A good number of supernova progenitor systems have now been identified, and Table 6.2 provides a list of those systems that reside within 1 kpc of the Sun at the present epoch. None of the systems listed in Table 6.2 pose any actual or immediate threat to the biosphere or life on Earth; they are simply reasonably well-understood (in some cases) stellar systems that will eventually end their days by undergoing supernova disruption. Stars such as Betelgeuse (α Ori), and HD 168625 are included in the list because they are supergiants of luminosity class Ia and must accordingly be massive, and because they are believed to be in a relatively advanced evolutionary state. The binary systems T Pyxidis and RS Ophiuchi are included since they are

Table 6.2 Supernova candidate systems within 1 Kpc of the Sun

System	D(pc)	Characteristics	SN Type (?)
IK Pegasi	46	A8m + DA	Ia
Betelgeuse (α Ori)	197	M2 Iab	II
WR 11 (γ² Vel)	260	WC8 + O7.5 III-V	Ic
RS Ophiuchi	> 600	M2 III + D [RN]	Ia
WR 147 (AS 431)	650	WN8 + B0.5 V	Ib
HD 168625	675	B6 Ia	II
WR 146 (HM19–3)	720	WC6 + O8	Ic
WR 144 (HM19–1)	790	WN4	Ib
WR 42b (SMSP 10)	870	WN4	Ib
WR 142 (sand 5)	950	WO2	Ic
WR6 (HD 50896)	970	WN4	Ib
T Pyxidis	~1000	? + D [RN]	Ia

Columns 1 and 2 provide the identification and distance to the candidate system. Column 3 indicates the system spectral type, and column 4 is our suggestion for the eventual supernova type that might be produced. Data for the Wolf-Rayet systems is from van der Hucht (2001). [RN] signifies that the system is a known recurrent nova

known recurrent novae, and the white dwarf component within these systems must therefore be undergoing some form of active accretion. IK Pegasi is included not because it has shown any recurrent nova-like activity, but because it contains one of the most massive white dwarfs ever recorded, weighing in at some 84% of the Chandrasekhar limiting mass. Several Wolf-Rayet stars are included in Table 6.2 since they correspond to an advanced stage of massive star evolution.

Column 4 of Table 6.2 indicates the possible supernova outcome for each system. Given the approximate correctness of the final supernova types, the data in Table 6.2 suggests that nine of the systems will eventually undergo disruption as Type II or Ib/c supernova, and three might undergo Type Ia disruption. Clearly, the distances (and order) to the systems listed in Table 6.2, at the point of their eventual supernova disruption, will be different to those given in column 2, and while IK Pegasi is currently the closest precursor supernova system to the Sun, Betelgeuse may well be the next closest supernova to physically occur—by this we mean spatially closest, rather than temporally, although we cannot be certain on either of these points since the exact evolutionary status, and more specifically the age and initial mass, of most of the systems listed in Table 6.2 are not precisely (if at all) known.

The potential number of stars within 1 kpc of the Sun that might eventually undergo Type II supernova disruption can be estimated from the number density of stars more massive than 8 $M_\odot$. This number essentially corresponds to the population of spectral type OV stars in the solar neighborhood, and taking for these stars a number density of 2.5×10^{-8} /pc^3 (Mihalas and Binney 1981), we arrive at a total of about 100 supernova precursor stars within 1 kpc of the Sun. Taking the lifetime of such stars to be <30 million years and adopting a characteristic space velocity of 25 km/s, then perhaps the subset of precursor OV stars formed within about 750 pc of the Sun (some 40 stars) might eventually detonate as a nearby (within a few tens of parsecs) supernova. These are Earth's potential future shock companions.

Supernova explosions pose a number of hazards for any nearby planetary systems, including various radiation hazards and a kinetic threat due to the rapidly expanding gas shell. Characteristically, the energy liberated during a supernova disruption is of order 10^{46} Joules, an amount equivalent to the rest mass energy of some 10^{29} kg of matter. Most of this energy is carried away by neutrinos, with the time-integrated luminosity of electromagnetic radiation accounting for of order 1% of the total energy available. If of order 1 $M_{\odot}$ of envelop material is ejected into space during the supernova disruption, then the typical observed ejection velocity of 10^4 km/s indicates that of order 1% of the total energy is entrained within the kinetic energy of the shell.

Detailed numerical calculations relating to the breakout of the shock front at the stellar surface (Klein and Chevalier 1978; Falk 1978) indicate that during the initial phase of supernova disruption, a hard-UV / soft X-ray luminosity of $L_X \sim 10^{38}$ J/s is emitted over a time interval of order 10^3 s. This relatively short duration pulse of energy is the first hazard posed to any nearby planetary systems. A longer duration hazard lasting many years and potentially several centuries will then develop due to the decay of the various radionucleids produced during the rapid neutron capture nucleosynthesis phase. This phase follows in the wake of the outward-propagating shockwave as it pushes through the progenitor star's outer envelope. Such radionucleid decays typically produce emission line gamma and X-ray radiation with luminosities of order 10^{30} J/s. The Crab Nebula, for example, has an X-ray (0.2 – 4 keV band) luminosity of 2.4×10^{30} J/s and a gamma ray luminosity (at 100 keV) of $L_{\gamma} \sim 5 \times 10^{28}$ J/s at the present epoch. Likewise, the Cas A remnant produces $L_X \sim 10^{31}$ J/s and L_{γ} (at 1.157 MeV) $\sim 5 \times 10^{28}$ J/s. In the Crab Nebula, the radiation is primarily due to the presence of a central pulsar, and the initial gamma and X-ray luminosities in its immediate environment would have been several orders of magnitude higher than 10^{30} J/s due to the initial predominance of ^{56}Co in the supernova envelope, but this would have rapidly decayed with a half-life of 77.27 days.

Given a prompt X-ray luminosity of L_X, and a shock breakout time of T_{sb}, the equivalent radiation dosage R (measured in Sieverts) at a distance D from the supernova will be:

$$R\,(Sv) = Q\omega \frac{L_X}{4\pi D^2}\ T_{sb}\left(\frac{A}{m}\right) \tag{6.2}$$

where m is the mass and A the surface area of the irradiated body, ω is a radiation weighting factor and $1 \geq Q \geq 0$ is a shielding factor. The shielding factor will be a complicated function of the density of any interstellar gas and dust between a planet and the supernova, as well as the composition and extent of any planet surrounding atmosphere. For the moment we do not address the scattering and absorption effects, but note that for the Earth it is estimated that of order 1% of the X-ray radiation

Fig. 6.1 Irradiation dosage at the top of Earth's atmosphere versus distance to the supernova. We adopt the following parameters for Eq. (6.2): $L_X = 10^{38}$ J, $T_{sb} = 10^3$ sec, $A/m = 0.0125$. The curves are labeled according to the shielding factor Q, and the dotted horizontal line indicates a lethal dosage level of 10 Gray. (Credit: Author)

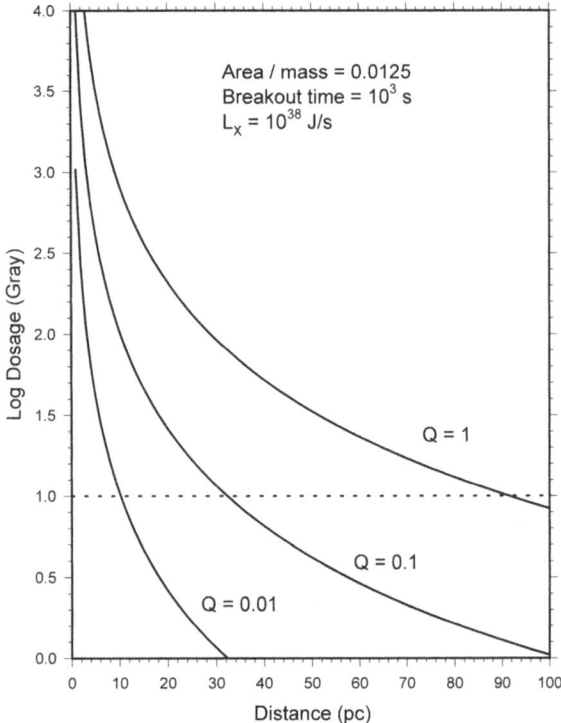

incident on the top of the atmosphere will find its way via scattering and secondary emission to the ground (Scalo and Wheeler, 2002). In contrast, more incident radiation will penetrate to the ground on Mars because of its less extensive atmosphere, and on the Moon there is clearly no atmospheric absorption at all—only substructure shielding will protect any humans who might eventually be living there. The ω term is introduced into Eq. (6.2) to allow for the fact that the biological effects will vary according to the wavelength and type of the incident radiation. For gamma and X-rays, however, ω is essentially equal to one. Figure 6.1 shows a series of representative dosage versus distance plots for typical values of $L_X = 10^{38}$ J, $T_{sb} = 10^3$ sec, assuming an area-to-mass ratio of 0.0125 (corresponding to that of a human being with $A = 1$ m^2 and $m = 80$ kg). A critical radiation radius, D_{crit}, can be derived from Eq. (6.2) on the basis that the gamma and X-ray radiation dosage exceeds some specified value. The generally accepted lethal dosage beyond which human death is inevitable is 8 Gray,[8] and accordingly this dosage is realized in the zero shielding ($Q = 1$) case when $D < 90$ pc. Allowing for 1% of any incident radiation to actually penetrate to the Earth's surface ($Q = 0.01$), the lethality distance

---

[8]The Gray relates to the absorbed dosage—the absorption of one Joule of radiation energy per kilogram of matter. The Sievert is also expressed in Joules per kilogram, but the kilogram in this case is specifically taken as being composed of human tissue.

for humans (here given dominance in our concerns) is reduced to $D < 10$ pc. Even a 1–2 Gray radiation dosage will result in human fatigue and a risk of reduced long-term health prospects through leucopenia, and in the 99% atmospheric shielding situation this requires $D < 30$ pc.

Data derived from atomic bomb testing experiments indicates that the acute lethal radiation dose for most plants ranges from ~10 to ~ 100 Gray (Woodwell, 1967; Terry and Tucker, 1968). Studies carried out at the Brookhaven National Laboratory by Woodwell (1967) further indicate that forest productivity is significantly downgraded for radiation dosages approaching that of lethality to humans. Pine forests are particularly sensitive to radiation damage, and exposure levels of just a few tenths of a Gray per day invariably result in tree death within 6 months. Oak trees are apparently somewhat more resilient than pines, but again dosages of a few Gray per day result in 100% mortality after several months. Grasses, mosses and lichens, in contrast to most plants and trees, are able to survive extremely high radiation dosages (in excess of several tens of Gray per day) for multiple years on end (Woodwell, 1967). Clearly, measurable radiation damage / lethality effects, whether related to humans, land/marine animals, insects, trees or plants, will vary considerably according to the dosage received, but in general it would also appear that for a supernova to have any discernable radiation damage effects, the critical distance cannot be any larger than 15 to perhaps 30 pc. Beyond this distance, any radiation effects by supernova upon land and marine biota will be essentially zero.

Exercise 6.3: Term Paper Topic Investigate a number of cases in which new technology, initially developed for military purposes, has been adapted for astronomical research. Examples include the CCD camera and satellite technology at optical and gamma ray wavelengths. See Neil deGrasse Tyson and Avis Lang's book, *Accessory to War: the unspoken alliance between astrophysics and the military* (W. W. Norton, 2018).

Having established an estimate for the critical distance corresponding to a specific radiation dosage, an estimate of the time interval T_{SN} between successive supernova events at the critical distance can be determined by dividing the area of the galactic disk, assumed to have a radius of 15 kpc, by the area of a disk of radius D_{crit}, and then dividing the ratio of areas by the galactic supernova rate of 0.02 supernovae (of all types) per year (van den Bergh and Tammann, 1991). If supernovae are assumed to occur at random locations within the galactic disk, the typical time interval for a supernova occurring with D_{crit} of the Sun will be

$$T_{SN} = \left(\frac{Area\,of\,galactic\,disk}{Area\,of\,critical\,SN\,disk} \right) \left(\frac{1}{galactic\,SN\,rate} \right)$$

$$= \left(\frac{15,000}{D_{crit}} \right)^2 \frac{1}{0.02} \quad \text{years.} \tag{6.3}$$

With $D_{crit} = 10$ pc, the time interval between successive events is expected to be of order $T_{SN} = 112$ Myr. This time interval caries a number of uncertainties and, for example, will be larger if one allows for the fact that most supernovae occur interior to the Sun's orbit about the galactic center (Lineweaver, Fenner and Gibson, 2004; Bartunov, Tsvetkov and Pavlyuk, 2006). The time interval will alternatively be shorter during the times of spiral arm crossing (Clark, McCrea and Stephenson, 1977; Leitch and Vasisht, 1998; Gies and Helsel, 2005), since during such epochs the Sun will encounter a greater number of massive, newly formed stars and young star clusters. The Sun encounters spiral arm structures at intervals of order 100 Myr, and each crossing takes about 10 Myr. Once again, the key point is that the supernova threat to the biosphere will vary from one epoch to the next, but it is unlikely to be shorter than ~ 75 Myr or longer than ~150 Myr.

In addition to high gamma and X-ray fluxes being produced during supernova disruptions, copious amounts of cosmic rays will also be generated. While it is generally accepted that the vast majority of cosmic rays with energies between 10^{10}–10^{15} eV (the so-called 'pre-knee' component in the cosmic ray energy spectrum) are produced in supernova events and within supernova remnants, it is far from clear what the actual flux and duration of the cosmic rays produced by any given supernova will be. Current observations suggest that cosmic rays carry away of order 10% of the supernova shell's kinetic energy (amounting, therefore, to some 10^{42}–10^{43} Joules), with active generation taking place on timescales of order 10^3 – 10^4 years.

The key hazards associated with an enhanced cosmic and gamma ray fluxes were outlined by Mal Ruderman (1974) and Gary Hunt (1978), and they primarily entail the generation of stratospheric nitric oxide (NO). The nitric oxide, importantly, initiates the catalytic destruction of ozone (O_3) and this accordingly weakens the atmosphere's ability to filter out harmful solar UV radiation. In this process, the incoming cosmic and/or gamma rays break down atmospheric molecular nitrogen (N_2) thereby enabling its two component nitrogen atoms to react with molecular oxygen (O_2) to form nitric oxide via the reaction: $N + O_2 \Rightarrow NO + O$. Importantly, the NO will begin to destroy atmospheric ozone via the reaction $NO + O_3 \Rightarrow NO_2 + O_2$ to produce nitrogen dioxide (NO_2). In turn, the NO_2 will react with atomic oxygen (O) to produce nitric oxide (NO): $NO_2 + O \Rightarrow NO + O_2$. This reaction is offset by the reaction $NO + N \Rightarrow N_2 + O$. The various reaction rates, however, see an overwhelming production of NO, leading to the rapid depletion of ozone. Accordingly, the atmosphere's ability to absorb the Sun's harmful UV radiation is compromised. Detailed calculations (Merlott et al., 2004) indicate that it might take between 5 – 10 years for the atmospheric ozone layer to recover from a significant, nearby, supernova-produced gamma ray hit.

Following Ruderman (1974), a typical supernova event situated some 15 pc distant from the Sun could produce a 50 times enhancement in the cosmic ray flux at the top of Earth's atmosphere for about 10^3 years. Such an enhanced flux would result in the significant reduction of stratospheric ozone (Hunt 1978; Thomas et al. 2005) and a concomitant period of distinct stress being placed upon the biosphere and surface life. Given the similarity of critical distances, the time interval between successive supernova events capable of producing significant ozone depletion in Earth's atmosphere will be about the same as that found for a lethal radiation dose and of order 100 Myr. Interestingly, a study of the cosmic ray exposure ages of iron meteorites by Shaviv (2003) has revealed evidence for a possible 143 ± 10 Myr periodicity in the cosmic ray flux during the past $\sim$700 My. Shaviv identifies the times of enhanced cosmic ray flux seen in the meteorite data with spiral arm crossing events (although we would comment that the correlation is not fully convincing), arguing that the variation is due to the Solar System being embedded within a more active supernova environment. There is some evidence to link the occurrences of ice ages with spiral arm passage events (Shaviv 2003; Gies and Helsel 2005), but it is far from clear how one might unambiguously disentangle this particular agent from the climate forcing induced by the (normally invoked) Croll-Milankovitch effect for glaciation cycles (Williams, and Pollard 2003).

The long-term, widespread environmental effects of a supernova explosion on the interstellar medium are illustrated at least in part by the Geminga supernova event that occurred an estimated 340,000 years ago. Indeed, it has been suggested that this particular supernova might be responsible for the otherwise puzzling low density of the interstellar medium (the so-called Local Bubble) in the solar neighborhood (Gehrels and Chen 1993; Salvati and Sacco 2008). The distance to the Geminga supernova at the time of its disruption is uncertain, but Pellizza et al. (2005) suggest a likely distance of between 90 and 240 pc from the Sun. They also argue that the formation location of the progenitor star was either within the Orion OB1 association (currently located between 240 and 340 pc from the Sun) or the Cassiopeia-Taurus OB association (presently located between 125 and 300 pc from the Sun). At a minimum distance of 90 pc from the Sun, the gamma and X-ray radiation dosage from the Geminga supernova at the Earth's surface would have been all but negligible (see Fig. 6.1). Likewise, the long-term ionization effects upon Earth's atmosphere by Geminga-produced cosmic rays would have been essentially negligible. These two results argue against the likelihood of any prolonged biospheric stress and/or mass extinction being associated with the Geminga event (this statement is made in contrast to the conclusions presented by Thomas et al. 2005; Ellis and Schramm 1995). Further, it is highly unlikely that the Local Bubble is entirely due to the Geminga supernova alone; rather, the Geminga event is simply the most recent supernova of the many that have acted to sculpt the local interstellar medium during the past 10^5 to 10^6 years.

The blast wave associated with the Geminga supernova would not reach the Solar System until at least $\sim$9000 years after the initial detonation, assuming a constant expansion velocity of 10,000 km/s (an upper limit value) and a (minimum) distance of 90 pc. As a supernova blast wave expands and propagates through the interstellar medium, it will loose both speed and energy, and if the encounter with the Solar System is going to have any noticeable effects, then at a minimum the ram pressure $P_{ram}(ISM)$ associated with the blast wave must be comparable to that of the solar wind, $P_{ram}(\odot)$. The ram pressure is related to the number density of particles n and their characteristic speed squared: $P_{ram}(ISM) \approx m_p\, n_{SN}\, V_\infty^2$, where m_p is the proton mass and n_{SN} is the number density of particles associated the expanding nebular, which is traveling at speed V_∞. The Solar wind ram pressure at a distance R from the Sun will be $P_{ram}(\odot) \approx m_p n_w V_w^2 (R_E/R)^2$, where $R_E = 1$ AU is the Earth's orbital radius. Characteristically, for the solar wind, $n_w \sim 7 \times 10^6$ m^{-3}, and $V_w \sim 4 \times 10^5$ m/s at $R = R_E$. For the ram pressure of the expanding nebular to be similar to that of the solar wind at 1 AU, the detonation distance must be closer than about 10–20 pc (Fields, Athanassiadou, Johnson, 2008). Only under such immediate detonation conditions might the supernova blast wave pose a direct kinetic (buck-shot-like) disruption threat to Earth's atmosphere and directly deposit radioisotope debris upon Earth's surface. In the case of the Geminga supernova, the minimum detonation distance of 90 pc excludes its expanding blast wave as being a meaningful agent for the onset of a mass extinction event and, for that matter, even having a noticeable effect upon Earth's atmosphere.

In contrast to the (as we have argued) overly interpreted Geminga situation, Benitez, Maiz-Apellaniz and Canelles (2002) have suggested that the active ^{60}Fe excess found in a core sample extracted from the ferromanganese ocean floor layers, deposited between three to six million years ago in the Pacific Ocean (Ocean Drilling Program sediment core 848, Leg 138), might have a supernova origin. This physical observation and its interpretation as a supernova-related phenomenon, is highly compelling. Specifically, Benitez et al. (2002) argue that the ^{60}Fe peak is due to the Solar System sweeping up the accumulated radionucleid debris from numerous supernovae located with the Scorpius-Centaurus association. The Sco-Cen associa-tion is presently situated about 150 pc away from the Solar System, but some 2–4 Myr ago it was considerably closer—indeed, segments of the association may have been as close as 50–100 pc. This distance, however, is still not close enough to reasonably explain significant ^{60}Fe accretion by the Earth. Rather, it is perhaps possible that a single close supernova (not inconceivably a high-velocity massive star ejected from the Sco-Cen association) occurred about three million years ago (Knie et al. 1999; Bishop and Egli 2011).

A full interpretation of the Pacific Ocean core sample is far from being complete, since no ^{60}Fe excess was found in a contemporaneous layer (Ocean Drilling Program sediment core 985, Leg 162) formed in the Atlantic Ocean (Fitoussi et al. 2008). This latter result weakens the argument that the ^{60}Fe was deposited globally over an extended period of time. Furthermore, Basu et al. (2007) have suggested that the ^{60}Fe excess detected in the Pacific Ocean core might be due to micrometeorite accretion and contamination, the ^{60}Fe being generated through galactic cosmic ray spallation of Ni while the meteoroids are unshielded in interplanetary space. With little doubt, the study of seafloor core samples is an exciting field to further pursue, with quantifiable data being (potentially) available, but it has not as yet produced a fully convincing argument for a clear and unambiguous supernova-related signal. However, the investigation of additional seafloor core samples continues, and recent work by Bishop and Egli (2011) has further identified a new approach to the study of ^{60}Fe deposition epochs via the analysis of single-domain magnetite crystals embed-ded within the microfossils of magnetotactic bacteria.

Among the various types of recognized supernovae are the rare, so-called hypernova subgroup thought to be responsible for producing some of the observed long-duration, high-energy gamma ray bursts (GRB). Such objects are thought to be produced by the core collapse of the most massive of stars that can possibly form (masses between 50–150 $M_\odot$) and which chance to be spinning rapidly (Woosley 2010; Woosley and Bloom 2006). The hypernovae are indeed rare, with perhaps 1 in 5000 (even 1 in 10,000) of supernova events producing such an outcome (Scalo and Wheeler, 2002; Fryer et al. 2007). Much has been made of the fact that GRBs tend to produce collimated jets of γ-ray emissions that, should one intercept the Earth, would cause significant biosphere degradation (primarily through the destruction of atmospheric ozone) even if situated of order 1 kpc away from the Sun. Since, however, GRBs appear to be much more prevalent in young, low-metallicity galaxies, it is quite likely that within our Milky Way galaxy this particular hazard is now all but negligible. We note however that Annis (1999) has argued that one of the key phase transitions any galaxy must pass through in order for life—and specifically intelligent life—to have a chance of evolving is that relating to the near cessation (or at least dramatic decline) in GRB activity. This being said, detailed numerical simulations by Gowanlock et al. (2011) following the time evolution of habitable planets within the Milky Way galaxy find that between 35 and 50% of the systems located 8 kpc from the galactic center are never sterilized by GRBs or nearby supernova.

The idea that nearby supernovae might have dramatic effects on Earth's atmosphere and thereby produce distinct extinction epochs was developed during the 1950s and 60s, predominantly by German paleontologist Otto Schindewolf (Albritton, 1989). Indeed, Schindewolf argued that the sudden onset of marine, insect and land vertebrate extinctions marking the end of the Permian some 250 million years ago was due to a supernova event. While Schindewolf proposed (without any direct evidence other than the abrupt extinction onset in the fossil record) that an extraterrestrial event produced the end Permian mass extinction, this particular episode in Earth history is now more generally associated with a time of global warming, possibly being brought about through the prolonged and massive basaltic outflows responsible for generation of the Siberian traps (Hallam, 2005). Terry and Tucker (1968) appear to be the first collaboration of a biochemist and astronomer to have considered the radiation flux and biological hazard associated with supernova. Using atomic bomb testing data relating to lethal dosages, these authors argued that supernovae closer than about 50 pc might indeed trigger extinction events on Earth, and that the extinction impact might reasonably be different for marine and land biota.

Exercise 6.4: Term Paper Topic Investigate the possible threats to life on Earth (mass extinctions) as posed by astronomical phenomenon. This topic can include not only supernova irradiation effects but those threats (among others) due to cometary and asteroid impacts—even hostile alien invasion could be included. You may wish to start by checking out my book, *Rejuvenating the Sun and avoiding other Global Disasters* (Springer, New York. 2008). See also David Raup's (excellent) book, *The Nemesis Affair: a story of the death of Dinosaurs and the ways of science* (W. W. Norton, New York, 1999).

Counter to what might at first be imagined, supernovae and GRBs were possibly less effective as evolutionary forcing and extinction agents in Earth's distant past when the atmosphere was less oxygen rich. Indeed, a detailed study by Martin et al. (2009) finds that the lethality distance was of order ten times smaller ($\sim$ 200 pc) in the Archean eon and Early Proterozoic era (3.8 – 2.5 billion years ago), when Earth's atmospheric oxygen content was some 10^{-5} times smaller than at present. The specific mechanism studied by Martin et al. (2009) relates to the UV flash emission (typically lasting about 10 s) derived from the interaction of primary γ- and X-ray radiation with atmospheric oxygen. Theses authors asses the effects of such a UV burst in terms of the daily equivalent of solar UV at the same wavelengths, and assume (following Thomas et al., 2005) that a factor of two or larger increase will be lethal to surface-dwelling biota (e.g., in the Archean epoch, cyanobacteria).

Importantly, Martin et al. (2009) note that the Archean biota would have already evolved to thrive in a higher background radiation environment (since at that time there was essentially no atmospheric oxygen and no ozone layer), and therefore any biosphere stress induced by the UV burst would be reduced. While the reemission UV burst, fluence would have been lower in the Archean for a given γ- and X-ray radiation flux than at the present time, the ground γ- and X-ray fluence would have been higher (again due to the lack of atmospheric oxygen to interact with), and this would have its own associated lethality, mutation and sterilization factors. Unfortunately, there is little observational data of any type with respect to the diversity of Archean lifeforms and/or the possible extinction events that might have occurred at that epoch. The full situation, thus remains to be resolved (Cockell 2002), possibly with the input from direct laboratory-based experiments. We note also that for the Earth to suffer from even a single GRB event at a stand-off distance of 200 pc during the Archean (the Earth then being between 1 and 2 billion years old), the supernova rate would need to be at least an order of magnitude higher than at the present time. The achievement of this latter condition is perhaps not entirely unreasonable since, following Annis (1999), the anticipated GRB rate was perhaps 4–5 times higher during the Archean eon.

A case—albeit not an entirely convincing one—has been made for the Ordovician-Silurian mass extinction some 450 million years ago being the result of a nearby GRB (Melott et al., 2004). Hallam (2004) has illustrates rather convincingly, however, that the Ordovician-Silurian marine extinction was much more likely a consequence of two distinct eustatic transitions. Likewise, Brakenridge (2011) has suggested that the Younger Dryas climate perturbation and the terminal Pleistocene extinctions, which occurred about 12–13,000 years before the present, were possibly a consequence of the Vela supernova detonation. In this latter case, the supernova is invoked to explain a rapid cooling of the climate and a concomitant rapid increase in the cosmogenic ^{14}C and ^{10}Be concentrations found within ice and marine core samples. While the argument is compelling, the problem is that we have no certain astronomical data to fully constrain the distance and age of the Vela nebula precursor at the time of its disruption. The estimated distance is of order 250 ± 30 pc, but the age could be anywhere between 11–16,000 years before the present (Cha et al. 1999). At best, the events may have occurred at about the same time, but no direct and/or overridingly compelling linkage has so far been demonstrated, and there are equally plausible non-supernova related explanations for all of the observational data.

It must be said that there is absolutely no unambiguous evidence within either the geological or paleontological record to indicate that the biosphere has ever suffered any extreme effects due to a GRB burst or nearby supernova event. This does not mean that none have occurred, but rather that there are no clear indicators of such events that everyone can agree upon. Indeed, finding distinct 'markers' or 'features' for a specific forcing mechanism is one of the inherent problems in interpreting past extinction and strong climate change episodes (Hallam 2004; Albritton 1989). It seems generally clear that unless specific GRB / supernova-related characteristics can be unambiguously identified, this particular astronomical forcing mechanism will never rise above being a possible 'when all other potentially more likely/more common terrestrial/astronomical explanations have been ruled out' biosphere affecting agent.

In a series of highly interesting research papers, Martin et al. (2009), Martin et al. (2010) and Penate et al. (2010) have begun to examine the consequences of how the UV flash associated with a GRB might affect food chains (that is, energy flow) within linked ecological systems. In this sense, the biosphere is viewed as a sum of interacting terrestrial and aquatic ecosystems. Such modeling is highly complex, with the possible outcomes of interactions being sensitive to the chosen vales of specific (and numerous) system constants (Wood et al. 2008). In essence, the sum of ecological systems readily lends itself to investigation under the umbrella of the Gaia hypothesis, first proposed by James Lovelock in the 1960s. The Gaia hypothesis, now greatly developed (see e.g. Harding 2006; Wood et al. 2008), essentially views the atmosphere, the oceans, the lithosphere, as well as all plant, forest, marine and land animals as one giant, continuously interacting positive feedback mechanism. To this end, Lovelock (see e.g. Lovelock 2006) considers the Earth to be a highly regulated system in which global stability (pertaining to the continued benefit of life) is maintained by feedback interactions between both its animate (plants, animals, insects, bacteria) and inanimate parts (e.g. lithosphere, ice coverage, atmospheric composition, carbon cycle and so on).

The simplest toy model for how Gaia essentially works is that demonstrated by Daisyworld (see e.g., Lovelock 1989; Wood et al. 2008; Beech 2009). In this model, a planet's temperature is regulated against an increasing stellar luminosity via a series of plant organisms with temperature-sensitive growing characteristics and different reflectance albedos. Much more complex ecological models have now been constructed (see e.g. Harding 2006; Wood et al. 2008) and it seems generally clear that the stronger and more diverse the various linkages between living ecosystems and their specific environments, the more stable and robust against change the overall system is—up to a point.

Global stability, however, does not exclude the possibility of local extinctions and dramatic changes in climate. Penate et al. (2010) have examined the consequences of a GRB-generated UV burst upon the survivability, with depth in the ocean, of phytoplankton. The key idea behind their study is that planktonic species reside at the starting point of the oceanic food chain and are responsible for the capture of CO_2 (a gas strongly linked to global warming; see e.g. Beech 2009), as well as the generation atmospheric oxygen. From this perspective, what is particularly interesting about the Gaian approach to a supernova-caused planktonic 'kill-off' is that not only could it result in the extinction of higher marine species (dependent upon plankton as food), but it could also produce significant atmospheric/climate perturbations. Lovelock and Kump (1994) have discussed this latter effect in a model that reveals the failure of a climate regulation model because of direct changes in the distribution of marine algae. Here the key point is that the alga is normally responsible for the production of dimethyl sulphide[9] (DMS = $(CH_3)_2S$), and this gas influences the cloud formation rate and cloud albedo—both of which in turn affect atmospheric temperature. By extending the ecological/Gaian modeling approach advocated by Penate et al. (2010), and as exemplified by Lovelock and Kump (1994), it should (at least in principle) be possible to identify causal linkages between specific animal extinction episodes and concomitant atmospheric/climate changes. This approach further provides an opportunity to shift the supernova model beyond the 'possible extinction mechanism, when all other explanations fail' category that it currently occupies.

Exercise 6.5: Term Paper Topic Investigate the origin, adaptation and controversies concerning James Lovelock's Gaia Hypothesis. In your essay, highlight the importance of positive and negative feedback mechanisms. Also, consider what impact the Gaia hypothesis and the continuing increase in the Sun's luminosity bring to the present-day problem of global warming.

Irrespective of past influences, the future role of galactic GRBs as a biosphere forcing agent is likely to be very small if not zero. Not only will their rate continue to decline as the galaxy ages, but the timescale between such events is typically of order the remaining lifetime of Earth's biosphere, estimated to be between 1.5 and 2.5 billion years (Beech 2008, 2009; Franck, Bounama and von Bloh, 2006; Caldeira and Kasting 1992; Lovelock and Whitfield 1982; and recall Sect. 5.7). The essential controlling factor behind the ultimate demise of the biosphere is that of a heat death brought about by the Sun's increasing luminosity and the loss of Earth's oceans via the runaway moist greenhouse effect (Beech 2009; Fogg 1991; Kasting 1988; see also Fig. 6.4).

The expected time interval between GRB events occurring within 1 kpc to the Sun can be developed in a similar manner to that outlined for Eq. (6.3), and it will be of order.

$$T_{GRB} \sim 50 \times (15000/1000)^2 \times 5000 \times BF \sim 56 \times BF \text{ million year} \qquad (6.4)$$

---

[9]It is dimethyl sulphide that provides the distinctive and refreshing tang to seaside air.

In Eq. (6.4), it is assumed that 1 in 5000 supernovae produce a GRB-generating hypernova, and the BF term accounts for the beaming factor associated with the collimated emission jets. The beaming factor relates to the solid angle subtended by the gamma ray beams, and accordingly $BF = 4\pi/(2 \times 2\pi(1 - \cos\theta))$. The additional factor of 2 in the denominator accounts for the fact that there are two beams and θ is the half cone angle for each beam. Typically, $5 \leq \theta\ (°) \leq 15$ (Podsiadlowski et al. 2004), and accordingly we find $30 \leq BF \leq 260$. With this correction term, the expected time interval for GRB events within 1 kpc of the Sun will be at least $T_{GRB} \sim 2$ Gyr. On the basis of this likely timescale we might expect at worst that there will be no more than one closer-than-critical-distance GRB event during the remaining natural lifetime of the biosphere.

(☺) **Exercise 6.6** Derive the formula for the beam factor $BF = 4\pi/(2 \times 2\pi(1 - \cos\theta))$.

The LBV star η Carina has often been touted as a possible pre-hypernova candidate. The system is estimated to be of order 100 $M_\odot$ in mass, placing it at least in the Type Ic supernova progenitor category. There is a growing body of evidence, however, to indicate that η Carina is actually a binary system (Bednarek and Pabich, 2011), and this will likely reduce the mass estimate of the pre-supernovae star, making a hypernova outcome much less likely. The estimated distance to η Carina is about 2.5 kpc, and this— even if it does produce a hypernova GRB—places it well outside of the critical, direct beam hit distance range for which significant biosphere damage is likely to occur.

Many pages within the popular press have been filled with the apparent doom that can be expected from the Wolf-Rayet star WR 104 (V* V5097 Sgr). This highly interesting massive WC + O star binary system sports a remarkable colliding-wind pinwheel nebula (Tuthill, et al., 2008). Initial estimates suggested that the system spin axis was pointing within a few degrees of the Solar System, and this raised the question concerning a future directed supernova (possibly GRB) jet intercept. However, given that WR 104 is at least 2.5 kpc distant, and that recent observations (Hill, 2009) find a much higher inclination for the system's spin axis away from the Sun, the future threat posed by WR 104 appears to be ill founded, if not entirely negligible.

Exercise 6.7: Term Paper Topic Investigate the observational history and the ongoing investigation of the luminous blue variable star η Carina.

The nearby supernova disruption and alteration of Earth's biosphere is a classic low-probability, high-consequence problem. There is presently no immediate threat to Earth's biosphere and/or humanity from a nearby supernova, and this will likely remain so for at least the next many millions of years. This being said, the threat posed to Earth's biosphere through the detonation of a nearby supernova is real, inevitable and potentially life-threatening. If nothing else, we do have the happy circumstance that future humanity will at least see the threat coming, since pre-supernova stars (especially Type II and Type Ib/c progenitors) are highly visible even at great distances, thereby providing a long damage mitigation preparation timescale. Type Ia precursors in contrast are less easily detected at great distances, unless they exhibit a dwarf nova phase, although again, at close quarters they are readily recognizable.

At some level, nearby supernovae must have affected Earth's biosphere in the distant past. Yet the current geological record provides no airtight case for any biotic extinctions and/or climate change episodes directly linkable to a nearby supernova. There is great difficulty in recognizing supernova-specific extinction and climate change 'markers', there being a wealth of terrestrial as well as other astronomical forcing mechanisms that can produce similar such changes. While nearby supernovae can conceivably have drastic effects upon Earth's biosphere, the fossil record tells us that life has never been fully eradicated by any external (or terrestrial) forcing mechanism during at least the past ~3.5 billion years, and this further suggests that, if nothing else, supernovae are not strong catastrophe candidates (Tegmark and Bostrom 2005) in the solar neighborhood.

The Earth's biosphere will likely last for at least another 1.5–2 billion years (recall Sect. 5.7), and given such a lifetime, it is likely that before its demise the Earth might experience at most one GRB event closer than 1 kpc (and this, according to jet cone angle, may have no effect on the biosphere at all) and perhaps 20 supernova events within 10 pc. Of the latter, some 15 will likely be core collapse, Type II or Ib/c supernova. The supernova threat to Earth's biosphere is certainly non-negligible, but it is also not one of immediate concern—indeed, asteroid and cometary nuclei collisions are the preeminent astronomical cataclysm to be faced by humanity in the near-term future (see e.g. Beech 2008; Bonnet and Woltjer 2008; Bailer-Jones 2011). During the next several million years, it seems highly unlikely that a supernova closer than Betelgeuse, located at a minimum distance of some 150 pc away from us, will occur.

Although the main thrust of this review has been to look at the supernova threat to Earth, the discussion applies in a general sense to any habitable, or more to the point, inhabited planet within the galaxy. While the interval between successive supernova events will vary somewhat according to galactocentric radius, the reactions to and responses against the threat by any sufficiently advanced civilization may involve some form of large-scale engineering projects. Such projects might constitute large orbital shields with the purpose of protecting a planet from the UV radiation emitted from its parent star; the ozone shield (say) having been temporarily destroyed through the supernova-induced production of nitrous oxide. Given that the majority of Sun-like stars (assuming such stars to be a standard, or at least an indicator, of life as set by our own existence) in the galactic habitability zone (GHZ) are ~1 billion years older than the Sun (Lineweaver, Fenner and Gibson, 2004), then the search for extraterrestrial technology might reasonably be directed towards those regions close to (within 10—20 pc of) any pre-supernova star or young supernova remnant observed. Indeed, any of the stars listed in Table 6.2 would be candidates for such a search since they straddled the GHZ, which forms an annulus with inner and outer radii of about 7 kpc and 9 kpc respectively. Given the potentially advanced age of extraterrestrial civilizations located within the inner part of the GHZ, any associated long-lasting civilizations may have already experienced (and survived?) perhaps one GRB event and of order 10 nearby supernovae.

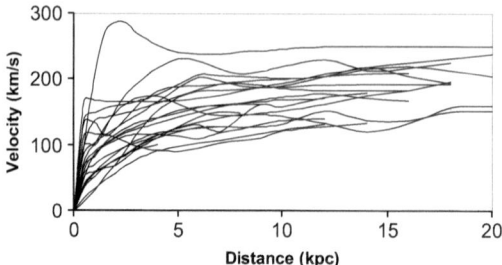

Fig. 6.2 Rotation curves for a series of spiral galaxies. The remarkable feature of these curves, which are quite typical of all spiral galaxies, is that the rotation rate does not tend towards zero at large distances from the galactic center, but instead remains constant with distance. Data from: V. Rubin et al., 1980. Rotation properties of 23 Sb galaxies. *Astrophysical Journal*, 261, 439–456. (Credit: Author)

6.3   Dark Matter Stars and Dark Energy

Recall exercise 3.23 concerning the characteristics of a large isothermal sphere. This exercise brought out the seemingly odd result that an object set in circular motion about the center of such a sphere would have the same orbital velocity irrespective of its radial distance from the center. The other situation (hinted at in the exercise) where this same result comes about is in relation to galaxy rotation curves. Such diagrams are a plot of the rotation velocities for the stars and interstellar gas clouds against distances away from the galactic center. A set of such curves is show in Fig. 6.2 What is remarkable about the observed rotation curves is that the rotation velocity does not decrease towards zero with increasing galactocentric distance; rather, the rotation rate remains near constant, with $V_{rot} \approx 200$ km/s. This is unexpected since most of the observed mass within a galaxy (the stars and the interstellar gas and dust) is located towards the galactic bulge and within the inner few kiloparsecs. Given the observed matter distribution, once the galactocentric distance exceeds some 5–10 Kpc, the rotation velocity would be expected to decrease. Specifically, it should decrease according to the inverse square root of the distance from the galactic center. This decrease has never been observed in any galaxy so far examined, and it appears to be a universal rule that in the outer regions of a galaxy, $V_{rot} \approx$ constant with distance.

   Since the speed with which a star or an interstellar gas cloud moves about the galactic center is determined by the matter interior to its orbit, something other than the normal matter (in stars and gas) must be keeping $V_{rot} \approx$ constant at large distances. This is taken to be the dark matter component of the galaxy. In this case, dark matter simply means that it is composed of some form of particle (although non-particle-specific possibilities have been suggested) that has a gravitational influence (a mass) but does not interact with electromagnetic radiation (which is why it is not directly detected with any kind of telescope). In the case of the Milky Way galaxy, the speed of the Sun in its galactic orbit indicates that there is something like 100 billion solar masses of gravitationally active material interior to its orbit, but only about five billion solar masses of this material can be accounted for in terms of stars and interstellar gas and dust.

Dark matter is real, and its gravitational contribution to stellar dynamics is measurable and inescapable.[10] But what is it, really? We currently do not know what dark matter is composed of, which is not to say that there aren't many possible theories to choose from. It is generally taken to be some form of presently unknown subatomic particle—one that indicates new physical theories beyond that of the standard model of particle physics. For all this lack of foundational explanation, however, it is possible to speculate on what various kinds of dark matter particles might exist, and how they might interact with each other.

(☺) **Exercise 6.8** Assuming circular orbits of radius a about the galactic center, use Kepler's 3rd law (Eq. (1.15)) to show that for a fixed (finite sized) central mass $m_1 >> m_2$, the rotational velocity V_{rot} will decrease as $a^{-1/2}$. *Hint:* For a circular orbit, $V_{rot} = 2\pi a / P$, where P is the orbital period.

6.3.1   Dark Matter Stars

The term 'Dark Star' has an ancient and modern meaning. Some two hundred plus years ago, it was taken to mean a star that was sufficiently small and compressed that the escape velocity at its surface was greater than that of the speed of light. Indeed, this was the speculation of John Mitchell in a letter to Henry Cavendish in 1783 and the idea is effectively a Newtonian version of a black hole. Mitchell's speculation was remarkable and far-sighted, and built upon the then-new determination that light actually traveled at a finite speed—a result first demonstrated by Danish astronomer Ole Romer in 1676. No such Dark Star as envisioned by Mitchell exists, however.

(☺) **Exercise 6.9** The formula for the escape velocity is given in Eq. (1.3). (1) Calculate how small a solar mass object would need to be in order to satisfy Mitchell's conditions for being a Dark Star. (2) What is the bulk density of such a Dark Star Sun, and how does this compare with the Sun's actual bulk density $\rho_\odot$ (recall exercise 3.4).

In the modern era, a Dark Star is an altogether different object to that envisioned by John Mitchell. Such stars are massive, large, highly luminous and powered by the annihilation of dark matter. The most recent models of Dark Stars build upon the possibilities derived from the (many) theories developed by particle physicists. The problem is highly complex and revolves around the identification of the lowest mass,

---

[10]Inescapable here means that dark matter must exist if Newtonian gravity holds true on very large scales. Alternative gravitational theories do exist, such as the Modified Newtonian Dynamics (MOND) model introduced by Mordehai Milgrom in the early 1980s, which can explain galactic rotation curves without invoking any dark matter component to the galaxy.

stable dark matter particle. The favored theory for producing various versions of dark matter particles is supersymmetry (usually just called SUSY). In this theory, every known standard particle has a much higher-mass-associated mirror companion, and the most promising SUSY candidate for the lightest supersymmetric dark matter particle is that of the neutralino. This is the massive supersymmetric partner to the W, Z and Higgs bosons, and if it actually exists—currently there is no experimental evidence that it does—then in addition to having the required properties to act as a dark matter particle, it is also its own antiparticle. It is this latter property that has resulted in the suggestion that Dark Stars, powered and stabilized by dark matter annihilation, might form in the very early universe; indeed, they would be the very first star-like objects. Such objects would look nothing like the stars we see in the universe today. This being said, it may be the case that Dark Stars, in spite of their implied early cosmological origins, still exist within the universe to this very day. The most recent study of Dark Star structure has been conducted by Katherine Freese (University of Michigan, Ann Arbor) and coworkers.

The first *real* stars (recall definition 2.1 and Sect. 4.4) to form in the universe—the so-called population III stars (recall Sect.1.5)—formed some 800,000 years after the Big Bang. These stars played a number of important functions in changing the properties of the cosmos. Firstly, they injected light (photons) into their otherwise dark surroundings, they re-ionized the universe, and they produced the first atomic elements beyond that of helium in the fledgling universe. These first stars are thought to form in dark matter halos with masses in the range 10^5 to 10^6 solar masses. Some 15% of the dark matter halo will be composed of a gas of hydrogen and helium, and this material will eventually form the corporeal body of an ordinary star. As the hydrogen gas within the dark matter halo cools, it begins to collapse and build up a steep density profile around a small core located at the center of the dark matter halo. The question asked by Freese and coworkers (2007, 2008a, 2008b) is what happens to the dark matter within the core region as the collapse of the ordinary matter proceeds. What their analysis found suggests that the dark matter density increases enough that the dark matter annihilation energy is sufficiently high to halt the collapse. Dark matter annihilation will generate energy at a rate Q_{DM} per unit volume of space per second of order

$$Q_{DM} = <\sigma v> \rho_{DM}{}^2/m_{DM} \qquad (6.5)$$

where $<\sigma v>$ is the dark matter annihilation cross section, ρ_{DM} is the dark matter density and m_{DM} is the mass of the dark matter particles. The characteristic value for the annihilation cross section is some 3×10^{-20} m^3 /s, and the characteristic dark matter particle mass is taken as $m_{DM} = 100$ GeV (1.8×10^{-31} kg)—although it should be noted that neither of these numbers is actually known for certain and/or even well constrained by current experimental data. In the early, low-density stages of the collapse, the energy produced through dark matter annihilation escapes from the protostar without any heating effect. However,detailed numerical calculations reveal that once the density of the protostar becomes sufficiently high, the dark matter annihilation energy can become trapped, resulting in a heating of the gas and the halting of the collapse. Freese and coworkers find that once the number density of the protostar reaches some 10^{19} m^{-3}, an equilibrium configuration with a radius of some 17 AU and a mass of 0.6 M$_\odot$ can come about. This extended structure is a Dark Star, and at this early stage its luminosity is about 140 L$_\odot$. As more matter falls onto the core, however, its surface luminosity will soon exceed the annihilation energy generation rate of the dark matter, and the contraction will once again set in. The contraction results in an increase in the dark matter density and produces a concomitant increase in the annihilation energy generation rate, leading once again to an equilibrium situation. The Dark Star thus evolves in a quasi-static manner, gaining mass and adjusting its size (radius) so as to maintain thermal equilibrium, with the energy radiated into space at its surface L being compensated for by the annihilation of dark matter within its interior.

$$L = L_{DM} = \int Q_{DM} \, dV \qquad (6.6)$$

A Dark Star is not a star in the sense that we have previously described such objects, since it is annihilation energy, not fusion energy, that is providing the internal heating. In some sense, they are delayed protostars. Only when the energy released by dark matter annihilation can no longer replenish the energy radiated into space at a Dark Star's surface will the cocooned ordinary matter collapse to produce a Population III star. In principle, a Dark Star will exist for as long as the inflow of dark matter continues. The canonical annihilation timescale will be of order

$$\tau_{DM} = \frac{m_{DM}}{\rho_{DM} < \sigma \nu >} \qquad (6.7)$$

which indicates $\tau_{DM} \sim 600$ million years if ρ_{DM} is taken as 10^{19} m^{-3}. This timescale is much longer than the collapse time for the ordinary matter (to form a Population III star) if there was no dark matter annihilation. The key potentially measurable outcome of Dark Star formation is hereby revealed in that the formation of such structures will delay the time (the age of the universe) at which the first population III stars will appear. The longer the Dark Star phase lasts, the older the universe will be before the first *bona fide* stars form.

To maintain thermal equilibrium, the luminosity L of the Dark Star must satisfy the condition that

$$L = 4\pi R^2 T_e^{\,4} = L_{DM} \qquad (6.8)$$

where R and T_e are the radius and effective temperature of the Dark Star. Freese and coworkers have produced a series of polytropic models for Dark Stars undergoing accretion. Setting the polytropic index to $n = 1.5$ and assuming that the interior of the Dark Star is fully convective, a series of mass-increasing models can be constructed. Just as with the determination of the Hayashi boundary (recall Sect. 5.1), the radius of a fully convective star is determined by the boundary condition $\kappa P = 2\,g/3$, where κ is the opacity, P the pressure and g is the surface gravity—this condition being derived from exercise 4.3 as the requirement that the optical depth outside of the photosphere is of order unity. Freese et al. use this condition to form a series of Dark Star models under the process of accretion.

The Dark Star model sequence is developed via an iterative process. For each Dark Star model mass, a radius R is assumed and the polytropic equation (3.46) is used to determine the run of the baryonic (that is ordinary matter) density. From this density distribution, the dark matter density is developed and the energy generated by dark matter annihilation is computed at each radius via Eq. (6.5) and (6.6). Starting at the center, the Lane-Emden equation for an $n = 1.5$ polytrope is integrated outwards to $r = R$, at which location the surface temperature T is deduced via Eq. (6.8). With T determined, the Stefan-Boltzmann Eq. (1.14) is used to determine the surface luminosity L. This luminosity is then compared to the luminosity L_{DM} due to dark matter annihilation—i.e. via Eq. (6.8). If $L_{DM} < L$, then a smaller radius R is implied and a new model computed; this new model will have a smaller luminosity L but a larger value of L_{DM} (since the model is now more compact). If $L_{DM} > L$, then a larger radius R is implied, so as to reduce the dark matter density distribution and accordingly the heating due to dark matter annihilation. In this manner, a back-and-forth approach is adopted with the radius guess varying up and/or down until the agreement that $L = L_{DM}$ is achieved.

Once Eq. (6.8) is satisfied (to some set accuracy limit), an equilibrium Dark Star model has been determined. Next, the mass of the model is increased (Freese et al. assume an accretion rate of 2×10^{-3} $M_{\odot}/yr$) and with a new initial guess for the radius, the next equilibrium model is computed. Since it is taken that a Dark Star is formed within an massive dark matter halo with a mass of order 10^{6} $M_{\odot}$, the Dark Star that develops might contain as much as ~1000 solar masses of ordinary matter. A Dark Star with this latter mass will form within 500,000 years (given the assumed accretion rate), and the models computed by Freese et al. indicate a final luminosity of several million $L_{\odot}$, a surface temperature of order 8000 K and a radius of order 10^{11} meters (~143 $R_{\odot}$). A Dark Star of total mass 1000 $M_{\odot}$ will contain about 0.3 solar masses of dark matter.

The final 1000 $M_{\odot}$ mass Dark Star has a very different structure to that of a Population III 1000 solar mass star formed in the absence of dark matter annihilation. The temperature and radius of a Dark Star are much lower and larger, respectively. Eventually, once the accretion of material stops, the heating via dark matter annihilation will no longer be able to keep a Dark Star in equilibrium and it must accordingly begin to collapse to produce a *bona fide* massive Population III star.

The picture of Dark Star formation as outlined by Freese and coworkers is dependent upon the currently unknown properties of dark matter. Not only this, but presently unarticulated mechanisms must eventually be invoked to limit the size (mass) of the Population III stars that are eventually produced. Studies of supernova progenitors indicate that once the mass of a star is greater than about 250 $M_{\odot}$ the final end state is that of direct collapse to a black hole with no significant ejection of chemically enhanced elements (recall Fig. 5.14). This latter effect is important since (recall) a chemical enrichment of the interstellar medium, out of which the next generation of Population II stars will eventually form, is required (Heger and Woosley, 2002; Heger et al. 2003).

(⊛) **Exercise 6.10: Term Project** Pat Scott (Stockholm University) and coworkers (2009) have made publicly available a Fortran95 computer code called *DarkStars*, which solves the equations of structure for low-mass stellar models accreting and generating energy via dark matter annihilation. The code is based upon the *EZ* platform (recall Exercise 4.23 and see the reference to Paxton (2004)) and allows for various dark matter capture and annihilation scenarios to be studied. Download the code and an appropriate compiler and produce a series of models for the Sun. Produce an HR diagram for a 100 solar mass Dark Star under the assumption that it is embedded within a (mass) range of dark matter halos.

6.3.2   Dark Energy

One of the pivotal scientific discoveries of the twentieth century was that the universe is expanding. This fundamental discovery was brought to the world's attention by Edwin Hubble in 1929, and the Hubble-Lemaitre law dictates that the speed V_{exp} with which a galaxy appears to be moving away from us (that is the Milky Way galaxy) is directly related to its distance D in such a way that $D = H_0 V_{exp}$, where H_0 is Hubble's constant at the present epoch. The distance units used in this formula are typically millions-of-parsecs (Mpc), while the velocity is expressed in units of km/s. Hubble's constant is accordingly expressed as $H_0 = 72 \pm 8$ km/s/Mpc, which, if one unravels the units, is really an inverse time. What this number tells us is that space is expanding by some 72 km every second over each unit length of one million parsecs. The implications of the Hubble-Lemaitre law are profound, specifically telling us that the Universe must have had a moment of origin (the Big Bang), and that the moment of the creation of the Universe was some $T_{Hubble} = H_0^{-1} =$ 13.6 billion years ago. While the Hubble-Lemaitre law revolutionized cosmological thinking in the last century, cosmologists in this century have been intrigued by the possibility that the universe is not just expanding, but expanding at an accelerating rate, with the acceleration apparently coming into effect some six billion years ago. The idea that the Universe is expanding at an apparently ever-increasing rate[11] remains controversial, and as of 2019 it is still far from being conclusively settled one way or another (see, Caldwell et al (2003)). For all this, however, most

---

[11] If the acceleration continues to increase exponentially, then space and everything in it (the atoms and sub-nuclear particles included) will be literally torn apart. The time of this so-called Big Rip is not that far distance in that it is (theoretically) set some 22 billion years into the future. The point, of course, is that since we don't know why the Universe is expanding at an accelerating rate in the first place, it can in principle stop accelerating at some time in the future—or it could also do something totally unexpected!

cosmologists are presently working under the pretext that the Universe is expanding at an accelerating rate, and therefore argue that this acceleration requires some form of driving energy behind it. This driving energy has been called Dark Energy, the term 'dark' being used since it has absolutely no agreed upon physical explanation at the present time.

The density of dark energy ρ_{DE} is related to the so-called cosmological constant Λ through the relationship

$$\rho_{DE} = \frac{\Lambda}{8\pi G} \tag{6.9}$$

where it is Λ that is deduced from the observations. In the presence of Dark Energy, the gravitational force F_g acting per unit mass over a spherical body of radius r and mass m is given by the relation

$$F_g = -\frac{Gm}{r^2} + \frac{\Lambda r}{3} \tag{6.10}$$

The additional term on the right-hand side of Eq. (6.10) acts against gravity, thereby reducing its attractive affect. Just as in our earlier development of the equation of hydrostatic equilibrium (recall Chap. 3), we now write

$$\frac{1}{\rho}\frac{dP}{dr} = F_g = -\frac{Gm}{r^2} + \frac{\Lambda r}{3} \tag{6.11}$$

Furthermore, we introduce a polytropic equation of state $P = K\rho^{\gamma}$, where $\gamma = (n+1)/n$, with n being the polytropic index and where K is a constant. Proceeding now by introducing the variables $r = \alpha\xi$ and $\rho = \rho_c\theta^n$, a slightly modified Lane-Emden equation (recall Eq. 3.46) is derived, with

$$\frac{1}{\xi^2}\frac{d}{d\xi}\left(\xi^2\frac{d\theta}{d\varsigma}\right) = -\theta^n + \Psi \tag{6.12}$$

where $\Psi = 2\rho_{DE}/\rho_c$, where ρ_c is the central density of the material body. Clearly, if Dark Energy disappeared, $\Psi = 0$ and the standard Lane-Emden equation would result. The question now is how the Ψ-term modifies any specific solutions to the Lane-Emden equation. We have described in Chap. 3 the zeroth-order, constant density model corresponding to $n = 0$, and if we make this substitution for n, the solution to the modified Lane-Emden Eq. (6.12) is:

$$\theta = 1 + \frac{\xi^2}{6}(\Psi - 1) \tag{6.13}$$

(☺) **Exercise 6.11** Verify that Eq. (6.13) is the correct solution to Eq. (6.12) when $n = 0$.

Accordingly, the surface of the material configuration, which recall is located at $\theta = 0$, is given by $\xi_1 = \sqrt{6/(1 - \Psi)}$. For any non-zero value of Ψ (that is, if Dark Energy really does exist), the radial configuration of a star will be larger than otherwise expected for its associated mass. We note that in principle, the closer Ψ is to unity, the larger the radius will become, such Dark Energy-pervaded stars bloating up like red giants. There is no indication that Ψ is anywhere near being close to unity at the present time, but, presumably if the Big Rip is going to occur, then the stars will grow ever larger in response to its approach.

(☺) **Exercise 6.12**   Find the solution to the modified Lane-Emden equation when $n = 1$ (see Merafina et al (2011)).

6.4   Stars in Other Universes

In Chap. 2, it was argued that one of the great triumphs of modern physics was the identification of various universal constants—pure numbers that literally define the way our Universe must work. There is currently no theory that explains why any of the universal constants must have the value that they do, and there is no specific theory that dictates that the so-called universal constants cannot change over time, varying in one way or another as the Universe ages (Uzan (2003)). Physicists have long worried about this situation, and rather than adopt a pragmatic approach and accept that the constants simply have the values that they are deduced to have, the current interpretation is that the fundamental constants can take on any value and that different combinations of the universal constants exist in other universes. Indeed, it is supposed by some researchers that there are an infinite number of other universes with effectively random values being adopted for the various fundamental constants, and that the various combinations of these constants will either support or negate the existence of stars and observers.

(☺) **Exercise 6.13: Term Paper Topic**   Mathematician and long-time *Scientific American* magazine columnist Martin Gardner (2001) has written that, "The stark truth is that there is not the slightest shred of reliable evidence that there is any universe other than the one we are in. No multiverse theory has so far provided any prediction that can be tested. As far as we can tell, universes are not plentiful". Find a copy of Gardner's article (from your local library - see also, Ellis (2008)) and then think hard about what he is saying.

The fact that we live in a universe with the specific universal constants that have been deduced, which in turn clearly allows for the formation of stars and planets and supernovae and life, can be described according to the Weak Anthropic Principle (WAP). A definition for the WAP has been provided by John Barrow and Frank Tippler (1986):

Definition Weak Anthropic Principle: *The observed values of all physical constants and cosmological quantities are not equally probable but they take on values restricted by the requirement that there exist sites where carbon-based life can evolve and the requirement that the Universe be old enough for it to have already done so.*

The WAP is not a predictive philosophy, but it does act to constrain some aspects of how the universe (that is physics) must work. It was seen in Chap. 5, for example, that the Chandrasekhar limiting mass for a white dwarf is

$$M_{CH} = const\,\mu_e^{-2} \left(\frac{hc}{Gm_H^{4/3}} \right)^{3/2}$$

(6.14)

Of interest in the present discussion is the value of the $(h\ c\ /\ G\ m_H^{4/3})$ term. In principle this term can take-on any value, big or small, according to the particular values adopted for c, h, m_H, and G in some particular universe. In our universe the Chandrasekhar limiting mass is 1.46 $M_\odot$. In a universe where the universal constants are the same as we find them, but in which G is half of our observed value, so the Chandrasekhar limiting mass would increase to 4.13 $M_\odot$. If G was twice as large as the value in our universe value so the limiting mass would be reduced to 0.52 $M_\odot$. The consequences of such variations in the gravitational constant are not easy to access since the variations will ripple through all areas of physical structure. Certainly, however, a change in the gravitational constant will have a direct affect upon the central temperature and luminosity of a star (see below), and a profound affect upon the mass range of stars capable of evolving into a final degenerate white dwarf state. For the larger limiting Chandrasekhar mass, for example, the production of Type Ia supernova would presumably be suppressed since a much greater mass exchange from a companion star would be required to push the white dwarf over its limiting mass. In contrast, the lower Chandrasekhar limiting mass would presumable enhance both Type II and Type Ia supernova. Fred Adams (University of Michigan - 2008) has investigated in some detail the effects of varying fundamental constants on stellar structure, and finds that while many universes (with different fundamental constants) will support the existence of stars, it is by no means a certainty that stars must exist in every conceivable universe (see below).

(☺) **Exercise 6.14: Term Paper Topic** Investigate the various forms of the Anthropic Principle (Weak, Strong, Participatory, Self-Sampling...) and discuss their philosophical implications.

Given that our existence as observers of the Universe is predicated upon the fact that supernovae have produced the appropriate complex atoms required to generate and sustain life (the carbon, oxygen, nitrogen...), any universe in which supernova production is suppressed or even invalidated will presumably not produce complex life (observers). Brandon Carter (1974) has examined this possibility and finds that the existence of supernova progenitor stars is sensitive to variations in the gravitational fine structure constant α_G as well as the fine structure constant α. Indeed, if the value of α_G was only slightly larger than observed (and/or the fine structure constant α just a little smaller), then all main sequence stars would be fully convective, and this would negate the formation of red giant stars and strongly affect the conditions leading to Type II supernova disruption. Carter's condition is partly based upon the Hayashi condition described in Sect. 5.1. The requirement is now that the surface temperature of a star must be smaller than the ionization limit of hydrogen in the outer envelope. This condition is set by the so-called Rydberg energy $E_R = \frac{1}{2} m_e c^2 \alpha^2$, where m_e is the mass of the electron and c is the speed of light. Carter finds that the surface temperature of a star will not exceed the ionization temperature if the following fundamental relationship holds true:

$$\alpha^{12} \left(\frac{m_e}{m_p} \right)^5 = 2.2 \times 10^{-39} \approx \alpha_G = 5.9 \times 10^{-39}$$

and, as seen, this 'coincident' condition is only just satisfied. What this inequality allows for is the ability of massive stars to have a radiative envelope, and that they are accordingly not fully convective within their interiors.

Since it enters into the determination of atomic structure, it is possible to look for variations in the fine structure constant α in the spectra of distant galaxies. The argument here is that since the light from distant galaxies was emitted when the universe was much younger, any changes in α over time should be measurable. Indeed, a research group of astronomers led by John Webb (University of New South Wales) has looked for variations in the fine structure constant in the spectra of over one hundred distant quasars, and they find (controversially, it should be pointed out) that not only is there evidence for a slight variation in α over the past 10–12 billion years, amounting to some $\Delta\alpha / \alpha \sim 6 \times 10^{-6}$, but that there is a spatial component to this variation as well. This latter variation has been called the Australian Dipole, and it is a particularly odd feature of the observations—Webb et al. (2011) argue that α decreases over time in one direction along the dipole axis and increases over time in the other. The overall variation in α is very small, and why it should show a spatial variation is entirely unclear, but these observations are either (1) wrong, and relate to some systematic error that has not been accounted for, or (2) they indicate profound new physics beyond the standard model. The implied change in α (if correct) is not enough to contravene Carter's 'coincidence' condition,[12] even in the very early universe, and it would appear that our universe (and the universal constants [sic] that define it) has always allowed for the possibility of stars to form.

The above being said, the fact that stars can exist and find an energy supply via fusion reactions is also determined by a remarkable coincidence between the fine

---

[12]The value of α would need to increase by more than 8.5% above its present-day value for Carter's 'coincidence' condition to be violated. This would have many additional consequences; recall exercise 6.5.

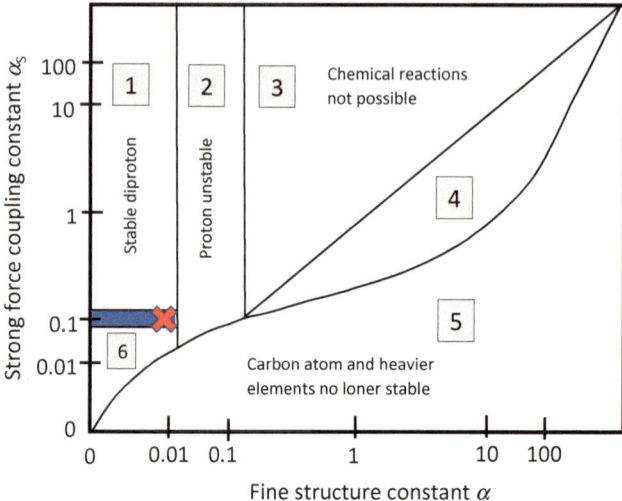

Fig. 6.3 Stability/insatiability domains for atoms and chemical reactions according to the relative size of the fine structure and the strong force coupling constants. The region shown in blue indicates where stars can form and generate energy via hydrogen fusion reactions, and in which the carbon atom (essential to our existence) is stable. The red cross indicates the values of α and α_S in our universe. The numbered regions are explained in the text. (Credit: Author—adapted from a diagram published by Luke Barnes)

structure constant α and the strong force coupling constant $\alpha_S = 0.1187$. This constant sets the relative strength of the strong nuclear force that holds atomic nuclei together and, with specific importance to the stars, determines the conditions under which nuclear fusion reactions can operate. Figure 6.3 shows a diagram (based upon the publications by Luke Barnes (2012), University of Sydney, Australia) in which the relative strength of the strong force coupling constant is compared against that of the fine structure constant. Once again, there is only a very small range of possible values for α and α_S under which stable stars can possibly exist. With reference to Fig. 6.3, region 1 is excluded. In this region, the diproton will have a stable bound state, which will dramatically speed up the hydrogen fusion process since the first and very slow P + P interaction in the PP chain in which a simultaneous inverse beta-decay must happen is no longer required (recall Fig. 4.6). The effect here will be to dramatically decrease the main sequence lifetime of a star. In region 2 the proton is unstable, while in region 3 the electron is unstable. In region 4, the typical energy of chemical reactions will be greater than the energy of nuclear reactions— for this range of values for α and α_S, the atoms will no longer maintain their identity during chemical reactions. In region 5, carbon and all other larger atoms are unstable, negating our existence as observers. In region 6 the deuteron is unstable, and accordingly the first reaction in the PP fusion chain will no longer proceed.

(☺) **Exercise 6.15: Term Paper Topic** Astrophysicist Fred Hoyle is justly famed for his prediction in the early 1950s that there must be something very special going on with the production of carbon atoms in the universe. The triple-alpha reaction in which three helium nuclei fuse to produce a carbon atom was discussed in Chap. 4, but it turns out that the process only works because of what is called a resonance level in the carbon atom. The fusion process runs as follows: 4He + 4He ⇒ 8Be, followed by 4He + 8Be ⇒ 12C, and it is the second process that is the fine-tuned peculiarity. The latter step in the process only works because 4He + 8Be has almost exactly the same energy as an excited state of 12C. Change the fine structure constant α by just a fraction and this resonance would disappear, and the universe would be largely devoid of carbon (carbon, of course, is essential to our existence). It is additionally important, as Hoyle noted, that there is no resonance that aids in the rapid destruction of carbon to form oxygen via the reaction 4He + 12C ⇒ 16O. Investigate the history of Hoyle's speculation and discuss the possibility of life evolving in a near carbon-free universe. Begin perhaps by finding a copy of Owen Gingerich's excellent book, *God's Planet* (Harvard University Press, Cambridge, 2014), and read through Chap. 3.

The Anthropic Principle—our existence as observers in the universe—can be applied to determine a maximum stellar mass rate (in our universe - Beech (1987)). It has been seen already (recall Sects. 4.5.4 and 5.4) that various forms of mass loss rate formula have been derived. The so-called Lucy-Solomon mass loss rate (Lucy and Solomon, 1970) for very high luminosity stars is based upon the radiation pressure acting upon specific spectral lines, giving a mass loss rate of $\dot{M} = -NL/c^2$, where L is the luminosity, $10 \leq N < 1500$ is a parameter accounting for the number of lines contributing to the mass loss and c is the speed of light. This mass loss rate will approach its maximum as $L \Rightarrow L_{EDD}$. Accordingly, the maximum mass loss rate for very massive stars is of order

$$\dot{M} = -N\left(1.6 \times 10^{14}\right)\left(M/M_{sun}\right) \text{ kg/s} \qquad (6.15)$$

For a 100 $M_\odot$ star, the maximum mass loss rate for $N = 1500$ is of order 2.4×10^{17} kg/s, or about 4×10^{-6} $M_\odot$/year.

Alternatively, one can argue that the minimum amount of energy required for mass loss to take place at a rate $\dot{M}$ is $GM\dot{M}/R$: this is the gravitational binding energy associated with a star of mass M and radius R. Additionally, the maximum amount of energy available for driving the mass loss will be λL, where L is the luminosity and $\lambda < 1$ is some numerical constant. In this manner, the maximum mass loss rate will be of order (Williams, 1967)

$$\dot{M} = -\lambda\left(\frac{L\,R}{GM}\right) \qquad (6.16)$$

Using expressions (1.24b) and (1.25b), the above mass loss rate for main sequence Sun-like stars is of order $\dot{M} = -2 \times 10^{15}(M/M_{sun})^{3.9}$ kg/s. In the case of the Sun, the solar wind mass loss rate amounts to about 2×10^{-14} M/year (or 1.3×10^9 kg/s), indicating that the actual mass loss rate of the Sun (at the present time) is well below ($\lambda \approx 6 \times 10^{-7}$) its maximum possible rate.

From a WAP point of view, the maximum stellar mass loss rate can be approached via the requirement that some stars must undergo a supernova disruption end phase. This requirement is set, recall, since it is through the nucleosynthesis taking place during supernova disruption that the heavy element abundance of the interstellar medium is enriched, and it is out of the metallicity-enhanced interstellar medium that later generations of stars and planets and human beings are made. If we define $< \dot{M} >_C$ to be the average critical mass loss rate such that a star of initial mass M_i will evolve into a final state of mass $M_f = M_{CH}$, the Chandrasekhar limiting mass, then for stars greater in mass than 8 $M_\odot$, the maximum average mass loss rate must satisfy the inequality $<\dot{M}> \ <<\dot{M}>_C$ if the star is to have a final mass greater than the Chandrasekhar limit and thereby be able to undergo supernova disruption. As long as this mass loss inequality holds true, an excess of white dwarfs is avoided and supernova events are not suppressed. Using Sect. 2.3 as our guide, the maximum average stellar mass loss rate can be expressed in terms of fundamental constants. Accordingly, the mass range of stars is described by the formulation $Q\alpha_G^{-3/2}m_p$ where $0.1 < Q < 50$, $\alpha_G = Gm_p^2/\hbar c\alpha_G$ is the fine structure constant and m_P is the proton mass. Furthermore, Bernard Carr and Martin Rees (1979) have shown in a classic research paper that the main sequence lifetime of a star is of order $t_{ms} = t_S{}^S/Q^2\alpha_G$, where $t_S \sim 10^{-23}$ seconds is the typical timescale of a strong nuclear interaction. The main sequence lifetime is the longest lived phase of any star, and its overall lifetime t_{total} to an ultimate end phase (white dwarf, neutron star or black hole) is of order $t_{total} = t_{ms} (1 + \varepsilon)$, where $\varepsilon < 0.5$. Accordingly, the maximum average mass loss rate for the most massive stars must be such that

$$< M >_C \approx \left(\frac{Q^3}{1 + \varepsilon} \right) \left(\frac{M_{pl}}{t_S} \right) \tag{6.17}$$

where M_{pl} is the Planck mass (recall Sect. 2.3). Substituting for values, Eq. (6.17) indicates that $< \dot{M} >_C \approx Q^3 10^{-8} M_\odot$/year—that is, the maximum average mass loss rate for the most massive stars cannot be greater than $\sim 10^{-3}$ $M_\odot$/year. This Anthropic maximum average mass loss is several orders of magnitude higher than the maximum Lucy-Solomon mass loss rate for a 100 $M_\odot$ star during its main sequence phase. The maximum WAP mass loss rate, however, is only about one order of magnitude higher than that observed for massive stars going through their advanced pre-supernovae stages of evolution. Indeed, the mass loss rate for some Wolf-Rayet stars is estimated to be as high as 10^{-4} $M_\odot$/year.

The Large Number Hypothesis was developed according to the initially unexpected, either very small or very large, comparisons between various combinations of the universal constants. Physicist Paul Dirac (1937, 1938) made great play of such combinations of pure number constants and noted that the ratio N_1 of the Hubble horizon radius of the universe (c / H_0, where H_0 is Hubble's constant) to the classical radius of the electron is of the same order of magnitude as the ratio N_2 of the electric to the gravitational force between an electron and a proton.

$$N_1 = \frac{c/H_0}{e^2/m_e c^2} \approx 6 \times 10^{39} \approx N_2 = \frac{e^2}{G m_P m_e} \approx 2 \times 10^{39} \qquad (6.18)$$

Arguing that the equality of N_1 and N_2 must hold at all times, one of the constants in Eq. (6.18) must accordingly change as the universe ages—recall that the age of the universe is given by the relation $t_0 = 1 / H_0$. Dirac argued that it is the gravitational constant that changes, and specifically that is must decrease as a function of time. Equally, one could argue that the gravitational constant G was truly constant and that the basic unit of charge e varied as a function of time. Such a change would of course have a direct affect upon atomic structure, fusion reactions and basic chemistry. Let us stick with Dirac's assumption that the gravitational constant is time varying. Accordingly, we have:

$$G = G_0(t_0/t) \qquad (6.19)$$

where G_0 is the gravitational constant at the present age of the universe $t_0 = 13.8$ billion years.

The near-exact equality of N_1 and N_2 in Eq. (6.18) is surprising and entirely unexpected, but what does it really mean? The answer lies at one of the two extremes: the coincidence is either telling us something profound and as yet not understood about the universe, or it means nothing, and the coincidence is just a pure and meaningless bit of numerology. Dirac argued that the large pure-number ratios that could be constructed from the various universal constants (i.e. N_1 and N_2) could be used to develop a new way of thinking about the universe. But no overall consensus on the topic exists. Many variants on the Large Number Hypothesis have been constructed since Dirac's early papers, and the issue of time varying universal constants refuses to go away.

Shortly after Dirac's papers on the Large Number Hypothesis first appeared, Edward Teller (1948) investigated the consequences of a varying gravitational constant on stellar properties. Teller's argument was built upon the argument that the luminosity of a star is determined according to the rate at which its radiant energy is lost via photon diffusion:

$$L = \frac{\left(\frac{4}{3}\pi R^3\right)\left(a T_c^4\right)}{\left(R^2/l c\right)} \qquad (6.20)$$

Using the results of Sect. 4.1, the mean free path $l = 1/\kappa \rho$ will vary as the temperature to the third power (assuming a Kramers-like opacity law). In this way,

$$L \sim R T^4 l \sim R T^7 \qquad (6.21)$$

Here the $\sim$ symbol simply means 'scales as'. Using Eq. (3.18), we also have that the temperature of a star scales as the ratio $T \sim G M / R$, and taking $M \sim R^3$ (from the bulk density formula), we derive the result that

$$L = \text{constant}\, G^7 M^5 \tag{6.22}$$

Equation (6.22) is just the mass-luminosity law for Sun-like stars, but in this case we have brought out the dependency upon the gravitational constant term. Having developed Eq. (6.22), Teller noted that the expression for the luminosity is not as such dependent upon any assumptions about the energy production mode within a star. Furthermore, what Eq. (6.22) reveals is that the luminosity of a Sun-like star is very sensitive to changes in the gravitational constant. For a constant mass, a 1% increase in the gravitational constant results in a dramatic 7% increase in the luminosity.

(☺) **Exercise 6.16** Use Eq. (6.19) to estimate how far back in time the gravitational constant would have been 10% larger than it is at the present time.

(☺) **Exercise 6.17** Think about what happens to Teller's argument if a different (non-Kramers-like) opacity law is adopted.

There are anthropic consequences associated with a varying gravitational constant. Indeed, it is straightforward to show that the simplest interpretation of Dirac's hypothesis must be wrong. The issue is a direct result that, as indicated by Eq. (6.20), the Sun's luminosity must have been higher in the past, and this means that Earth's temperature must also have been higher in ancient times. Given that the Earth orbits the Sun at some circular orbital distance a, its orbital velocity V_E will vary as

$$\frac{V_E^2}{a} = \frac{GM_{sun}}{a^2} \tag{6.23}$$

Accordingly, if the gravitational constant G changes over time, a and V_E must change as well. Conservation of angular momentum dictates that $(a\,V_E)^2 = \text{constant}$, and accordingly from Eq. (6.23) we have that

$$(a V_E)^2 = a\,GM_{sun} = const \tag{6.24}$$

From this equality we find that the Earth's orbital radius will change in response to a change in the gravitational constant as $1/GM_{sun}$. That is, in the distant past, the Earth's orbital radius must have been smaller (since G was then larger under the Dirac scenario). The effect of reducing the Earth's orbital radius and increasing the Sun's luminosity on the Earth's surface temperature T can be determined by thinking of the Earth as a blackbody radiator, where the Earth is heated according to the solar energy flux at an orbital distance a, but at the same time the Earth radiates energy back into space from its surface in accordance with the Stefan-Boltzmann law. In this manner:

$$\left(\frac{L}{4\pi a^2}\right)\left(\pi\, R_E^2\right) = (1 - A)\left(4\pi R_E^2\right)\varepsilon\,\sigma T^4 \tag{6.25}$$

where R_E is the Earth's radius, A is albedo of Earth's atmosphere and ε is the emissivity. In deriving Eq. (6.23), it has been assumed that the Earth spins sufficiently rapidly that it radiates equally over its entire surface. This being noted, it is also the case that the size of the Earth (through the R_E value) is not actually a factor in the determination of its surface temperature. Using present-day values for the Sun's luminosity and the Earth's orbital radius (1 astronomical unit), Eq. (6.25) yields an equilibrium temperature for the Earth of $T_{now} = 278$ K (taking $\varepsilon = 0.6$, $A = 0.3$). Now, using Eqs. (6.22) and (6.24), with Eq. (6.25) we find that

$$T = T_{now}[G/G_0]^{5/4} \tag{6.26}$$

that is, the Earth would have been much hotter in the past since G was larger then. This, result runs in direct contrast to the geological and paleontological record.

(☺) **Exercise 6.18** Use Eqs. (6.19) and (6.26) to construct a diagram showing the temperature of the Earth as a function of time in Dirac's varying G cosmology. Your results should look something like Fig. 6.4.

Dirac's simplest cosmology based upon the Large Number Hypothesis, with the gravitational constant varying as the inverse age of the universe (via Eq. (6.19)),

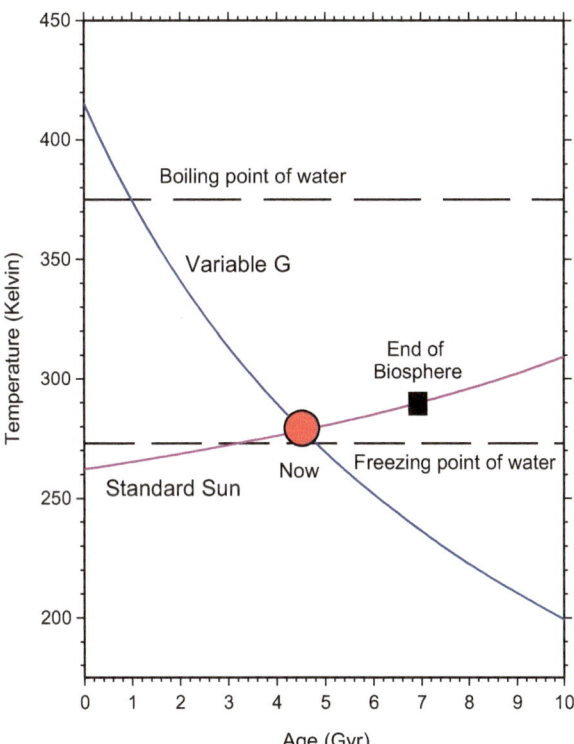

Fig. 6.4 The past temperature of the Earth based upon the assumption that the gravitational constant decreases as the inverse age of the universe (blue line). The red line shows the temperature variation of the Earth based upon the evolution of a standard solar model (recall Sect. 5.2). The filled square indicates the time, some 2.5 billion years in the future, when the biosphere will come to an end and the moist greenhouse effect will cause the loss of Earth's oceans. (Credit: Author)

cannot be correct. Other cosmologies based upon large number coincidences or simple assertion can and indeed have been constructed, and these have proved much more difficult to negate. Constraints on any change in the gravitational constant over time (in our corner of the universe) have been set by lunar-laser-ranging measurements, and these reveal that $\dot{G}/G = (4 \pm 9) \times 10^{-12}$ year^{-1}. Within the context of the present day many-worlds hypothesis, Dirac's time-varying gravitational constant should (presumably) be realized in some universes (somewhere), and presumably life as we know it might struggle to come into existence there. Indeed, within a Sun-Earth analog system in a Dirac varying G universe, the parent Sun would be well passed middle-age before the temperature of its companion Earth was even below the boiling point of water. It would thus be highly unlikely that the analog Earth would ever experience any associated water, life or wisdom—a sad universe indeed.

6.5 Answers to Exercises

Following are the answers to all exercises in this chapter. The exercises for which no answer is provided are those in which the student is asked to check a result or complete the algebra steps between two equations given in the text.

Exercise 6.6 The beam factor is derived from the area of a spherical cap on a unit sphere, and is given by the double integral:

$$\int_0^{2\pi} \int_0^{\theta} \sin\theta \, d\theta \, d\varphi$$

where 2θ is the apex angle of the cap cone. The 4π term in the BF is the solid angle subtended by the entire unit sphere, i.e. in our case when the cone angle opens up to $90°$ with $\theta = \pi/2$.

Exercise 6.9 (1) $R \approx 3$ km. (2) $\rho = 1.8 \times 10^{19}$ kg/m$^3 \sim 10^{16} \rho_{\odot}$.

Exercise 6.16 The time would be $t = 1.25$ billion years ago. This is in the Mesoproterozoic era, when photosynthesis first began to evolve in plants.

References

Bibliography for the Section 6.1.1

Babu, G. J., & Feigelson, E. D. (1996). *Astrostatistics*. Chapman and Hall.
Caes, C. J. (1988). *Studies in Starlight: understanding our universe*. Tab Books.

Deutsch, A. J. (1962). In T. Page (Ed.), *Stars and galaxies: Birth, aging and death in the universe*. Prentice-Hall.

Deutsch, D. (2011). *The beginning of infinity: Explanations that transformed the world*. Allen Lane.

Eddington, A. S. (1935). *The nature of the physical world*. Mayflower Publishing Co.

Golub, L., & Pasachoff, J. M. (2001). *Nearest star; the surprising science of our sun*. Harvard University Press.

Impey, C. (2010). *How it ends: From you to the universe*. W. W. Norton and Co.

Macpherson, H. (1923). *The romance of the modern astronomy*. Seeley, Service and Co.

McCrea, W. H. (1950). *Physics of the sun and stars*. Hutchinson's University Library.

Meadows, A. J. (1967). *Stellar Evolution*. Pergamon Press.

Newcomb, S. (1832). *Astronomy for everybody*. New Home Library.

Noyes, R. W. (1982). *The sun: Our star*. Harvard University Press.

Ryan, S. G., & Norton, A. J. (2010). *Stellar evolution and Nucleosynthesis*. Cambridge University Press.

Sagan, C. (1980). *Cosmos*. Ballantine Books.

Searle, A. (1875). *Outlines of astronomy*. Ginn Brothers.

Young, C. (1899). *A text-book of general astronomy*. Ginn and Company.

Zeilik, M., & Gaustad, J. (1990). *Astronomy: The cosmic perspective*. Wiley and Sons.

Zirin, H. (1988). *Astrophysics of the sun*. Cambridge University Press.

References for the Section 6.1.2

Adams, F. (2010). The birth environment of the sun. *Annual Reviews in Astronomy and Astrophysics, 48*, 47–85.

Arny, T. (1990). The star makers: A history of the theories of stellar structure and evolution. *Vistas in Astronomy, 33*, 211–233.

Bastian, N., Covey, K. R., & Meyer, M. R. (2010). A universal stellar initial mass function? A critical look at variations. *Annual Reviews in Astronomy and Astrophysics, 48*, 339–389.

Dole, S. H. (1964). *Habitable planets for man* (p. 101). Rand Corporation.

Gowanlock, M. G., Patton, D. R., and McConnell, S. M. (2011). A model of habitability within the milky way galaxy. http://arxiv.org/pdf/1107.1286.pdf.

Henry, T. J., Jao, W.-C., Subasavage, J. P., Beaulieu, T. D., Lanna, P. A., Costa, E., & Mendez, R. A. (2006). The solar neighborhood XVII. Parallax results from the CTIOPI 0.9 m program: 20 new members of the RECONS 10 parsec survey. *Astronomical Journal, 133*, 2360–2371.

Lineweaver, C. H., Fenner, Y., & Gibson, B. K. (2004). The galactic habitability zone and the age distribution of complex life in the milky way. *Science, 303*, 59–62.

Mihalas, D., & Binney, J. (1981). *Galactic astronomy: Structure and kinematics*. W. H. Freeman and San Francisco.

Parker, E. N. (2000). The physics of the sun and the gateway to the stars. *Physics Today (June), 53*, 26–31.

Rocha-Pinto, H. J., & Maciel, W. J. (1996). The Metallicity distribution of G dwarfs in the solar Neighbourhood. *Monthly Notices of the Royal Astronomical Society, 279*, 447–458.

Spiegelhalter, D., Pearson, M., & Short, I. (2011). Visualizing uncertainty about the future. *Science, 333*, 1393–1400.

Tayler, R. J. (1989). The sun as a star. *Quarterly Journal of the Royal Astronomical Society, 30*, 125–161.

Wyatt, M. C., Clarke, C. J., & Greaves, J. S. (2007). Origin of the Metallicity dependence of exoplanet host stars in the Protoplanetary disc mass distribution. *Monthly Notices of the Royal Astronomical Society, 380*, 1737–1743.
Yoon, J., Peterson, D. M., Kurucz, R. L., & Zagarello, R. J. (2010). A new view of Vega's composition, mass, and age. *The Astrophysical Journal, 708*, 71–79.

References for the Section 6.2.1

Albritton, C. C. (1989). *Catastrophic episodes in earth history*. London: Chapman and Hall.
Annis, J. (1999). An astrophysical explanation of the great silence. *Journal of the British Interplanetary Society., 52*, 19–22.
Bailer-Jones, C. A. L. (2011). Bayesian time series analysis of terrestrial impact cratering. *Monthly Notices of the Royal Astronomical Society, 416*, 1163–1180.
Bartunov, O. S., Tsvetkov, D. Y., & Pavlyuk, N. N. (2006). Sternberg Asytonomical institute supernova Calalogue, and radial distribution of supernovae in host galaxies. *Highlights in Astronomy, 14*, 316–323.
Basu, S., Stuart, F. M., Schnabel, C., & Klemm, V. (2007). Galactic-cosmic-ray-produced He3 in a ferromanganese crust: Any supernova Fe60 excess on earth? *Physical Review Letters, 98*, 141103–141107.
Bednarek, W., & Pabich, J. (2011). High-energy radiation from massive binary system eta Carina. *Astronomy and Astrophysics, 530*, A49.
Beech, M. (2008). *Rejuvenating the sun and avoiding other global disasters*. New York: Springer Publishing.
Beech, M. (2009). *Terraforming: The creating of habitable worlds*. New York: Springer Publishing.
Beech, M. (2011). The past, present and future supernova threat to Earth's biosphere. *Astrophysics and Space Science, 336*, 287–301.
Benitez, N., Maiz-Apellaniz, J., & Canelles, M. (2002). Evidence for nearby supernova explosions. *Physical Review Letters, 88*, 81101–81105.
Bishop, S., & Egli, R. (2011). Discovery prospects for a supernova signature of biogenic origin. *Icarus, 212*, 960–966.
Bizzarro, M., Ulfbeck, D., Trinquier, A., Thrane, K., Connelly, J. N., & Meyer, B. S. (2007). Evidence for a late supernova injection of 60-Fe into the protoplanetary disk. *Science, 316*, 1178–1181.
Bonnet, R.-M., & Woltjer, L. (2008). *Surviving 1,000 centuries – Can we do it?* New York: Springer Publishing.
Brakenridge, G. R. (2011). Core-collapse supernovae and the younger Drayas / terminal Rancholabrean extinctions. *Icarus, 215*, 101–106.
Caldeira, K., & Kasting, J. F. (1992). The life span of the biosphere revisited. *Nature, 360*, 721–722.
Cha, A. N., Sembach, K. R., & Danks, K. R. (1999). The distance to the vela supernova remnant. *Astrophysical Journal, 515*, L25–L28.
Clark, D. H., & Stephenson, F. R. (1977). *Historical Supernovae*. Oxford: Pergamon Press.
Clark, D. H., McCrea, W. H., & Stephenson, F. R. (1977). Frequency of nearby supernovae and climatic and biological catastrophes. *Nature, 265*, 318–319.
Cockell, C. (2002). Photobiological uncertainties in the Archaean and post Archaean world. *International Journal of Astrobiology, 1*, 31–38.
Ellis, J., & Schramm, D. N. (1995). Could a nearby supernova explosion have caused a mass extinction? *Proceedings of the National Academy of Sciences, 92*, 235–238.

Falk, S. W. (1978). Shock steepening and prompt thermal emission in supernovae. *Astrophysical Journal, 225*, L133–L126.

Fields, B. D., Athanassiadou, T., & Johnson, S. R. (2008). Supernova collisions with the heliosphere. *Astrophysical Journal, 678*, 549–562.

Fitoussi, C., et al. (2008). Search for supernova-produced 60-Fe in a marine sediment. *Physical Review Letters, 101*, 121101–121104.

Fogg, M. (1991). Terraforming as part of a strategy for interstellar communications. *Journal of the British Interplanetary Society, 44*, 183–192.

Franck, S., Bounama, C., & von Bloh, W. (2006). Causes and timing of future biosphere extinctions. *Biogeosciences, 3*, 85–92.

Fryer, C. L., et al. (2007). *Constraints on type Ib/c supernovae and gamma-ray burst progenitors* (Vol. 119, pp. 1211–1232). Publications of the Astronomical Society of the Pacific.

Gehrels, N., & Chen, W. (1993). The Geminga supernova as a possible cause of the local interstellar bubble. *Nature, 361*, 706.

Gies, D. R., & Helsel, J. W. (2005). Ica age epochs and the Sun's path through the galaxy. *Astrophysical Journal, 626*, 844–848.

Gowanlock, M. G., Patton, D. R., and McConnell, S. M. (2011). A model of habitability within the milky way galaxy. http://arxiv.org/pdf/1107.1286.pdf.

Hallam, T. (2004). *Catastrophes and lesser calamities – The causes of mass extinctions*. Oxford: OUP.

Harding, S. (2006). *Animate earth: Science, intuition and Gaia*. Vermont: A Science Writers Book, Chelsea Green Publishing.

Hill, G. A. (2009). WR 104: Are we looking down the gun barrel of a future GRB? *Bulletin of the American Astronomical Society, 41*, 475.

Hunt, G. E. (1978). Possible climatic and biological impact of nearby supernovae. *Nature, 271*, 430–431.

Kasting, J. F. (1988). Runaway and moist greenhouse atmospheres and the evolution of earth and Venus. *Icarus, 74*, 472–494.

Klein, R. I., & Chevalier, R. A. (1978). X-ray bursts from type II supernovae. *Astrophysics Journal, 223*, L109–L112.

Knie, K., Korschinek, G., Faestermann, T., Wallner, C., Scholten, J., & Hillebrandt, W. (1999). Indication for supernova produced 60-Fe activity on earth. *Physical Review Letters, 83*, 18–21.

Leitch, E. M., & Vasisht, G. (1998). Mass extinctions and the Sun's encounters with spiral arms. *New Astronomy, 3*, 51–56.

Lineweaver, C. H., Fenner, Y., & Gibson, B. K. (2004). The galactic habitability zone and the age distribution of complex life in the milky way. *Science, 303*, 59–62.

Lovelock, J. E. (1989). The ecoposiesis of dairy world. *Journal of the British Interplanetary Society, 42*, 583–586.

Lovelock, J. E. (2006). *The revenge of Gaia*. London: Allen Lane.

Lovelock, J. E., & Kump, L. R. (1994). Failure of climate regulation in a geophysiological model. *Nature, 369*, 732–734.

Lovelock, J. E., & Whitfield, M. (1982). Life span of the biosphere. *Nature, 296*, 561–563.

Martin, O., Galante, D., Cardenas, R., & Horvath, J. E. (2009). Short term effects of gamma ray bursts on earth. *Astrophysics and Space Science, 321*, 161–167.

Martin, O., Cardenas, R., Guimarals, M., Horvath, J. E., & Galante, D. (2010). Effects of gamma ray bursts in Earth's biosphere. *Astrophysics and Space Science, 326*, 61–67.

Melott, A., et al. (2004). Did a gamma-ray burst initiate the late Ordovician mass extinction? *International Journal of Astrobiology, 3*, 55–61.

Mihalas, D., & Binney, J. (1981). *Galactic astronomy: Structure and kinematics* (p. 229). San Francisco: W. H. Freeman and Company.

Pellizza, L. J., Mignami, R. P., Grenier, I. A., & Mirabel, I. F. (2005). On the local birthplace of Geminga. *Astronomy and Astrophysics, 435*, 625–630.

Penate, L., Martin, O., Cardenas, R., & Agusti, S. (2010). Short-term effects of gamma ray bursts on oceanic photosynthesis. *Astrophysics and Space Science, 326*, 211–217.

Podsiadlowski, P., Mazzali, P. A., Nomoto, K., Lazzati, D., & Cappellaro, E. (2004). The rates of hypernovae and gamma-ray bursts: Implications for their progenitors. *Astrophysics Journal, 607*, L17–L20.

Reynolds, S. P., et al. (2008). The youngest supernova remnant: G1.9+0.3. *Astrophysical Journal Letters, 680*, L41–L44.

Ruderman, M. A. (1974). Possible consequences of nearby supernova explosions for atmospheric ozone and terrestrial life. *Science, 184*, 1079–1081.

Salvati, M., & Sacco, B. (2008). The Milagro anticenter hot spots: Cosmic rays from the Geminga supernova? *Astronomy and Astrophysics, 485*, 527–529.

Scalo, J., & Wheeler, J. C. (2002). Astrophysical and Astrobiological implications of gamma-ray burst properties. *Astrophysical Journal, 566*, 723–737.

Shaviv, N. J. (2003). The spiral structure of the milky way, cosmic rays and ice age epochs on earth. *New Astronomy, 8*, 39–77.

Tegmark, M., & Bostrom, N. (2005). Astrophysics: Is a doomsday catastrophe likely? *Nature, 438*, 754.

Terry, K. D., & Tucker, W. H. (1968). Biological effects of supernovae. *Science, 159*, 421–423.

Thomas, B. C., et al. (2005). Gamma-ray bursts and the earth: Exploration of atmospheric, biological, climatic and biogeochemical effects. *Astrophysical Journal, 634*, 509–533.

Tuthill, P. G., Monnier, J. D., Lawrance, N., Danchi, W. C., Owocki, S. P., & Gayley, K. G. (2008). The prototype colliding-wind pinwheel WR 104. *Astrophysical Journal, 675*, 698–710.

van den Bergh, S., & Tammann, G. (1991). Galactic and extragalactic supernova rates. *Annual Reviews of Astronomy and Astrophysics, 29*, 363–407.

van der Hucht, K. (2001). The VIIth catalogue of galactic wolf-Rayet stars. *New Astronomy Reviews, 45*, 135–232.

Williams, D. M., & Pollard, D. (2003). Extraordinary climates of earth-like planets: Three-dimensional climate simulations at extreme obliquity. *International Journal of Astrobiology, 2*, 1–19.

Wood, A. J., Ackland, G. J., Dyke, J. G., Williamns, H. T. P. and Lenton, T. M. (2008). Daisy world: A review. Reviews in Geophysics, 46, RG1001.

Woodwell, G. M. (1967). Radiation and the patterns of nature. *Science, 156*, 461–470.

Woosley, S. E. (2010). Models for gamma-ray burst progenitors and central engines. http://arxiv.org/pdf/1105.4193.pdf.

Woosley, S. E., & Bloom, J. S. (2006). The supernova gamma-ray burst connection. *Annual Reviews of Astronomy and Astrophysics, 44*, 507–556.

References for the Section 6.3.1

K. Freese, P. Gondolo, and D. Spolyar. (2007). The effect of dark matter on the first stars: A new phase of stellar evolution. http://arxiv.org/pdf/0709.2369.pdf.

K. Freese, P. Bodenheimer, D. Spolyar, and P. Gondolo. (2008a). Stellar structure of dark stars: A first phase of stellar evolution resulting from dark matter annihilation. http://arxiv.org/pdf/0806.0617.pdf.

K. Freese, D. Spolyar, and A. Aguirre. (2008b). Dark matter capture in the first stars: A power source and limit on stellar mass. http://arxiv.org/pdf/0802.1724.pdf.

Heger, A., & Woosley, S. E. (2002). The nucleosynthetic signature of population III stars. *Astrophysical Journal, 567*, 532–543.

Heger, A., Fryer, C. L., Woosley, S. E., Langer, N., & Hartman, D. H. (2003). How massive single stars end their life. *Astrophysical Journal, 591*, 288–300.

B. Paxton, (2004). It's EZ to evolve ZAMS stars: A program derived from Eggleton's stellar evolution code. http://arxiv.org/pdf/astro-ph/0405130.pdf.

P. Scott, J. Edsjo, and M. Fairbairn, (2009). The DarkStars code: A publicly available dark stellar evolution package. http://arxiv.org/pdf/0904.2395.pdf.

References for the Section 6.3.2

R. Caldwell, M. Kamionkowski and N. Weinberg. (2003). Phantom energy and cosmic doomsday. https://arxiv.org/pdf/astro-ph/0302606.pdf.

M. Merafina, G. S. Bisnovatyi-Kogan, and S. O. Tarasov. (2011). Polytropic configurations with non-zero cosmological constant. https://arxiv.org/abs/1102.0972.pdf.

References for the Section 6.4.1

F. Adams. (2008). Stars in other universes: Stellar structure with different fundamental constants. PDF available at - http://arxiv.org/pdf/0807.3697.pdf.

L. Barnes. (2012). The fine-tuning of the universe for intelligent life. PDF available at –http://arxiv.org/pdf/1112.4647.pdf.

Barrow, J., & Tippler, F. (1986). *The anthropic cosmological principle*. Cambridge: Cambridge University Press.

Beech, M. (1987). WAP arguments for a maximum stellar mass loss rate. *Speculations in Science and Technology, 11*, 233–235.

Carr, B., & Rees, M. (1979). The anthropic principle and the structure of the physical world. *Nature, 278*, 605–612.

B. Carter. (1974). Large number coincidences and the anthropic principle in cosmology. In: *Confrontation of cosmological theories with observational data*. Proceedings of the Symposium, Krakow, Poland, September 10–12, 1973. Dordrecht, D. Reidel Publishing.

Dirac, P. (1937). The cosmological constants. *Nature, 139*, 323.

Dirac, P. (1938). A new basis for cosmology. *Proceedings of the Royal Society of London A, 165*, 199–208.

Ellis, G. (2008). Opposing the multiverse. *Astronomy and Geophysics, 49*, 2.33–2.35.

Gardner, M. (2001). Multiverses and blackberries. *Skeptical Inquirer, 25*, 5.

Lucy, L. B., & Solomon, P. M. (1970). Mass loss from hot stars. *Astrophysical Journal, 159*, 878–893.

Teller, E. (1948). On the change of physical constants. *Physical Review, 73*, 801–802.

Uzan, J.-P. (2003). The fundamental constants and their variation: Observational and theoretical status. *Reviews of Modern Physics, 75*, 403–455.

J. Webb et al. (2011). Indications of a spatial variation of the fine structure constant. http://arxiv.org/pdf/1008.3907.pdf.

Williams, I. P. (1967). Maximum mass loss from stars. *Monthly Notices of the Royal Astronomical Society, 136*, 341–346.

Appendix

The 4th Order Runge-Kutta (RK4) Integration Scheme

There are many numerical integration schemes for finding the solution to ordinary differential equations (ODEs), but one of the most straightforward and accurate is that developed by German mathematicians Carl Runge and Martin Kutta in the early twentieth century. The RK4 scheme applies to the solution of a first order ODE when the following initial information is available:

$$\frac{dy}{dx} = f(y, x), \text{ and } y(x_0) = y_0$$

where $y(x)$ is the dependent variable to be determined for successive values of x, and $f(y, x)$ is some specified function. By providing a specific starting value $y(x_0) = y_0$, this makes for an initial value problem. In the RK4 scheme, the value of $y(x)$ at the point $x + h$, where h is some specified step size, is approximated by a weighted average of the values of $f(x, y)$ at four points in the interval $x + h$. The formula for this is given as:

$$y(x + h) = y(x) + (k_1 + 2(k_2 + k_3) + k_4)/6$$

where

$$k_1 = h f(x, y)$$
$$k_2 = h f(x + \tfrac{1}{2}h, y + \tfrac{1}{2}k_1)$$
$$k_3 = h f(x + \tfrac{1}{2}h, y + \tfrac{1}{2}k_2)$$
$$k_4 = h f(x + h, y + k_3)$$

With this scheme in place, once $y(x + h)$ has been determined, $y(x + 2h)$ can be evaluated, and so on until some final value of x has been obtained.

© Springer Nature Switzerland AG 2019
M. Beech, *Introducing the Stars*, Undergraduate Lecture Notes in Physics,
https://doi.org/10.1007/978-3-030-11704-7

The RK4 is easily implemented in any programming language, and a brief pseudocode is given below:

```
h = 0.001 (set the step size)
x = x0 (the initial value of x)
y = y0 (the initial value of y(x0))
J = 0 (initiate the while-do-loop counter)
While J < 100 do (perform 100 integration steps)

    gosub RK4(x, y, h)
    gosub Output (x, y)
    J = J + 1

Endo
End.
Subroutine RK4(x, y, h)
Begin

    FK1 = func(x, y)
    FK2 = func(x + h/2, y + h*FK1/2)
    FK3 = func(x + h/2, y + h*FK2/2)
    FK4 = func(x + h, y + h*FK3)
    y = y + h*(FK1 + 2(FK2 + FK3) + FK4) / 6
    x = x + h

End
Return
Function func(x, y)
Begin

    func = f(x, y) (this is the function being integrated)

End
Return
Subroutine Output (x, y)
Begin

    Write out x and y (chose some final print-out format)

End
Return
```

In order to integrate the Lane-Emden equation, an extra step needs to be taken when developing an RK4 scheme. Recall from Eq. (3.42) that we are looking for the numerical solution to an equation of the form

$$\frac{1}{\xi^2}\frac{d}{d\xi}\left(\xi^2\frac{d\theta}{d\xi}\right) + \theta^n = 0 \tag{A.1}$$

which is to be integrated from $\xi = 0$ to the surface $\xi = \xi_1$ where $\theta(\xi_1) = 0$. The first step in solving for (A.1) is to reduce the equation to a first order ODE. Expanding the differential term in the equation, we have:

$$\left(\frac{2d\theta}{\xi d\xi} + \frac{d^2\theta}{d\xi^2}\right) + \theta^n = 0 \tag{A.2}$$

Accordingly, we now introduce the substitution that

$$z = \frac{d\theta}{d\xi} \tag{A.3}$$

which is easily integrated step by step, under the boundary condition that $\theta = 1$ at $z = 0$, as

$$\theta(z+h) = \theta(z) + zh \tag{A.4}$$

With substitution (A.3), Eq. (A.2) is transformed into a first order differential equation that can be integrated via the RK4 scheme:

$$\frac{dz}{d\xi} = -2\frac{z}{\xi} - \theta^n \tag{A.5}$$

In this case we have (with respect to the pseudocode above)

$$\text{func} = f(z,\xi) = -2z/\xi - \theta^n \tag{A.6}$$

and the boundary condition is that $z = d\theta / d\xi = 0$ at $\xi = 0$. There is technically a problem lurking in Eq. (A.6) at $\xi = 0$, in the sense that z / ξ is indefinite. Starting the integration with $z /\xi = -1/3$ when $\xi = 0$ solves this problem (see below).

Once you have developed your code for the numerical solution of the Lane-Emden equation, compare your results against the numbers given in Table 3.1, and against the analytic solutions when $n = 0$, 1 and 5. In order to achieve a good numerical match, you will have to experiment with the step size h—if h is set too big, the results will not be very accurate, whereas if h is set very small, the code will take a long time to run since there will be many more steps that it has to work through. A step size of $h = 0.001$ will be fine for what needs to be computer in the exercises. You will also have to experiment with the number of passes through the 'While-Do' part of the code; the J counter will eventually need to be set to a value much larger than the 100 that is shown. Rather than using a loop-counter, one could alternatively set up an end condition based upon the attainment of $\theta = 0$, which will occur when $\xi = \xi_1$.

According to what is required, there are various techniques available for solving the Lane-Emden equation. Procedures other than RK4 can be developed (see e.g. Vanani and Aminataei, 2010) and a power series solution is also available (see e.g. Mohan and Al-Bayaty, 1980; Roxburgh and Stockman, 1999; Rohue, 2008.). Indeed, for small ξ, the first five terms of the power series solution are:

$$\theta_n = 1 - \frac{1}{6}\xi^2 + \frac{n}{120}\xi^4 - \frac{n(8n-5)}{15120}\xi^6 + \frac{n(122n^2 - 183n + 70)}{3265920}\xi^8 + \cdots \quad (A.7)$$

From Eq. A.7, the only obvious analytic solution is that for $n = 0$, where all but the first two terms vanish. The numerical dodge mentioned with respect to the RK4 method where $z/\xi = -1/3$ when $\xi = 0$ is also revealed in A7. Keeping just the first two terms when ξ is very small, we have: $z = d\theta / d\xi = -\xi/3$. A number of approximate analytic solutions to the Lane-Emden equation, when $n = 1.5$ and 3 have been developed from the power series expansion A7 (see e.g. Beech, 1987; Fowler and Hoyle, 1964). The following approximations apply within the central regions:

$$\vartheta = \exp(-\xi^2/6) \quad \text{when } n = 1.5 - \text{see Beech (1987)}$$
$$\vartheta = \text{sech}(\xi/\sqrt{3}) \quad \text{when } n = 3 - \text{see Beech (1987)}$$
$$\vartheta_n = \left[1 - \left(\frac{n}{120} - \frac{1}{72}\right)\xi^4\right]\exp(-\xi^2/6) - \text{see Fowler and Hoyle (1964)}$$

References

Beech, M. (1987). An approximate solution for the polytrope n = 3. *Astrophysics and Space Science*, *132*, 393–396.

Fowler, W., & Hoyle, F. (1964). Neutrino processes and pair formation in massive stars and supernovae. *Astrophysical Journal Supplement*, *91*, 201–319. See specifically their equation C79.

Mohan, C., & Al-Bayaty, A. R. (1980). Power-series solutions of the Lane-Emden Equation. *Astrophysics and Space Science*, *73*, 227–239.

Rohue. K. (2008). Computing the coefficients for the power series solution of the Lane-Emden equation with the Python library *SymPy*. https://arxiv.org/ftp/arxiv/papers/1409/1409.2008.pdf

Roxburgh, and Stockman. L. (1999). Power series solutions of the polytropic equations. *Monthly Notices of the Royal Astronomical Society*, *303*, 466–470.

Vanani, S. K., & Aminataei, A. (2010). On the numerical solution of differential equations of Lane-Emden type. *Computers and Mathematics with Applications*, *50*, 2815–2820.

Index

© Springer Nature Switzerland AG 2019
M. Beech, *Introducing the Stars*, Undergraduate Lecture Notes in Physics,
https://doi.org/10.1007/978-3-030-11704-7